People and Culture in Construction

Also available from Taylor & Francis

Communication in Construction: Theory and Practice
A. Dainty *et al.*

Hb: ISBN 0–415–32722–9
Pb: ISBN 0–415–32723–7

OSH in Construction Project Management
H. Lingard *et al.*

Hb: ISBN 0–419–26210–5

Managing Diversity & Equality in Construction
A. Gale *et al.*

Pb: ISBN 0–415–28869–X

Communication in Construction Teams
C. Gorse *et al.*

Hb: ISBN 0–415–36619–4

Risk Management in Projects 2nd ed.
M. Loosemore *et al.*

Hb: ISBN 0–415–26055–8
Pb: ISBN 0–415–26056–6

Information and ordering details

For price availability and ordering visit our website **www.tandfbuiltenvironment.com**
Alternatively our books are available from all good bookshops.

People and Culture in Construction

A reader

Edited by

Andrew Dainty, Stuart Green and Barbara Bagilhole

Routledge
Taylor & Francis Group

LONDON AND NEW YORK

First published 2007
by Taylor & Francis
This edition published 2015
by Routledge
2 Park Square, Milton Park, Abingdon, Oxfordshire OX14 4RN

Simultaneously published in the USA and Canada
by Routledge
711 Third Avenue, New York, NY 10017

*Routledge is an imprint of the Taylor and Francis Group,
an informa business*

First issued in paperback 2015

Typeset in Times New Roman by
Integra Software Services Pvt. Ltd, Pondicherry, India

British Library Cataloguing in Publication Data
A catalogue record for this book is available from the British Library

Library of Congress Cataloging in Publication Data
People and culture in construction : a reader / [edited by] Andrew Dainty,
  Stuart Green, and Barbara Bagilhole.
    p.   cm. — (Spon research)
  Includes bibliographical references.
  ISBN 978-0-415-34870-6 (hardback : alk. paper)   1. Building—
  Superintendence.   2. Construction industry—Personnel management.
  3. Construction industry— Customer services.   4. Corporate culture.
  I. Dainty, Andrew.   II. Green, Stuart, 1958–   III. Bagilhole, Barbara,
  1951–
  TH438.P383 2006
  338.4'7624—dc22   2006024093

ISBN 978–0–415–34870–6 (hbk)
ISBN 978–1–138–97810–2 (pbk)
ISBN 978–0–203–64091–3 (ebk)

Contents

Tables

Figures

Contributors

The editors

Andrew Dainty is Professor of Construction Sociology at Loughborough University's Department of Civil and Building Engineering. A renowned researcher in the field of human resource management and organisational behaviour in the construction industry, he has published widely in both academic and industry journals. He holds a number of research grants from the EPSRC, ESRC and various government and European agencies, as well as advising a wide range of contracting and consultancy firms on human and organisational issues. Professor Dainty is co-author of *HRM in Construction Projects* (2003) and *Communication in Construction* (2006), also published by Taylor & Francis. He has recently co-edited special issues of both *Personnel Review* and *Construction Management and Economics* on construction labour and equality issues.

Stuart Green is Professor of Construction Management in the School of Construction Management and Engineering at the University of Reading. He is also Director of the Innovative Construction Research Centre (ICRC) where he is responsible for a multi-million pound research programme in collaboration with industry. Previously he worked in contracting for several years before gaining design experience with an engineering consultancy. Professor Green has published widely in a range of international journals and has extensive experience of large multi-institutional, multi-disciplinary research projects. He is a chartered civil engineer and chartered builder. He is well known throughout the UK construction industry as a dissenter to accepted notions of 'best practice'. Previous topics of critique include lean construction, partnering and construction process improvement.

Barbara Bagilhole is Professor of Equal Opportunities and Social Policy and Associate Dean Research in the Faculty of Social Sciences and Humanities at Loughborough University. She has taught, researched and published extensively in the field of Equal Opportunities and Diversity legislation and policies across the areas of gender, race, disability, sexual orientation, religious belief and age. Recent research has looked at women and science, engineering and technology, and other women in non-traditional occupations. Her last book published was *Women in Non-traditional Occupations: Challenging Men* (2002).

Author biographies

Andrew Agapiou gained a PhD in Construction Labour Resource Planning from Loughborough University, and is currently Academic Co-director of the Centre for the Built Environment at the University of Strathclyde. He has published widely, is a reviewer for many international journals, conferences and funding agencies and has led or collaborated on several funded research projects in the area of labour resources for the construction industry. He is also one of the co-organisers of the ESRC/EPSRC funded Trans-disciplinary Research Seminar Series concerning People and Culture in Construction jointly hosted by Strathclyde and Glasgow Caledonian Universities.

J. Craig Barker is Professor of Law at the University of Sussex. Having received his doctorate from the University of Glasgow in 1991, he spent some time in legal practice with the leading Scottish corporate law firm, Dickon Minto W.S. Before taking up his Chair at Sussex he spent ten years teaching at the University of Reading specialising in International Law and Employment Law. He has published numerous articles and books including, most recently, a book entitled *The Protection of Diplomatic Personnel* (Ashgate, 2006).

Andrew Caplan is Senior Research Consultant with the Centre for Ethnic Minority Studies, Royal Holloway College, University of London, the college which awarded him his doctorate in 1971. He has more than 30 years experience in London education as a lecturer in history and humanities. When the ILEA was abolished in 1990 he was Senior Inspector for Multi-Ethnic Education, and in the 1990s he held senior education posts in LEAs, including Chief Officer with the London Borough of Lambeth. In addition, Dr Caplan has served over 25 years with the Open University as an Associate Lecturer in the Arts Faculty.

Simon A. Burtonshaw-Gunn is a visiting professor in the School of the Built Environment at the University of Salford. He has experience of construction partnering and collaborative working and undertook research in this topic for his PhD under the supervision of Professor Bob Ritchie. Professor Burtonshaw-Gunn holds a full-time consultancy within a major UK defence company and has represented it on a number of research projects. Since 2000 he has facilitated numerous initial, interim and close-out multi-company partnering workshops and has been a member of the Association of Partnering Advisors since 2001.

Paul Chan is a lecturer in the School of the Built Environment at Northumbria University. Dr Chan has been actively researching in the area of skills and competencies, with particular emphasis on the deployment of skills for the effective implementation of project processes. His research interests cover the area of human resources management and development, which stem from his earlier research into training participation in the UK construction industry and his PhD research on construction labour productivity. He has also taught project management and team development and facilitation, and is keen on the personal development planning of students.

Linda Clarke is Professor of European Industrial Relations at the Westminster Business School, University of Westminster, undertaking research on training, skills, labour relations and diversity in employment in Europe, particularly in the construction sector. She is on the presidium of the European Institute for Construction Labour Research. In addition to numerous articles, Professor Clarke's major publications include *Women in Construction* (co-editor, 2004); *The Dynamics of Wage Relations in the New Europe* (joint editor, 2000); *A Blueprint for Change: Construction Skills Training in Britain* (with C. Wall, 1998); and *Building Capitalism* (1992).

Joanna Cullinane is a principal lecturer in HRM at the University of Greenwich. Before joining Greenwich, she worked for the University of Glamorgan and the University of Waikato (New Zealand). Outside of academia, Dr Cullinane previously worked as a HRM officer in a large city council. Her main research interests are in intra-personal conflict and in general areas such as public sector management, unionisation, health and safety, personal grievances, internal marketing, cross-cultural HRM, change in the voluntary sector and the relationship between employment relations and shifts in wider political economy.

Ann de Graft-Johnson is a senior lecturer in Planning and Architecture at the University of the West of England. Ann is an architect with special expertise in equal opportunities and diversity issues. She undertakes research primarily in relation to gender, culture and equal opportunities issues. She is also a practising architect. She was a member/director of Matrix Architects Feminist Co-operative, which actively worked to redress the balance in relation to women's marginalised role in decision-making processes which affect the built environment. She was also a founder member of the Society of Black Architects (SOBA).

Janet Druker is Senior Pro-Vice Chancellor at Canterbury Christ Church University. Formerly Head of Research and Organisation with the Union of Construction Allied Trades and Technicians (UCATT), she has also worked for the Universities of Greenwich and East London. She holds a PhD from Warwick University and is a fellow of the Chartered Institute of Personnel and Development. Professor Druker's research interests are concerned with changes in work and employment, in both the UK and internationally. She has published extensively on these issues in relation to the construction industry. Ongoing research is concerned with the problems and experiences of small- and medium-sized enterprises.

Chris Forde is a senior lecturer in Industrial Relations in the Work and Employment Relations Division, Leeds University Business School. Dr Forde's interests centre around the nature and development of contingent employment arrangements, particularly agency work; the relationship between alternative forms of work organisation, the external environment and performance; the social and economic consequences of restructuring and redundancies; the employment experiences of migrant workers, asylum seekers and refugees; and the relationship between deunionisation and wage inequality.

Paul W. Fox is an assistant professor in the Department of Building and Real Estate, The Hong Kong Polytechnic University. Prior to becoming an academic, Dr Fox gained over 10 years' experience in national and international contracting. His research MSc was a study of the Hong Kong construction industry using Systems Theory (University of Salford, 1989). His PhD (QUT, 2003) on construction industry development identified significant cultural factors. Together with Professor Lu You-Jie of Tsing-Hua University, Beijing, he co-authored an ILO Working Paper on the labour aspects of the Chinese construction industry. For over 10 years he has taught a postgraduate subject on Managing People to Hong Kong and Mainland Chinese students.

Clara Greed is Professor of Inclusive Urban Planning at the University of the West of England. She has written over 10 books on town planning, urban design and education and practice issues in the built environment professions. Her main interests are gender mainstreaming, equality, accessibility and the social aspects of urban design. Currently, Professor Greed is undertaking EPSRC funded research with colleagues at UCL on the role of public toilets in creating accessible city centres. She is a member of the RTPI and the RICS and fellow of the CIOB.

Georg Herrmann is a project manager for the East Thames Housing Group. He was until recently a Research Fellow at the University of Westminster with a particular research interest in labour market issues in Europe, especially in the construction sector. Dr Herrmann obtained an MA and PhD in Modern History at Munich University, Germany and an MBA at Imperial College, London.

Stephen Ison is a senior lecturer within the Department of Civil and Building Engineering, Loughborough University. An economist with 20 years' experience in Higher Education, his research is in the area of applied economics and policy, in particular regional economics, labour markets and transport. In recent years, Dr Ison has undertaken a number of attitudinal surveys, most recently in terms of ascertaining key stakeholder opinions on traffic-related issues in urban areas and the likely effectiveness and acceptance of various policy options. In addition to survey work, his research has involved the use of detailed case studies and in-depth interviews. He is currently supervising research in the area of construction labour markets and skills. He has published widely in the area of applied economics.

Ammar Kaka is Professor of Construction Economics and Management in the School of Built Environment, Heriot-Watt University. Professor Kaka has developed a Dynamic Cash Flow Forecasting model, mapping projects' financial management processes, and an occupancy cost prediction model for buildings. He has also undertaken research projects aimed at developing best practices for the design and procurement of sustainable and economic housing for the twenty-first century. Externally, he is also a member of several Commissions of Building Economics and Construction Management, and is presently a member of the EPSRC review college and the 'Gardiner & Theobald Procurement Forum'.

Christian Koch is an associate professor based within the Department of Civil Engineering, Technical University of Denmark. He researches change in construction with a particular focus on production and site issues, such as work organisation, organisational culture, quality and production management. Dr Koch's other interests include ICT and knowledge management. Christian Koch has published widely both within and outside of the construction research community.

David Langford is Professor of Construction Management at Glasgow Caledonian University's School of the Built and Natural Environment. He has been a construction researcher for over 25 years and has held research contracts on Construction Management from the research councils (EPSRC and ESRC), the European Union, the UK government, the Scottish government, private sector organisations, the BRE and overseas research organisations. Much of this work has related to human issues in construction. He has published widely, and human and labour issues have featured strongly in his output. He recently co-edited a special edition of *Personnel Review* featuring labour research on the construction industry.

Karolina Lorenz is a teaching assistant and PhD student at the Institute for Building Informatics at Graz University of Technology. She studied civil engineering, specialising in construction management and economics, at Graz University of Technology, University of Calgary and University of Innsbruck. In 2003 she undertook her master's study on cultural differences in the construction industry under the supervision of Professor Marton Marosszeky at UNSW in Sydney.

Steven McCabe is a senior lecturer in the School of Property, Construction and Planning, University of Central England. His first employment in the industry was as a site engineer in the late 1970s. After attending UMIST he worked as a surveyor and estimator in both the private and public sectors. A suggestion by an ex-tutor led to work as a visiting lecturer at Birmingham Polytechnic in 1987. Since then he has completed an MSc and doctorate in construction management at the University of Birmingham. His PhD researched quality management in large contractors. Dr McCabe's publications include conference articles, book chapters and two textbooks on the application and effects of quality management, benchmarking and human resource management in construction.

Robert MacKenzie is a senior lecturer in Industrial Relations in the Work and Employment Relations Division, Leeds University Business School. His research interests are concerned with the regulation of the employment relationship and industrial restructuring. Dr MacKenzie's work in the construction and telecommunications industries has focused upon the use of subcontracting as a means of regulating the supply of labour, and the implications this holds for skills and labour reproduction. Other research interests include the social and economic impact of restructuring and redundancy in the Welsh steel industry. More recently he has conducted research on the social and labour market experience of migrant workers, asylum seekers and refugees.

Sandra Manley is a principal lecturer in the School of Planning and Architecture, University of the West of England. She teaches urban design and planning and has a special interest in the design of inclusive environments and the involvement of community in urban design and regeneration projects. She has written articles on these subjects, including a recent contribution to the *Universal Design Handbook*, and has spoken at conferences both in the UK and in USA. Sandra was a key member of the team that developed the combined architecture and planning degree at the University of the West of England.

Marton Marosszeky is the Multiplex professor of Engineering Construction Innovation in the School of Civil and Environmental Engineering at the University of New South Wales, Sydney, Australia. His primary research interests are in the area of construction process improvement in relation to quality, safety and production using Lean and TQM concepts. Professor Marosszeky is a co-author with John Oakland of *Total Quality in the Construction Supply Chain*.

Michael Pye is a senior lecturer in HRM at the University of Hertfordshire. In his youth, he worked as labourer in the construction and demolition industries before becoming a nurse. He has spent most of his working life to date in the health sector in a variety of roles, including a ten-year period in various forms of 'non-standard' employment, as he made the move from healthcare to academia. His main area of research interest is in the working lives of frontline clinical managers and the organisation of health and sickness work.

Ani Raidén is a lecturer in Human Resource Management/Organisational Behaviour at the University of Glamorgan Business School. She completed her PhD at Loughborough University in 2004 developing a strategic employee resourcing framework (SERF) for construction organisations. Prior to becoming an academic she worked as an HR transformation executive at Eircom, a large telecommunications company in the Republic of Ireland, and as a personnel assistant at Hampshire Ambulance NHS Trust in the UK. Dr Raidén's research interests focus on employee resourcing, employee involvement and flexibility within project-based environments.

Bob Ritchie is Professor of Risk Management, Lancashire Business School, University of Central Lancashire. His research specialisation in risk management includes applications in supply chains and SMEs. He is an author of numerous best-selling international texts and journal articles in the field and is a journal editor.

Neil Ritson is a senior lecturer at Cumbria Business School, University of Central Lancashire. He qualified as a psychologist and then moved into business management. He has been an employee relations adviser with Exxon-Mobil and the Engineering Employers Federation, and a management consultant. He has taught in a variety of universities including St Andrews, Newcastle, and the OU. He has published numerous articles on Employee Relations, HRM and Strategy in major academic and industry journals.

Katherine Sang is a researcher based in Loughborough University's Department of Civil and Building Engineering. She is currently researching the psychological well-being of architects, with a particular focus on gendered differences within the profession. Her previous research includes studies into the experiences of women working in the sector and in the relationship between skills and productivity within construction. She has a MRes in civil engineering and a MA Hons in history and psychology, both from the University of Dundee.

Acknowledgements

First and foremost, we would like to thank all of the individual authors for their contributions to this book. The critical and insightful perspectives provided by these authors is a tribute to the quality of the scholarship upon which their contributions are based.

This book has its origins in a transdisciplinary Economic and Social Research Council (ESRC) funded research seminar series. This series of critical seminars aimed to bring together leading researchers from a range of different disciplines in order to facilitate an understanding of the employment and cultural issues facing the construction industry. Many of the chapters which follow this introduction are based on presentations previously made at the ESRC-funded research seminars. We would like to express our thanks both to the chairs of these seminars (who in addition to ourselves included Professor Irena Grugulis, Dr Peter McDermott, Dr David Bartram and Professor David Farnham) and to all of the 270 people who participated in the series as speakers or delegates. The lively and critical discussion which characterised each event has helped to shape many of the contributions within this book.

Finally, we would like to thank Dr Rachel Stewart for her efforts helping to edit the contributions contained within this book.

The following figures are reproduced with kind permission from the copyright holder:

Figure 11.1 © John Wiley & Sons, Inc.

Figure 11.2 © Geert Hofstede

Figure 12.2 © European Foundation for the Improvement of Living and Working Conditions, 2005, Wyattville Road, Loughlinstown, Dublin 18, Ireland. Original language English, first published in 2004.

Table 11.1 from *Diagnosing and Changing Organizational Culture*, Cameron and Quinn, © John Wiley & Sons, Inc., 2005. Reprinted with permission of John Wiley & Sons, Inc.

Part I

The construction employment context

1 People and culture in construction

Contexts and challenges

Andrew Dainty, Stuart Green and
Barbara Bagilhole

1.1 Introduction

The construction industry is one of the largest, complex and most people-intensive sectors. In many developed countries, however, construction is also one of the most highly criticised sectors with regard to its employment practices and industrial relations climate. Underlying this criticism lays a fundamental tension. Too often construction relies upon informal and casualised employment practices which provide low barriers to entry for those wanting to work within the sector. Yet it also maintains an ingrained exclusionary culture which militates against the entry of those who cannot conform to its norms and stereotypes. Those that do choose to work in the industry experience a workplace environment character-ised by structural fragmentation, a wide diversity of employment practices and an endless succession of short-term projects. Diverse groups of people who are brought together for short periods of time are expected to rapidly establish co-operative working relationships while frequently being engaged on entirely different terms and conditions. Tensions are exacerbated by the need to deliver buildings under evermore-stringent time and cost constraints. With contractors being increasingly removed from the physical task of construction, the organisa-tion of production receives little management attention. Human resource issues too often lie outside the remit of project managers who neither know nor care about the employment status of many operatives on the project for which they are responsible. What results is an employment relations climate characterised by separation, conflict, informality and a reluctance to embrace change – issues which are frequently cited as being at the root of many of the sector's enduring problems.

Recently, there appears to have been an acknowledgement that the way in which people are managed and developed is limiting the industry's ability to improve its performance. As such, the industry has been challenged to address its poor performance on people management and cultural issues. Unfortunately, the accepted improvement agenda too often abounds with simplistic exhortations that the construction industry must 'change its culture'. Indeed, as will be examined later in this chapter, the discourse of culture change within the industry appears strangely disconnected from the broader defining literature. This book begins to address this failing by presenting a broad range of critical and empirical insights on construction employment and culture within a single volume. Although eclectic in terms of their coverage and philosophical standpoints with regard to industry practice, together they offer a set of different lenses through which the current failings of the industry can be examined critically. The authors, who are all prominent researchers within their respective fields, share a common approach in that they all aim to be critical but pragmatic. Rather than seek simple normative performance-enhancing solutions to ingrained problems, they explore the implications of the industry's approaches for both research and practice. By locating salient issues within a wider theoretical framework, they provide the reader with a set of thought-provoking perspectives on the challenges confronting the construction sector and how they might be viewed differently in the future.

This introductory chapter attempts to chart the broad contextual terrain of the industry's labour market, its employment practices and its apparent inability to embrace reform. It should be read as a precursor to the more in-depth treatment such issues receive within the book. It briefly reviews the mutually supporting contributions of the later chapters, which together reveal the interplay of structural and cultural factors which have recursively shaped the industry's employment practices and cultural climate. The coverage is primarily focused on the UK, although not exclusively. International comparisons are invariably useful in challenging assumptions that there 'is no other way of operating'.

1.2   Defining characteristics of the construction industry

The construction sector presents a complex, problematic and yet fascinating context within which to explore the creation, enactment and impact of employment practices. As such, it is initially important to consider the characteristics of the sector and the ways in which these shape its employment context and culture.

The construction industry: Boundary definitions and defining characteristics

In seeking to understand 'people and culture' in construction, it is initially important to be aware of the defining characteristics of the sector. The construction industry differs from other sectors in a number of important respects that are fundamentally important in shaping employment practices and cultural recipes. But precisely what constitutes the 'construction industry' is in itself subject to a range of different boundary definitions. It is also possible to argue that the construction industry is not a single industry, but several separate sub-industries (Ive and Gruneberg, 2000). Skills and equipment vary significantly across different types of specialist firms, which in consequence require very different capital structures. For example, there are more differences than commonalities between painting and decorating firms and specialist piling contractors. Likewise, large international contractors are very different from local jobbing builders. In the light of this significant diversity, there are few generalisations that can be made with any confidence about the 'construction industry'. Nevertheless, it is possible to identify a range of characteristics that in combination comprise a unique industry context.

Scope of construction sector

Pearce (2003) distinguishes between narrow and broad definitions of the construction sector. The narrow definition focuses on on-site assembly and the repair of buildings and infrastructure as performed by contractors. This interpretation of the construction sector broadly follows the Standard Industrial Classification (SIC) system that allocates economic activity to different divisions in accordance with a generic coding system (Office for National Statistics, 2003). The statistics routinely quoted to indicate the size of the construction are those produced by the Department of Trade and Industry (DTI) in accordance with SIC Division 45. On this basis, the UK construction sector in 2004 had a provisional annual output of £102.363 billion, with a total workforce of 2,216,000 (DTI, 2005a). The size of the workforce alone justifies construction as a worthwhile context for social research, irrespective of any desire to 'improve productivity'. Understanding is, however, too often subjugated to an overriding objective of improving performance, thereby reducing the construction workforce to 'human capital' which must be exploited to that end. The aspirations and lived realities of over two million workers deserve better attention from the construction research community. Sometimes we perhaps need to remind ourselves that the capitalist system is supposed to serve the interests of society, and not the other way around.

It is not difficult to argue that construction activity is significantly larger than suggested by the figures indicated above. As Pearce (2003) observes, the SIC 45 category excludes those involved in self-build, direct labour and the 'informal' sector, the latter being estimated at £10 billion. The narrow definition of construction also excludes those involved in professional services, including those who provide design and engineering services. An increasing number of professionals are also involved in the delivery of facilities management services. Furthermore, a significant amount of construction activity frequently fails to be reported because it is not the 'principal activity' of the organisation filing the return (Briscoe, 2006). Consider, for example, large client organisations which employ their own construction professionals to provide design, construction and facilities management services for their own property portfolio. Such 'in house' activity is invariably omitted from the official statistics. The recent phenomenon whereby large 'contractors' reclassify themselves as service organisations again serves to under-represent the extent of construction activity. On the basis of his narrow definition of construction, Pearce (2003) suggests that construction contributes in the region of 5 per cent of UK gross domestic product (GDP). However, if Pearce's broad definition is adopted, the boundary extends beyond on-site activity to include quarrying of raw materials, manufacture of building materials, sale of construction products and professional services. On this basis, the contribution of the construction industry to GDP doubles to 10 per cent.

Structure of construction sector

The construction sector is highly fragmented and dominated by small firms. According to the Small Business Service (DTI, 2005b) the UK construction sector comprises 114,510 SMEs (excluding sole traders). In conjunction with 'sole proprietors' (i.e. self-employed), these firms collectively account for 66.1 per cent of private sector turnover. Furthermore, they account for 82.9 per cent of the workforce. Even more strikingly, of all sectors of the economy, construction has the highest percentage of enterprises with no employees, constituting 86.6 per cent of the total. A 'typical' firm of the construction sector therefore comprises a self-employed 'owner-manager'. Collectively, these sole proprietors account for 23.8 per cent of private sector turnover (DTI, 2005b). The population of small firms is of course subject to continuous change as new firms are established and existing firms go out of business; figures should therefore always be treated with caution (Briscoe, 2006). Nevertheless, the dominance of small firms within the construction sector cannot be denied, with lasting implications for employment practices and career structures. Indeed, clichés about fragmentation hardly do justice to the myriad of small firms that is forever moving from project to project. An institutionally embedded reliance on subcontracting means that main contractors frequently employ very few operatives directly. In such circumstances, notions of 'human resource management' are the exclusive preserve of the construction industry's managerial staff.

Among all the usual caveats regarding construction industry statistics, the extent of self-employment continues to be especially contentious. On the basis of a different sampling methodology, DTI (2005a) suggest that 38 per cent of the industry's workforce is self-employed. This compares with a recorded high point of 46.3 per cent in 1996. The subsequent reduction is widely accredited to a clamp-down on tax evasion by the Inland Revenue over the last decade (see Briscoe *et al.*, 2000). The confusion of employment status is conflated, however, with a widespread culture of tax evasion, thus rendering statistics on the levels of self-employment especially unreliable. Official sources are held to consistently underestimate the number of self-employed workers by at least 200,000 (Cannon, 1994; Harvey, 2003). The statistical picture is further clouded by the high number of transient migrant workers, especially in the south-east of England. What is clear is that the construction sector accounts for an unusually high proportion of non-standard forms of employment, with significant implications for career structures, training, and health and safety. Notwithstanding recent talk about a 'return' to direct employment, current indications suggest that the percentage of self-employed is rising once again.

Cyclical demand and structural flexibility

The demand for construction is highly cyclical (Hillebrandt, 2000). Capital investment is the first thing to suffer during any economic downturn. Non-transportable, durable goods are especially sensitive to the business cycle as they cannot be transported to other locations where demand is greater (Bosch and Philips, 2003). The demand for residential housing is also highly sensitive to interest rates, which again links demand to the business cycle. The memory of previous cycles of boom and bust remains writ large in the industry's collective memory. In response, construction firms continue to place a premium on structural flexibility, that is the ability to expand and contract in accordance with fluctuations in demand. The strategy of structural flexibility goes some way towards explaining the industry's high reliance on subcontracting and self-employment (Winch, 1988). It also helps explain the widespread reluctance to invest in training and the collapse of the apprentice system. However, the culture of self-employment has undoubtedly been voluntarily embraced by many operatives. In this respect, the government incentivisation of self-employment during the period 1980–95 undoubtedly played an important part in shaping today's employment patterns (Harvey, 2003).

Project-based industry

The construction industry is frequently cited as the epitome of a project-based industry. The working lives of most construction operatives are characterised by an endless progression from project to project. By and large, unique project organisations are established for every significant new construction project. Cherns and Bryant (1984) coined the term 'temporary multiple organisation' (TMO) to describe the project team created on every project. Each TMO comprises a complex and temporary set of inter-organisational relationships, governed by project-defined interactions (Bresnen *et al.*, 2004). Simply put, construction projects tend to be constructed by a combination of firms and individuals most of whom will not have worked together before and are not likely to work together again. Each project brings together a range of different cultural recipes and employment regimes. Managers are involved in an endless process of trading-off the objectives of the firm with those of the project. For most of the workforce, any commitment to the project is mediated by a permanent sense of uncertainty regarding which project they will be working on next. The competitive tendering system combines with cyclical demand to mean that few contractors can be confident about steady workloads. Such factors militate against stable employment regimes and any long-term investment in the 'psychological contract' between employers and employees (Dainty *et al.*, 2005). In part, the lack of stability in the construction labour market is attributable to the physical characteristics of the industry's products.

In the case of new work, the industry's products tend to be large, heavy and expensive (Hillebrandt, 1988). Once built, they tend to be fixed in location. Furthermore, most products are assembled to a unique specification. The fact that each project is different, both in terms of the product and in terms of the people involved, makes it difficult to achieve the degree of repetition and routinisation achieved in other industries (Bresnen and Marshall, 2001). However, the characteristics of construction projects are common across the world, and they do not in themselves necessarily translate into casualised employment practices. Indeed, among developed countries the UK occupies an unusually extreme position in terms of its deregulated and casualised labour market.

But the argument that construction is a project-based industry is often over-stated. Too often this is used as an excuse not to tackle deep-rooted problems regarding employment patterns in the sector. Supporting arguments tend to be biased towards large firms and major projects, neither of which are representative of the industry as a whole. It is frequently forgotten that around 45 per cent of industry output (on the basis of the narrow definition) is routinely accounted for by repair and maintenance. Much of this work is routine and is funded from operational rather than capital budgets. While reliable statistics are difficult to obtain, the repair and maintenance sector is even more dominated by small firms and single traders. On this basis, construction could be construed as a service industry, in that productive activity takes place at the point of consumption (Bosch and Philips, 2003). It is also apparent that the lived realities of the workers that maintain the nation's existing property stock receive even less attention than those involved in new construction. And the skills required for repair and maintenance are less prone to solution through supposed technical fixes such as pre-assembly and modularisation. Little enough is known about 'people and culture' in the context of new construction; even less is known about 'people and culture' in the context of repair and maintenance.

1.3 Construction industry's employment practices

It should be made clear that self-employment and a high reliance on subcon-tracting have always been evident within the construction sector and that both have perfectly legitimate roles to play. The issue is rather the degree to which industry relies on such recipes, and the extent to which the destructive side-effects can be alleviated. Problems of insecure work in construction are hardly new (cf. Tressell, 1914/1957), but the level of self-employment is inexorably linked with the government's willingness to promote labour market regulation. Of particular note since the mid-1970s is the dramatic reduction in directly employed labour by contractors (and subcontractors) in favour of extensive outsourcing. The few large construction firms which operate in the UK are virtually all exemplars of hollowed-out 'flexible' firms (cf. Atkinson, 1984), with very few directly employing operative labour. This situation has been exacerbated by decades of government policy which has done little to regulate industry employment, either through direct labour market governance or via public sector procurement influence. Of particular concern has been the way that contractors have systemi-cally sought to avoid statutory responsibilities through *bogus* self-employment, with significant implications for job security, training and health and safety.

The fractured workforce that emerges from the above is diverse and includes unskilled, craft, managerial, professional and administrative workers. These groups operate as part of a largely itinerant labour force, working in informal, *ad hoc* groups to complete short-term project objectives in a variety of workplace settings (see Cox and Thompson, 1997). This in turn creates additional pressures to ensure flexibility in both employment and working arrangements (Bresnen *et al.*, 1985; Yaw and Ofori, 1997; Belout, 1998; Loosemore *et al.*, 2003).

The adverse implications of industry fragmentation and casualisation are particularly apparent in the area of skills. Employers who do not offer any long-term job security are, understandably, reluctant to invest in developing their employees' skills for fear the relationship will not last long enough for them to realise that investment. Full-time permanent workers, who are already highly educated and highly skilled, are far more likely to receive training than their insecure and unskilled colleagues (Cully *et al.*, 1999; Forde *et al.*, 2005). Within the UK, construction and engineering construction are the only sectors for which the maintenance of a legally enforceable levy and grants system to support training investment has remained necessary (Druker and White, 1996). The Construction Industry Training Board (CITB) was one of the few bodies to survive when the British government dismantled most of the collective, sectoral arrangements for developing skills in the 1980s. This is just one example of many where the industry has retained aspects of a training infrastructure long since discarded by other sectors. Skills are clearly too important to leave to the vagaries of the free market.

The skills issue has grown in prominence within the UK in recent years, mainly in recognition of the low-skills equilibrium from which the country suffers (Keep and Mayhew, 1998; CEC, 2000 also known as Lisbon Strategy) and the alleged productivity gap between the UK and the US/mainland Europe (Bloom *et al.*, 2004). Research has consistently shown that the industry continues to encounter recruitment problems at both operative (Agapiou, 2002; Dainty *et al.*, 2005) and managerial/professional levels (Dainty and Edwards, 2003). Skills deficiencies are also apparent in terms of the *quality* of skills available to employers (Bloom *et al.*, 2004: 3). This may, in part, be attributable to the qualification structure which now underpins construction training. A number of writers (e.g. Callender, 1997; Grugulis, 2003) have criticised the National vocational qualification (NVQ) framework for its lack of academic rigour and dilution of technical content. In recent years there has been wider acknowledgement that skills development stems from a combination of formal education and work-based experience (Ford, 1992; Bloom *et al.*, 2004). This demands that employers are engaged in providing an appropriate training experience for their employees. However, as Grugulis (2003: 470) states, employers generally respond to immediate and hence short-term skills needs, with few planning skill formation and development over more than a two-year period. The unfortunate legacy of such employment decisions and the industry's continued reliance on outsourcing is an enduring paucity of high quality skills within the labour market.

1.4 Workforce diversity in the construction sector

An obvious implication of the industry's approach towards its workforce is the detrimental effect that it has on its external image and the knock-on effect that this has on recruitment. Within the UK there is a predominance of men in both craft and professional positions, and most of the women who embark on careers within the industry are engaged in ancillary or supporting roles (Fielden *et al.*, 2001). The industry's labour market is therefore segregated both horizontally and vertically by sex. A similar situation is also apparent in terms of the representation of ethnic minority workers. Only 2 per cent of the construction workforce comprises black and Asian people, compared with an economically active population of 5 per cent (Ansari *et al.*, 2002). This statistic defines construction as *the* most white-male dominated of all major UK industrial sectors. Although figures for the employment of disabled people are limited, evidence from Newton and Ormorod (2005) indicates a similarly low level of representation within the industry's larger companies, despite construction being one of the most likely industries to disable its workers. 'Macho' cultural images undoubtedly help to reproduce the perception that construction is unsuitable for groups currently underrepresented within the sector, which in turn reinforces the homogeneity of its workforce. For those minorities who do decide to enter construction, research evidence points to it providing a problematic context within which to develop a career (Dainty *et al.*, 2000; Ansari *et al.*, 2002).

Demographic change and workplace reform have rendered workforce diversity and equal opportunities – two of the most pressing challenges facing contemporary construction organisations. Restricting recruitment to under half of the population (white, non-disabled men) is likely to severely restrict organisational growth and development, while treating those from underrepresented groups differently is discriminatory and morally indefensible. In recent years, the sector has been placed under increasing pressure to diversify its workforce and to address its inequitable workplace environment. Arguments have tended to emphasise the business, ethical and legislatively defined drivers for the industry to embrace both equality and diversity, although there are important differences between the concepts. Whereas equality of opportunity tends to be externally initiated and assumes assimilation, diversity emphasises pluralism in its acknowledgement of the distinct cultural perspectives, influences and contributions of different groups (Kandola and Fullerton, 1998). Proponents of diversity suggest that the industry is under-utilising the full range of skills and talents in the population: that it should be possible for organisations to increase their efficiency and effectiveness by projecting a more pluralistic self-image and that a diverse workforce leads to a better-informed, more innovative and adaptable organisation which is closer to its customers (Bagilhole, 1997). The popularity of diversity as a concept represents a shift in emphasis from legislation in support of equal opportunities, towards a voluntarist policy of diversity supported by an espoused 'business case'. However, equality of opportunity remains central to any civilised society and to the quality of life for those who work within the industry.

Some of the reasons for the industry's lack of success in attracting a more diverse range of new entrants are most likely rooted in its unpopular image. Just as engineering is considered by many to be a dull, uncreative activity, associated with the so-called 'old-economy' (Malpas, 2000), construction is seen as a tough, heavy and dirty sector, ideas which have also affected wider perceptions of its cultural image as a 'macho' masculine domain (Gale, 1994). The decline of heavy manufacturing in the UK and the rise of service industries have combined to erode the privileged occupations of unionised white working class men. Societal changes since the Second World War and the advent of a post-industrial economy have undermined workplace certainties and have eroded opportunities for semi-skilled male workers with low levels of educational achievement. In some respects, the construction sector remains one of the last bastions of the traditional male working class that feels increasingly alienated by social change. The propagation of the industry's cultural image as a 'macho' domain therefore protects the vested interests of a social group that sees itself as being bypassed by post-modernist notions of the 'knowledge economy'. Such complexities, however, defy simplistic interpretation, and need to be understood within the broader context of societal change.

1.5   Barriers to change

The backcloth presented in this chapter thus far suggests a problematic context for improving the ways in which people are managed within the sector. It is clear that trite calls for the industry to 'change its culture' belie the nature of the concept, and the extent to which it is intertwined with the structural context within which it is enacted. Culture is a complex and multifaceted phenomenon which develops through ongoing social interaction within particular contexts (Meek, 1988; Barthorpe *et al.*, 2000). Change therefore is difficult to engender and the process is likely to be lengthy. This is not to infer that the culture of the industry is immutable, but attempts to *impose* cultural change usually result in resistance unless they take account of the structural landscape within which the industry's culture is manifested (see Bresnen and Marshal, 1998). The emphasis on people issues within cultural change has been extremely limited within the construction management research community. The overriding focus has been on research *for* management, rather than research *of* management. The popularity of techniques such as lean production and business process re-engineering are symptomatic of this attitude, as they reflect fashions in mainstream management which are themselves based upon a traditional culture of prescription and control (Richardson, 1996). This focus on project, organisational and outturn perform-ance requirements at the expense of human needs inevitably leads to employee dissatisfaction, reduced commitment, industrial conflict, increased turnover, more accidents, de-professionalisation, recruitment problems and a continued poor public image (Loosemore *et al.*, 2003). Even managerialist concepts of human resource management (HRM) remain seriously under-researched and there is little evidence of cross-disciplinary learning from other relevant fields (CRISP, 2002).

As was suggested above, the engineering orientation of the construction research community serves to reinforce the contentious assumption that 'culture' is an organisational variable that is subject to conscious manipulation. Indeed, the entire discourse of culture change within the construction industry appears strangely disconnected from the broader defining literature (e.g. Smircich, 1983; Willmott, 1993; Legge, 1994; Ogbona and Harris, 2002). The 'management of change' literature is replete with examples of failed attempts to change culture which have focused on the content of change programmes at the exclusion of under-standing the context and process of change (Pettigrew and Whip, 1991). Thus, a more nuanced understanding of construction culture and recognition that it is mutually constituted with its structure are needed if change is to be properly understood and responded to (see Seymour and Hill, 1996). To assume that structural imperatives can be transgressed by appealing to a new common culture oversimplifies the ingrained nature of industry practice described within this chapter.

Given the supposed severity of the skills crisis and the obvious failings of the industry's approach towards people management, it is reasonable to assume that such issues would provide a focus for industry improvement activity in both research and practice. In practice terms, the overt response to industry-wide skills shortages and fragmentation has been a series of government-backed reports, exhorting the sector to address its lamentable performance on people management issues as part of a wider 'performance improvement agenda' (Latham, 1994; Egan, 1998). Most recently, the industry's 'Strategic Forum' has laid down chal-lenging targets for the improvement of its people management practices within its 'Accelerating Change' report (Strategic Forum for Construction, 2002). Unfortunately, this agenda consists largely of simplistic exhortations to the construction industry to address its past failings with little acknowledgement of the aforementioned labour market constraints which impede change. In the meantime, most firms within the industry have acted to alleviate the skills shortage in the tried-and-tested manner: by increasing their reliance on migrant workers. The casualised structure of the construction labour market means that rapid increases of migrant workers can be absorbed entirely below the radar of official statistics. British construction firms are currently especially adept at freeloading on the training regimes of Eastern European countries.

Flows of labour between poorer and richer economies have, of course, always been a feature of the construction sector (ILO, 2001). In previous eras, the UK was heavily reliant on Irish labour. Migrant workers have always added much to the UK construction sector and to the dynamism of society at large. However, too often they are employed on different terms and conditions from the domestic workers whom they are working alongside. It is also worth making the point that British workers continue to work overseas in significant numbers, especially professional and technical groups. Nevertheless the extent to which the UK can morally continue to rely on less developed economies to train its construction operatives is debatable.

An alternative strategy adopted by the industry has been the attempt to mitigate its people's needs by redesigning jobs to reduce the skills required, or to shift them to other places within the construction process. Indeed, suggestions for 'innovating away' the skills shortage have formed popular discussion topics for many industry commentators and policy bodies in recent years. Concerns about skills development are often emasculated by a managerialist discourse that mobilises opaque concepts such as dynamic capabilities (Teece *et al.*, 1997), core competencies (Prahalad and Hamel, 1990) and organisational learning (Chan *et al.*, 2005) as routes to enhance the productive capacity of organisations. Yet, despite the rhetorical claims to enhance flexibility and continuous improvement, a lack of any proven relationship with performance remains (Scarbrough, 1998). There is also a noticeable lack of interest in the way the enactment of such ideas shapes the lived reality of the workforce. Attempts to circumvent skills needs by shifting the locus of production away from the site may actually exacerbate skills gaps in the future (Braverman, 1974). Moreover, given the very limited market penetration of offsite construction methods and materials (especially in maintenance and refurbishment – see Goodier and Gibb, 2005), the impact of technical solutions on skills shortages appears limited.

1.6   Fresh perspectives on people and culture in construction

This book has its origins in a transdisciplinary Economic and Social Research Council (ESRC) funded research seminar series. This series of critical seminars aimed to bring together leading researchers from a range of different disciplines in order to facilitate an understanding of the employment and cultural issues facing the construction industry. Many of the 16 chapters which follow this introduction are based on presentations made at the seminars, together with invited contributions from international researchers whose insights either enhance or act as counterpoints to the UK-based contributions. Each chapter offers either a critical or an empirically informed insight into the ways in which the industry manages people, or helps to foster understanding of the culture of the sector and its impact on those that work within it. Although all 16 chapters are very different, their underlying point of connection is that each offers a fresh insight into the ways in which the industry's employment practices mutually reinforce each other. The net effect is to create the cultural climate which so many see as representing the most fundamental barrier to change within the industry.

The following chapters are divided into three broad parts. Part I examines the employment context of the industry's labour market. It examines the emergence of the industry's informal and casualised employment practices alluded to above, charting the legislative and employer-driven issues which have led to the problematic employment climate which underpins many of the industry's failings. Comparative insights are provided both from continental Europe and from further afield to demonstrate the impact that more enlightened practices can have on employment within the sector. Following this contextual backdrop to employment within the industry, Part II of the book examines the corollaries of the employment culture from the perspective of the people that work within it. These chapters highlight the ways in which the structure and culture mutually reinforce each other to ingrain the issues emerging from the earlier chapters. Part III of the book presents a collection of chapters which examine critically the likely impact of industry trends on employment issues and cultural change. Consideration is also given to the implications of any change in the sector's employment relations climate given the problematic context presented earlier in the book. Some of these chapters adopt a forward-looking orientation, examining the likely impact of the industry's current developmental trajectory on its future employment practices and culture. A summary of the individual chapter contents is provided below.

Part I: The construction employment context

This section examines the employment context of the industry's labour market. It charts the legislative and employer-driven issues which have created the current employment climate that underpins many of the industry's failings. Comparisons with other countries demonstrate how such a situation is not an inevitable result of construction activity, but of particular policy decisions, industry structures and employer characteristics which shape industry culture and practice.

In Chapter 2, Chris Forde and Robert MacKenzie report on survey-based research which has examined the implications of high levels of contingent labour in the sector. Their work reveals how, regardless of which contractual form is used, labour shortages have driven a reliance on outsourcing and self-employment within the sector, which has in turn led to further casualisation of the employment relationship. The implications of this self-fulfilling spiral are examined in relation to recruitment and training. Employers appear to abdicate responsibility for ensuring the recruitment and development of a suitably qualified workforce. The authors suggest that the use of contingent labour fails to offer a basis for maintaining the sector's long-term skills profile. Indeed, recent reports suggest a paucity of self-employed labour in skilled construction trades. The damaging legacy of contingent employment is shown in relation to its impact on its productive capacity, safety and training standards.

In Chapter 3, Ani Raidén, Michael Pye and Joanna Cullinane explore some of the implications of the contingent employment revealed by Forde and MacKenzie in Chapter 2. They argue that the sector's history of atypical employment practices and its reliance on small firms offer a unique arena within which to examine employment relations and the legal, social and psychological dimensions of the employer/employee relationship. They use a range of secondary data sources to examine whether the employment culture of the sector effectively represents a multi-tier contracting arrangement or a misinterpreted 'contract' of employment. They suggest that the concept of multi-tier contracting masks a reality of workers being starved of employment benefits and job security. This, they argue, feeds the low-cost, low-skill, low-productivity culture which pervades the sector. It therefore has fundamental implications for the competitiveness of the industry, particularly in comparison to more regulated construction sectors in mainland Europe (see Chapters 6 and 7).

Several chapters explore the employment and labour market governance implications of the casualised employment picture painted by the earlier contributions.

In Chapter 4, Craig Barker examines the difficulties that self-employment and casualisation present from an employment law perspective. Key legal cases and tests developed by the courts to deal with labour-only subcontracting and an over-reliance on agency workers have done little to help the position of those working under such contracts. Legal precedent is complex and inconsistent, which reflects the nature of the sector with regard to employment issues. Barker concludes that the legal process is likely to have little impact on the problems of self-employment and casualisation. Ultimately, the industry must decide whether its interests are served by a continued reliance on large-scale casualisation.

The industrial relations implications of the industry's labour market context are particularly significant. In Chapter 5, Janet Druker provides a fascinating exposition around the apparent tension between two apparently contradictory perspectives. The issue of concern is the concurrent reliance of the sector on self-employment on the one hand, and collective bargaining as a mechanism to regulate wage rates on the other. She argues that employers maintain this seemingly paradoxical situation as a form of risk management, reducing employee voice and involvement, while simultaneously maintaining the machinery necessary to ensure sector-wide multi-employer collective agreements. This appears symptomatic of the industry's focus on damage limitation and cost containment which are the hallmarks of its risk management approach. If maintained, such an approach effectively undermines the involvement of the workforce in mutual interdependence and trust, an approach advocated as part of the Egan agenda for harnessing skills, realising potential and raising productivity.

This section is completed by two chapters which provide international perspectives on employment regulation and its implications for the culture of the industry. In Chapter 6, Linda Clarke and Georg Herrmann compare the divisions of labour in the British and German construction industries. They reveal that the divisions between professional and operative labour are more entrenched in Britain than in Germany, whereas the vertical skills structure is far less permeable. The British model of weak institutional linkages, individualisation of the employment relationship and managerial prerogative is contrasted with Germany's highly regulated labour market and collective employment relations. The British model reflects a preoccupation with firms, sites and costs rather than employment relations. This reinforces the divide between trade operatives and professions which, in turn, hampers the productive effort. The chapter's conclusions not only have resonances with all of the preceding contributions, but also reveal a potential way forward for the industry by showing the benefits that labour force integration and collective regulation can bring.

In Chapter 7, Christian Koch extends the themes of both international comparison and maintaining positive industrial relations raised in Chapters 5 and 6, although this time in relation to adapting to industrial change in the context of the Danish construction industry. It focuses on explaining how the Danish construction workers' trade union has helped to shape the adoption of lean principles, thereby showing how they can contribute to change. Although he argues that the Danish union's experiences should be viewed as embedded in the context within which they are played out, such approaches can still be seen as an attempt to construe and re-construe basic values of egalitarianism, fraternity and solidarity, all of which can be seen from the previous chapters to be lacking in the UK construction sector.

Part II: Implications for people management practices and culture

The second section of the book examines the implications of the climate revealed in Part I on the employment practices and culture of the sector. These chapters highlight the ways in which the structure and culture of the industry mutually reinforce an employment climate, which in turn militates against change within the sector.

Chapter 8 examines the ways in which the employment culture of the industry detrimentally affects those working within it. Katherine Sang, Andrew Dainty and Stephen Ison explore the experiences of those working within the industry in relation to the concept of psychological well-being. They draw upon a range of secondary sources to examine how the nature of the industry's products, coupled with the cultural norms which define the nature of work and commitment within the sector, lead to high workloads, long hours, stress and psychological burnout. The consequences of these symptoms include absenteeism, low levels of performance and a decline in performance and productivity. The authors suggest that the industry should change its reliance on lowest bid contracting and contingent labour if the sector is to be rendered a more pleasant and attractive place to work. Recommendations for showing greater concern for employee needs and for recognising employee voice in project deployment decisions chime with the sentiments discussed in Chapters 6 and 7 with regard to employee relations.

Chapters 9 and 10 explore the implications of the people management practices of the sector in relation to the position of under-represented groups within the sector. In Chapter 9, Andrew Caplan examines the structural reasons for the under-representation and under-achievement of minority ethnic people in the industry. Drawing upon the findings of four research projects, he suggests that fragmentation and an ingrained culture have been largely responsible for the industry escaping the scrutiny to which other sectors have been subjected. The result is that institutionalised racism acts as both a barrier to entry and an inhibitor of progression should access to the industry be achieved. He argues that challenging the current status quo will demand change on many fronts. These include client influence through procurement practices and leadership direction at senior and middle management levels within the industry. Ultimately, however, the key relationships for ethnic minority construction workers are with their white colleagues. Addressing ingrained and institutionalised attitudes will require a lengthy transition.

Chapter 10 switches attention to the position of women in the industry. Ann de Graft-Johnson, Sandra Manley and Clara Greed explore how the architectural profession can begin to close the gender gap in architectural practice. They argue that deep structures of discrimination within the industry have proved resilient to change. They contend that for the architectural profession these problems are not UK specific, but that women face particularly acute difficulties given that the prevailing climate is one of a bullying, 'macho' culture with poor work-life balance for those within the profession. They cite the experience of Finland as an example of how a more appropriate balance of men and women can be achieved. Here, an anonymous bidding system has led to a higher ratio of architectural competitions being won by female architects. As with Chapters 6 and 7, it seem that the UK has much to learn from its European competitors in terms of how to manage people in a fair and egalitarian way.

This section is again completed with some international perspectives on the cultural differences between different sectors. In Chapter 11, the points of comparison are provided by the German, Austrian and Australian industries. Karolina Lorenz and Marton Marosszeky explore the differences between designers and contractors from Austria and Germany with those from Australia. The study reveals significant differences in the way in which the industry is regulated and in the way in which organisations are managed. These differences have had marked effects on the cultural climate which emerges within the two countries' construction industries. Trades unions are more powerful in Australia and influence the ways in which labour-related issues are negotiated. As such, health and safety issues receive higher priority within the Australian industry, as does the documentation of management procedures. However, project specific contingent employment remains the norm within Australia, whereas the Austrian employment context is organisationally determined. The findings presented within this chapter emphasise that the employment relations practices and cultural climate which define the UK construction industry are not so much a product of the nature of construction work as rooted in the approach towards industrial relations and employment. It shows the importance of institutional influences which provide the governance structure for the industry within different countries.

Part III: Critical perspectives on construction employment and change

The final section of the book presents a collection of chapters which examine critically the likely impact of current industry practices on people and employment issues from a variety of different perspectives. Attention is also given to the implications of any change in the employment relations climate of the sector given the problematic backdrop presented in the preceding chapters. Consideration also extends to the impact the industry's current developmental trajectory may have on its future employment practices and culture.

In Chapter 12, David Langford and Andrew Agapiou draw upon a range of data sources in order to explore the impact of the enlargement of the European Union on the construction labour market. They argue that accession is likely to increase the mobility of Eastern European labour, but that this will bring with it risks in terms of the prevalence of 'informal' employment, wage levels, health and safety, and productivity. Clearly, a large influx of foreign workers may ultimately have an effect on the culture of the industry and its employment, but whether the levels of migration are likely to reach the levels necessary to induce a cultural shift is a moot point. This chapter deals with a highly topical issue for the UK industry as it struggles to cope with high levels of demand and a declining indigenous workforce.

In Chapter 13, Simon Burtonshaw-Gunn and Bob Ritchie examine the impact of the growing practice of supply chain management within the industry on the sector's cultural norms, behaviours and practices. They identify that a willingness to engage in problem resolution and continuous improvement is weak within the industry. Their research into partnering and supply chain management shows that while clients and contractors recognise the need to change the nature of their relationships, there is much less clarity on how this can be achieved. They further suggest that those involved in supply chain management activities need to engage in new patterns of behaviour, especially in the development of trust, which underpins the success of long-term relationships. Given the fragmentation and cultural separation of the various parts of the supply chain discussed in Chapters 1–6, realising the full benefit of supply chain management seems some distance away.

Chapter 14 examines the potential value of involving employees in performance and productivity improvements. Paul Chan and Ammar Kaka report on case study research which examined two separate projects with a view to establishing the most appropriate ways of ensuring engagement with performance improvement approaches. In both cases they note how supervisors adopt a benign paternal role in relation to their relations with operatives. They suggest that the development of a strong bond between white-collar and blue-collar workers represents a cornerstone of a positive employment relationship, but that ways of achieving such a bond tend to be inimitable and hence must be developed in response to particular project circumstances. What is clear, however, is that managers must accept that operatives possess tacit knowledge that could unlock productivity improvements and the need to tap into this knowledge appropriately. This recognition demands efforts to break down some of the barriers alluded to in Chapters 5 and 6 in order to overcome years of division and disconnection between the industry's occupational constituencies.

Chapter 15 broadens the horizons of the text to examine the engineering construction sector's approach to industrial relations. Neil Ritson examines the implications of client management of contract maintenance within the Oil and Gas sector. Although this sector has many parallels with construction maintenance and repair, the structure of the sector is such that managers are able to create a single union agreement for the contractor's various workforces. This policy was adopted as a deliberate attempt to engender change in the endemic culture on a refinery site. Ritson found that a kind of managerial ideology prevailed in pursuit of labour-based savings as opposed to capital investment. The implications for the less stable environment of the construction industry are discussed. Although a national agreement is more suited to the construction sector's geographically dispersed context, the argument in favour of single union agreements deserves serious consideration.

The final two chapters address both the issue of cultural change and the susceptibility of the industry to adapting modernising work practices with questionable implications for people working within the sector. In Chapter 16, Paul Fox examines the extent to which a construction 'culture' really exists. He argues the existence of an industry-wide culture which pervades many different countries and contexts and which has become accepted within the prevailing discourse surrounding industry improvement and change. He argues that the ability to change this culture in a way which improves the industry is to some extent determined by the maturity of the individual country concerned: those with a weaker institutional context will require greater governmental intervention. Fox suggests, however, that directed cultural change may not be possible in an industry such as construction; instead, the role of research is to monitor cultural evolution in order to foster understanding of how to respond better to the demands of the future.

In the final chapter, Steven McCabe casts a critical eye over the industry improvement rhetoric which has pervaded the sector over the last few years. In Chapter 17, the origins of organised labour in construction are charted, with a particular emphasis on the rise and fall of trade unionism within the sector. The argument is that the industry's present problems are rooted in two centuries of poor treatment of people working within the industry. This provides a backcloth for an examination of post-Egan initiatives such as 'Respect for People'. These, it is argued, represent little more than an attempt by employers to deal with the threat of increased power for workers. McCabe's highly thought-provoking chapter could have easily been positioned at the beginning of this text to frame the issues in need of change within the sector. However, making this the concluding chapter of the book underscores the scale of the problem facing the industry if it is to develop its people management practices in a way which has meaning for those who work within it.

Summary

The preceding description of the chapters contained within this book raises a number of recurring issues worthy of discussion which may point towards new foci for industry practice and research in the future. Probably the most significant recurring theme concerns the fragmented nature of the industry and the resultant tendency to outsource and rely on contingent labour. Despite early optimism that networks of small firms and independent workers would facilitate specialisms free from bureaucratic regulation (Castells, 1996), the reality of outsourcing and subcontracting has been much less glamorous. It creates many problems for organisations attempting to exert control and monitor quality over processes that they no longer manage directly (see Grugulis *et al.*, 2003). The plethora of different contractual arrangements which govern the operation of project teams on site further fracture an already fragmented production process. They also mean that the responsibility for skills development is devolved, often repeatedly, until it rests with the individual unskilled worker who is least likely to have information, resources or inclination to embark on a lengthy training programme. It is unsurprising, given the trend towards outsourcing and self-employment in recent years, that skills shortages in the British construction industry are so severe (Hillage *et al.*, 2002). It is even less surprising that repeated calls for the industry to improve its people management practices are ignored. The most vocal and most publicised requests come from official reports and industry reviews that suggest how improvements in people management practices could impact on efficiency, productivity and cost-effectiveness. Few are effective, however, with many resulting in little more than rhetorical exhortations to change. It seems that in this sector neither markets nor morality helps to improve people management practices (Dainty *et al.*, 2007). It is against this background that the chapters within this book were commissioned.

And so where do the collective insights provided by this book leave our understanding of the industry's employment practices and culture and how can it be changed in the future? Clearly, it is difficult to draw firm conclusions from a book such as this. The chapters mobilise a variety of different discourses and approach issues from different frames of reference. These are themselves rooted in socially constructed perspectives on the industry and the ways in which people work and interact which may, or may not, accord with those of the reader. None the less, if we return to the critical but pragmatic standpoint from which we started, it is important that the insights provided are used to shape future action if the employment practices of the industry are to be improved. If it is accepted that organisational culture is critical to organisational success, and HRM is critical in managing and changing cultures (see Redman and Wilkinson, 2006: 21), then it is reasonable to suggest that there are better ways of managing people which could, over time, achieve the cultural change seen as so crucial by industry commentators.

What is clear from the contributions provided within this book is that mitigating the industry's skills challenge and inducing cultural change will require a concerted and joined-up effort over a sustained period. Several issues have emerged which offer directions for future efforts in both research and practice in this regard. First, there needs to be a much better understanding of the industry's employment practices and their impacts on the workforce. All too often the perspectives of those that are at the receiving end of the industry's attempts to improve its performance, those that work in the industry, receive scant consideration from the advocates of change. A second issue concerns the need to establish ways of preventing the abdication of responsibility for skills development on the part of employers. Without a genuine and sustained commitment to workforce investment, construction careers and occupations will remain a last resort for many of those whom the industry needs to attract. A third issue concerns the need to bridge the gulf between the operative and the managerial workforce. Much of the attention has been focused on the managerial constituency within the industry at the expense of the productive majority. There needs to be a far greater emphasis on understanding the perspective of construction operatives and how they utilise skills within their occupational roles. Such an understanding is essential if the tacit knowledge of the workforce is to be tapped and employee relations improved (see Cully *et al.*, 1999; Stasz, 2001). Ultimately, this will require a dismantling of the institutionalised structures which maintain existing intra-sectoral divisions and which have been shown to stymie change within the sector. A final crucial issue concerns the need to re-focus industry attention on people issues and away from performance outcomes. The overwhelming emphasis on project, organisational and outturn performance requirements at the expense of human needs has played an important part in shaping the problems reviewed in this book. Such problems include employee dissatisfaction, reduced commitment, industrial conflict, propensity for accidents and the poor public image for which the industry has become renowned. In no small respect, the construction industry is shaped by the policies of its leaders and the regulatory frameworks (or lack of them) set by the government. For too long the industry agenda has emphasised short-term 'efficiency' with little recognition of the long-term side effects. We see this book as a first step at redressing the imbalance in research and industry priorities.

Although this chapter has signposted the reader towards key issues facing the sector, it is incumbent on the individual to make sense of the contributions from their own perspective. A key feature of this book is that it is *not* driven by a desire to find ways of improving the industry's performance. Rather, the reader is encouraged to assimilate with their own experiences and perceptions the multiple perspectives and discourses provided and to draw their own conclusions as to what they might mean for future employment practice. However, it is hoped that the thought-provoking chapters presented in this book will stimulate further discussion and research into the complex interaction between people and the environment in which they work within the construction sector. It provides a theoretical and empirical basis for the study of people and culture issues, and for challenging some of the industry's ingrained assumptions.

References

Agapiou, A. (2002) Perceptions of gender roles and attitudes toward work among male and female operatives in the Scottish construction industry, *Construction Management and Economics*, 20, 697–705.

Ansari, K.H., Aujla, A., Caplan, A., Chowdhury, H., Dainty, A.R.J., Gilham, J., Jackson, J., Kathrecha, P. and Mann, P. (2002) *Retention and Career Progression of Black and Asian People in the Construction Industry*. Report commissioned by the CITB, Royal Holloway, University of London.

Atkinson, J. (1984) *Emerging UK Work Patterns, in Flexible Manning – The Way Ahead*, IMS Report, No. 88, Institute of Manpower Studies, Brighton.

Bagilhole, B. (1997) *Equal Opportunities and Social Policy: Issues of Gender, Race and Disability*, London: Longman.

Barthorpe, S., Duncan, R. and Miller, C. (2000) The pluralistic facets of culture and its impact on construction, *Property Management*, 18(5), 335–351.

Belout, A. (1998) Effects of human resource management on project effectiveness and success: Toward a new conceptual framework, *International Journal of Project Management*, 16(1), 21–26.

Bloom, N., Conway, N., Mole, K., Möslein, K., Neely, A. and Frost, C. (2004) *Solving the Skills Gap*, summary report from a CIHE/AIM management research forum.

Bosch, G. and Philips, P. (2003) Introduction, in G. Bosch and P. Philips (eds), *Building Chaos: An International Comparison of Deregulation in the Construction Industry*, London: Routledge, pp. 1–23.

Braverman, H. (1974) *Labour and Monopoly Capitalism*, New York: MRP.

Bresnen, M. and Marshall, N. (1998) Partnering strategies and organisational cultures in the construction industry, in W. Hughes (ed.), *Proceedings*, ARCOM 14th Annual Conference, University of Reading, 9–11 September 1998, Vol. 2, pp. 465–476.

Bresnen, M. and Marshall, N. (2001) Understanding the diffusion and application of new management ideas in construction, *Engineering, Construction and Architectural Management*, 8(5/6), 335–345.

Bresnen, M.J., Wray, K., Bryman, A., Beardsworth, A.D., Ford, J.R. and Keil, E.T. (1985) The flexibility of recruitment in the construction industry: Formalisation or recasualisation? *Sociology*, 19(1), 108–124.

Bresnen, M.J., Goussevskaia, A. and Swan, J. (2004) Embedding new management knowledge in project-based organizations. *Organization Studies*, 25(9), 1535–1555.

Briscoe, G. (2006) How useful and reliable are construction statistics? *Building Research & Information*, 34(3), 220–229.

Briscoe, G., Dainty, A.R.J. and Millett, S.J. (2000) The impact of the tax system on self-employment in the British construction industry. *International Journal of Manpower*, 21(8), 596–613.

Callender, C. (1997) *Will NVQs Work? Evidence from the Construction Industry*. Report by the Institute of Manpower Studies (IMS) for the Employment Department Group.

Cannon, J. (1994) Lies and construction statistics, *Construction Management and Economics*, 12(4), 307–313.

Castells, M. (1996) *The Rise of the Network Society*, Oxford: Blackwell.

CEC (2000) *Commission Staff Working Paper: A Memorandum on Lifelong Learning*, Brussels: European Commission.

Chan, P., Cooper, R. and Tzortzopoulos, P. (2005) Organisational learning: Conceptual challenges from a project perspective. *Construction Management and Economics*, 23, 747–756.

Cherns, A.B. and Bryant, D.T. (1984) Studying the client's role in construction, *Construction Management and Economics*, 2, 177–184.

Cox, A. and Thompson, I. (1997) 'Fit for purpose' contractual relations: Determining a theoretical framework for construction projects, *European Journal of Purchasing and Supply Management*, 3, 127–135.

CRISP (2002) *Culture and People in Construction – A Research Strategy*, London: CRISP.

Cully, M., Woodland, S., O'Reilly, A. and Dix, G. (1999) *Britain at Work: As Depicted by the 1998 Workplace Employee Relations Survey*, London: Routledge.

Dainty, A.R.J and Edwards, D.J. (2003) The UK building education recruitment crisis: A call for action. *Construction Management and Economics*, 21, 767–775.

Dainty, A.R.J., Bagilhole, B.M. and Neale, R.H. (2000) A grounded theory of women's career under-achievement in large UK construction companies. *Construction Management and Economics*, 18, 239–250.

Dainty, A.R.J., Ison, S.G. and Briscoe, G.H. (2005) The construction labour market skills crisis: The perspective of small-medium sized firms, *Construction Management and Economics*, 23, 387–398.

Dainty A.R.J., Grugulis, I. and Langford, D.L. (2007) Understanding construction employment: The need for a fresh research agenda, *Personnel Review*, 36(2) (in press).

Druker, J. and White, G. (1996) *Managing People in Construction*, London: Institute of Personnel and Development.

DTI (2005a) *Construction Statistics Annual*, London: Department of Trade and Industry.

DTI (2005b) *Small and Medium-sized Enterprise (SME) Statistics for the UK*, London: Department of Trade and Industry.

Egan, J. (1998) *Rethinking Construction: The Report of the Construction, Task Force,* London: Construction Industry Council.

Fielden, S.L., Davidson, M.J., Gale, A.W. and Davey, C.L. (2001) Women Equality and Construction, *Journal of Management Development*, 20(4), 293–304.

Ford, G.W. (1992) Integrating technology work organisation and skill formation, in M. Costa and M. Easson (eds), *Australian Industry: What Policy*? Sydney: Pluto Press.

Forde, C., MacKenzie, R. and Robinson, A. (2005). Firm foundations? Contingent labour and employers' provision of training in the UK construction industry. Presented at Second International Conference on Training, Employability and Employment, 21–23 September, Monash University Centre, Prato.

Gale, A.W. (1994) Women in non-traditional occupations: The construction industry. *Women in Management Review*, 9(2), 3–14.

Goodier, C.I. and Gibb, A.G. (2005) *The Value of the UK Market for Offsite*, Loughborough University, report for *BuildOffsite*, London.

Grugulis, I. (2003) The contribution of National Vocational Qualifications to the growth of skills in the UK. *British Journal of Industrial Relations*, 41(3), 457–475.

Grugulis, I., Vincent, S. and Hebson, G. (2003) The rise of the 'network organisation' and the decline of discretion, *Human Resource Management Journal*, 13(2), 45–59.

Harvey, M. (2003) Privatization, fragmentation and inflexible flexibilization in the UK construction industry, in G. Bosch and P. Philips (eds), *Building Chaos: An International Comparison of Deregulation in the Construction Industry,* London: Routledge, pp. 188–209.

Hillage, J., Regan, J., Dickson, J. and McLoughlin, K. (2002) *Employers Skill Survey 2002*. In Research Report, Department for Education and Skills, Nottingham.

Hillebrandt, P.M. (1988) *Analysis of the British Construction Industry*, London: Macmillan.

Hillebrandt, P.M. (2000) *Economic Theory and the Construction Industry*, 3rd ed., London: Macmillan.

ILO (2001) *The Construction Industry in the Twenty-First Century: Its Image, Employment Prospects and Skill Requirements*, Geneva: International Labour Office.

Ive, G.L. and Gruneberg, S.L. (2000) *The Economics of the Modern Construction Sector*, Basingstoke: Macmillan.

Kandola, R. and Fullerton, J. (1998) *Diversity in Action. Managing the Mosaic*, London: Institute of Personnel and Development.

Keep, E. and Mayhew, K. (1998) *Was Ratner Right? Product Market and Competitive Strategies and Their Links with Skills and Knowledge.* Employment Policy Institute Economic Report, 12(3).

Latham, M. (1994) *Constructing the Team*, London: HMSO.

Legge, K. (1994) Managing culture: Fact or fiction, in K. Sisson (ed.), *Personnel Management: A Comprehensive Guide to Theory and Practice in Britain,* Oxford: Backwell, pp. 397–433.

Loosemore, M., Dainty A.R.J. and Lingard, H. (2003) *Managing People in Construction Projects: Strategic and Operational Approaches,* London: Taylor & Francis.

Malpas, R. (2000) *The Universe of Engineering: A UK Perspective.* Available at: http://www.engc.org.uk/publications/pdf/Malpas_report.pdf.

Meek, V.L. (1988) Organisational culture: origins and weaknesses, *Organisation Studies,* 9(4), 453–473.

Newton, R. and Ormorod, M. (2005) Do disabled people have a place in the UK construction industry? *Construction Management and Economics,* 23(10), 1071–1081.

Office for National Statistics (2003) *UK Standard Industrial Classification of Economic Activities 2003,* London: Stationery Office.

Ogbonna, E. and Harris, L.C. (2002) Organizational culture: A ten year, two-phase study of change in the UK food retailing sector, *Journal of Management Studies,* 39(5), 673–706.

Pearce, D. (2003) *The Social and Economic Value of Construction: The Construction Industry's Contribution to Sustainable Development,* London: nCRISP, Davis Langdon Consultancy.

Pettigrew, A. and Whipp, R. (1991) *Managing Change for Competitive Success,* Oxford: Blackwell.

Prahalad, C.K. and Hamel, G. (1990) The core competence of the corporation. *Harvard Business Review,* May–June, 79–91.

Redman, T. and Wilkinson, A.J. (2006) *Contemporary Human Resource Management: Text and Cases,* 2nd edn, Harlow Essex: Pearcon.

Richardson, B. (1996) Modern management's role in the demise of sustainable society, *Journal of Contingencies and Crisis Management,* 4(1) March, 20–31.

Scarbrough, H. (1998) Path(ological) dependency? Core competencies from an organisational perspective. *British Journal of Management,* 9, 219–232.

Seymour, D. and Hill, C. (1996) The first-line supervisor in construction: A key to change? *Proceedings of the 13th Annual ARCOM Conference,* Vol. 2, 655–664.

Smircich, L. (1983) Concepts of culture and organizational analysis, *Administrative Science Quarterly,* 28(3), 339–358.

Stasz, C. (2001) Assessing skills for work: Two perspectives. *Oxford Economic Papers,* 3, 385–405.

Strategic Forum for Construction (2002) *Accelerating Change,* London: Rethinking Construction.

Teece, D., Pisano, G. and Shuen, A. (1997) Dynamic capabilities and strategic management. *Strategic Management Journal,* 18(7), 509–533.

Tressell, R. (1914/1957) *The Ragged Trousered Philanthropists,* Moscow: Foreign Languages Publishing House.

Willmott, H. (1993) Strength is ignorance: Slavery is freedom: Managing culture in modern organisations, *Journal of Management Studies,* 30(4), 515–552.

Winch, G. (1998) The growth of self employment in British construction. *Construction Management and Economics*, 16(5), 531–542.

Yaw, A.D. and Ofori, G. (1997) Flexibility, labour subcontracting and HRM in the construction industry in Singapore: Can the system be refined? *The International Journal of Human Resource Management*, 8(5), 690–709.

2 Concrete solutions?

Recruitment difficulties and casualisation in the UK construction industry

Chris Forde and Robert MacKenzie

Abstract

There has been long-standing interest in issues around the nature of employment in construction. Historically, high levels of contingent labour use in the sector, including the use of subcontractors, agency workers and self-employed workers, have led to debate over the linkages between casualisation and labour reproduction. Recent skills shortages and recruitment difficulties in the sector have intensified this debate further. This chapter examines the relationship between recruitment difficulties, the use of contingent labour and skills reproduction in the construction sector. Drawing on results from an original survey of construction employers conducted in 2002, the chapter examines recruitment difficulties among construction employers, their responses to these difficulties and their use of contingent labour. The chapter also examines the relationship between contingent labour and training. The results of the survey point to overall labour shortages in the construction sector, regardless of contract type, and suggest that further casualisation of the employment relationship and the increased use of contingent labour in the UK construction industry offer no long-term solution to labour and skills shortages in the sector.

Keywords: casualisation, subcontracting, agency labour, self-employment, recruitment, training and skills shortages.

2.1 Introduction

There has been long-standing interest in issues around the nature of employment in construction. Employers in the construction industry have historically relied on a range of contract forms – including direct employment relationships, subcontracting arrangements, self-employment and agency labour – to regulate the supply of labour. These have often been labelled 'contingent' forms of employment (see, for example, Barker and Christensen, 1998) to reflect the job insecurity or unpredictability typically associated with such arrangements for workers. The high levels of 'contingent' labour in the industry have led to concerns about the implications for the reproduction of skills in the industry. The high levels of skills shortages and recruitment difficulties experienced in recent years by many construction employers have intensified this debate, and led to a closer examination of the linkages between contingent labour usage, recruitment and labour reproduction. Industry bodies, such as the Construction Industry Training Board (CITB), have argued that the long-term reliance on contingent labour may be contributing to long-term skills shortages in the sector. Unions representing workers in the sector have also pointed to the negative consequences of the use of contingent labour, arguing that the culture of casualisation has undermined industry-wide initiatives, including the implementation of safety and training standards. In contrast, others have argued that the use of contingent labour allows firms effectively to meet their demands for production skills, since these can be bought in from the market as and when they are needed.

The chapter examines in more detail the linkages between recruitment difficulties, the reproduction of skills and the use of contingent labour in the construction industry. Drawing on results from an original survey of construction employers conducted in 2002, recruitment difficulties experienced among construction employers and their responses to them are explored. This exploration reveals the importance of both 'internal' and 'external' responses to these difficulties, the former involving the upgrading of the skills of existing staff, and the latter including the increased use of contingent labour. The survey, in line with other studies, points to the widespread use of contingent labour in the sector. It also identifies that many employers have increased their use of these contingent contract forms in recent years, although this has not generally reflected a deliberate strategy on the part of construction employers. With many construction employers also experiencing difficulties recruiting contingent labour, the problem appears to be a more general shortage of labour in the industry, regardless of contract type. The survey also reveals that the extent and scope of training for contingent labour is often more limited than that offered to directly employed staff. While many firms appear to be seeking recourse to contingent labour as a means of overcoming perceived labour shortages, these results suggest that further casualisation of the employment relationship and the increased use of contingent labour in the UK construction industry offer no long-term solution to the problem.

2.2 Background

Interest in the linkages between recruitment difficulties, skills shortages and labour reproduction in the construction industry in the UK is not new. Over a century ago, evidence from the 1890–94 Royal Commission on Labour, for example, revealed concerns among employers, unions and the government about labour shortages and skills deficiencies in the sector (Royal Commission, 1892). More recently, national surveys have pointed to the high concentration in the construction sector of recruitment difficulties that stem from skills shortages, compared with other sectors (Hillage *et al.*, 2002). In 2002, 15 per cent of construction employers reported a 'skill shortage' job vacancy (defined as a hard-to-fill vacancy resulting from low numbers of applicants with required skills, a lack of applicant work experience or a lack of required qualifications) compared to 8 per cent of all establishments. The same survey revealed that only a minority of establishments reporting such vacancies had responded by increasing or expanding trainee schemes, with the more commonly adopted responses being to increase salaries, job advertising or expanding recruitment channels (ibid.).

Concerns over skills reproduction have been heightened because of the construction industry's structural characteristics. Firms within the industry have historically used the panoply of contract forms – including direct employment relationships, subcontracting arrangements, self-employment and agency labour – to meet their labour requirements. Contingent labour is commonly utilised, as in other sectors, for the 'traditional' reasons of meeting peaks in demand, for one-off tasks, or to provide specialist skills. The greater reliance on contingent contract forms in construction may reflect the uncertain nature of demand in the sector, and the greater need for specialist skills at particular points in the production process. Yet it has also been argued that the 'casualisation' of employment in construction reflects employers' attempts to avoid the costs associated with some basic employment rights that accrue to employees but not to 'workers'. Under this argument, a large proportion of the increase in self-employment that occurred from the 1960s onwards would be seen as 'bogus' or 'false' self-employment, in that workers may have been compelled to accept classification as self-employed in order to get a job (Harvey, 2001; UCATT, 2005). Reported levels of usage of some of these contingent forms of labour in the construction industry are high compared to other sectors. The 2001 UK Labour Force Survey revealed that around 35 per cent of the construction workforce was in self-employment, a decline from the figure of 45 per cent in 1996, but still three times the level of self-employment in any other sector (Harvey, 2001). Subcontracting arrangements have also been used extensively alongside or as alternatives to direct labour (Bresnen and Fowler, 1994; Druker and White, 1997: 133), and 57 per cent gross output in the construction sector involved the buying in of materials and subcontractor services (Dainty *et al.*, 2001).

Some writers have suggested that the extensive use of these contingent forms of labour can have a deleterious effect on skills reproduction. Training for workers on contingent contracts tends to be dominated by induction training at the outset of employment and by training courses pursued by individuals outside of work (Green, 1999). In the construction industry, self-employed workers have demonstrated a lower propensity than other groups to pursue training (Gospel and Fuller, 1998). The ongoing reliance upon subcontractors and self-employed workers in the industry may also act as a deterrent to offering apprenticeships, and may impact on the capacity of firms to provide training in-house (ibid.). The use of self-employed labour is also thought to perpetuate reliance on craft-based training within UK construction, and Clarke and Wall (2000) suggest that this has generated less transferable and lower levels of skills compared to the broader, industry-based training approach apparent in such countries as Germany and the Netherlands.

Certainly, industry bodies have identified connections between the use of these contract forms and skills shortages, arguing that reliance on contingent labour may have detrimental consequences for the sector's long-term skill and training needs (CITB, 2002). According to Green and May (2003), the continued use of casual employment practices raises serious questions about the industry's ability to deliver the 'high-quality construction' emphasised within the performance-improvement agenda. Recent reports have argued that the extensive use of contingent labour offers firms 'little opportunity' to invest in training (CITB, 2003a). Workers on contingent contracts are said to be less likely to receive training than their directly employed counterparts, and this training is often of a much more limited scope than that offered to directly employed workers (CITB, 2003b). Differences in training provision have been on the basis that contingent workers, often being employed to undertake a specific task or to provide specialist skills, may be more skilled than directly employed staff, and that the relatively short period of time for which these workers are employed militates against training provision (ibid.).

However, the poorer training provision that is offered to contingent workers is only part of the problem. The CITB has argued that firms that use extensive amounts of contingent labour are less likely to offer certain training initiatives to all their staff, including both contingent labour and those on directly employed contracts. This is notably the case with provision of industry-based apprenticeships, which in turn is problematic given the central role afforded by the CITB to such apprenticeships in meeting the industry's long-term skills requirements (CITB, 2003a). Apprenticeships are seen as a key means of attracting new workers into the industry, alongside the provision of ongoing training for existing workers (CITB, 2001). At least part of the problem of the industry's skills shortages, then, is seen to be its reliance on contingent labour.

This view can be contrasted with a government report of 2000 endorsed by the CITB, in which this reliance on contingent labour is seen as relatively unproblematic in terms of the consequences for training and skills reproduction. The report contends that firms can use contingent labour to 'buy in' the production skills that they need rather than offering such training to directly employed staff (Department for Education and Employment (DfEE, 2000: 10)). The DfEE report argues that many construction companies now see the key skills to be provided in-house as being those that relate to 'conceiving, scheduling and managing projects' (ibid.). This report – downplaying the importance of production skills and seeing the use of contingent labour as a means of meeting production skill requirements in the industry – would appear to offer a quite different diagnosis of the industry's skills shortages, and the means by which these shortages might be overcome.

These observations point to the complexities of the relationship between the use of contingent labour, training, skills shortages and recruitment difficulties, within a sector that has historically relied on the range of contract forms to regulate the supply of labour. Recent developments in the industry have added further complexity to the structures of employment. These include the wider use of external employment agencies and changes to the tax regime, the latter appearing to impact in particular on the use of self-employment in the industry (Harvey, 2001). Furthermore, recent years have seen the rapid growth of large construction management firms, which co-ordinate and manage a large proportion of construction projects, but often employ few workers directly. The links between the use of contingent labour within the resultant complex structures of employment have been the focus of some attention within the construction management literature. The use of subcontractors in particular has been seen as posing a challenge to the implementation of integrated project delivery programmes of the kind espoused in recent government reviews of the sector (see Dainty *et al.*, 2001). It is also increasingly recognised that the casualised and frag-mented nature of employment in the industry may have limited the success of 'top down' policy initiatives that have attempted to overcome skills shortages in the sector (see Dainty *et al.*, 2005). Given these complexities of employment structures in the industry, and the historical reliance on contingent labour, the construction sector thus offers an interesting context in which to explore the relationship between recruitment difficulties, skills shortages and the use of contingent labour.

2.3   Methodology

The majority of data presented in this chapter are drawn from a national postal questionnaire of construction employers, distributed in 2002. This technique has been supplemented with ongoing interview-based fieldwork for purposes of triangulation. While the majority of results are drawn from the survey, some observations from the interviews are also included, where pertinent. The decision was made to focus specifically on the non-residential construction and civil engineering segment of the industry. This section of the industry has a broader importance in terms of its linkages with other sectors, such as the utilities industries, and with general economic activity, in terms of road construction, bridge and tunnel construction. The rationale for this decision was to allow an insight into employment practices in this vital infrastructure industry, while avoiding the sample being dominated by the numerically superior housebuilding and residential construction firms. Using Dun and Bradstreet's MarketPlace UK dataset, a population of 8038 units was identified, from which a sample of 1780 firms was generated. In total, 188 usable questionnaires were returned, providing a response rate of 10.6 per cent, which is in line with other studies of the construction sector and so represents a meaningful basis for analysis.

2.4   Results

Recruitment difficulties and solutions

A large proportion of respondents (54.3 per cent) reported suffering problems with recruitment of direct staff over the past five years. Firms were asked how they had responded to these problems with recruitment, selecting all responses used from a list of 10 options. The three most cited responses were increases in pay (66 per cent), increased use of subcontractors (60 per cent) and upgrading skills of existing staff (52 per cent). Firms were then asked to rank the three most important responses; the results are summarised in Table 2.1. The three most important responses were the same as above. Only a small minority of firms, less than one in seven, reported that an increase in apprenticeships was one of their three most important responses to recruitment difficulties (14.9 per cent). Even fewer firms ranked an increase in other forms of entry-level training among the three most important responses (7.4 per cent).

These responses suggest a duality in firms' responses to the problem of shortages of suitably skilled workers. 'Internal' responses are oriented around investment in the skills of their existing workforce and 'external' responses based on the increased use of contingent workers, mainly in the form of subcontractors, but also agency workers to some extent. The results indicate that employers are making some investment in training to overcome recruitment difficulties. However, employers are affording a relatively limited role to entry-level training as a means of overcoming recruitment difficulties. Coupled with the CITB's concerns that new workers are not being attracted into the industry, the small amount of entry-level training suggests that the pool of available labour in the industry is not being increased, thereby potentially exacerbating longer-term problems with labour reproduction within the industry. The results are also in

Table 2.1 Most important responses to recruitment shortages with direct staff

	Respondents (%)
Increase pay	55.3
Upgrade skills of existing staff	36.2
Increase use of subcontractors	36.2
Change recruitment methods	27.7
Increase non-wage benefits	13.8
Increase use of self-employed workers	25.5
Increase use of agency staff	23.4
Expand geographical catchment area	16.0
Increase apprenticeships	14.9
Increase career development opportunities within the firm	9.6
Increase other entry-level training	7.4
Number	94.0

Notes
• Table shows the percentage of those firms experiencing recruitment shortages who cited a particular reason as one of the three most important responses to such a shortage.
• 102 firms (54.3%) reported recruitment shortages for direct staff. The table is limited to the 94 firms who provided ranking information on the importance of the responses they adopted.

keeping with Gospel and Fuller's (1998) assertion that the availability of contingent labour may act as a deterrent to the provision of apprenticeships within firms.

The extent of use of contingent labour

In terms of the extent of construction employers' reliance on contingent labour, the findings from this survey were in keeping with other studies. The overwhelming majority of firms in the sample (83.5 per cent) reported utilising some form of contingent labour, with the greater part of the sample (60 per cent) making use of more than one form (see MacKenzie *et al.*, 2006). When these figures were disaggregated by contract type, the most widely used form of contingent labour proved to be subcontractors: 79.7 per cent of the sample utilised labour from this source, compared with 47.3 per cent of respondents who used self-employed labour and 35.6 per cent who used agency labour. The importance of subcontracting, revealed in these figures, arguably reflects the long-standing tradition for such arrangements within the construction sector; however, the presence of other forms of contingent labour is far from insignificant, often utilised in conjunction with subcontract arrangements.

For many firms, contingent labour constituted a large proportion of their workforce. For almost nine out of ten firms using contingent labour, such employees accounted for more than 10 per cent of total workforce. For over one-third of firms (34.9 per cent), half or more of the total workforce was made up of contingent labour. The importance of subcontracting can be seen in Table 2.2, which disaggregates employers' use of contingent labour by contract type. For nearly one-quarter of those firms using subcontract labour, 50 per cent or more of their workforce was employed on a subcontract basis. There appeared to be less extensive use of self-employed labour, with the distribution evenly split between users with less than 10 per cent of their workforce being engaged on these terms and those with more than 10 per cent self-employed. The use of agency workers was predominantly low scale, with four out of five users located in the lowest bracket of less than 10 per cent of the total workforce coming from this source. Only among a very small minority of users of agency labour (2.4 per cent) did such employees account for half or more of the total workforce, indicating less willingness by employers to rely on this form of labour to constitute the majority

Table 2.2 Construction employers' use of contingent labour

	Low (<10% of total workforce)	Medium (10 to <50% of total workforce)	High (50% or more of the total workforce)	Number
Subcontract labour	13.3	62.6	24.3	107
Agency labour	80.5	17.1	2.4	41
Self-employed labour	50.0	43.3	6.7	60
Total contingent labour	12.7	52.4	34.9	126

of the workforce. This finding perhaps hints at a difference in attitude towards different types of contingent labour: the use of subcontractors reflects their long-standing centrality to the production process within the construction industry, while the more recent growth of agency employment is based on their use in a more short-term, contingent fashion.

To assess how patterns of usage of contingent labour had developed recently, data were collected on changes in the amount of work undertaken by contingent labour over the previous five years, and whether the range of tasks performed by contingent labour had altered. In terms of the amount of work undertaken by contingent labour, increases were particularly apparent among users of subcontract workers. Of the firms utilising labour from this source, 51.7 per cent reported an increase over the last five years in the amount of work undertaken by subcontract workers. A higher proportion of firms who had experienced difficulties in recruiting directly employed labour (60.2 per cent) had increased the amount of work undertaken by subcontractors, providing further evidence of a link between recruitment difficulties and the use of contingent labour. While a majority of subcontract users reported an increase in the amount of work allocated, among other contract forms trends were more mixed. Of those firms using agency workers, more had increased the amount of work allocated to this type of labour than had decreased it, but the largest grouping reported no change. For users of self-employed labour, the largest grouping (37.8 per cent) had decreased the amount of work allocated, compared to 26.7 per cent who had increased and 35.4 per cent who reported no change. These figures may reflect legislative changes that have impacted upon the way self-employed workers are taxed. The finding is consistent with preliminary results from qualitative research, which has indicated that a fall in the availability of self-employed workers has led firms to shift towards other contract forms. In terms of the range of tasks performed by contingent labour, the majority of respondents reported no change across all three categories of labour. There were similar trends to the changes in the amount of work allocated, in that increases in the range of tasks performed were largest for subcontract workers and, perhaps for the reasons cited above, decreases were largest for self-employed workers. As can be seen from these figures, slightly more firms reported an increase in the range of tasks performed by agency staff than reported a decrease.

To ascertain whether changes in the patterns of use over recent years reflected any strategic change in labour utilisation policy, respondents were also asked whether they had made a conscious shift between the use of direct labour and contingent contracts (Table 2.3). The largest grouping of respondents (39.9 per cent) reported no conscious shift between direct and contingent labour. Where firms had made a conscious shift the results were interesting. Although a significant number of firms (20.7 per cent) reported making the conscious shift from the use of direct labour to contingent labour of some form, a greater number (37.2 per cent) had gone in the opposite direction, moving from contingent to direct labour, with 2.1 per cent reporting a shift in both directions over the previous five years. Further analysis of these data revealed some interesting trends, however. Of

Table 2.3 Changes in use of contingent labour

	Increased	Stayed same	Decreased	Number
Amount of work				
Subcontract	51.7	40.0	7.3	145
Self-employed	26.7	35.4	37.8	107
Agency	35.2	43.8	21.0	129
Range of tasks				
Subcontract	31.5	61.6	6.8	145
Self-employed	15.1	58.7	26.2	107
Agency	19.8	65.1	15.0	129

those firms who had reported a conscious shift from contingent to direct labour, 81 per cent also reported that they had increased the amount of work allocated to subcontractors over the last five years, 74 per cent had increased the amount allocated to agency labour and 42 per cent had increased the allocation to self-employed labour. Despite a conscious attempt to move away from contingent labour, many firms had in fact increased the amount of work allocated to contingent forms. It seems that many firms using contractors are not making a conscious choice to use contingent labour, but are perhaps being compelled to use these labour forms as an alternative to direct labour.

Recruitment and training of contingent labour

As noted above, many firms had experienced difficulties in recruiting direct labour. However, problems with labour supply are not restricted to the recruitment of direct labour: 24.6 per cent of firms that made use of subcontract labour reported they had experienced problems hiring sufficient numbers of subcontract workers. Similarly 22.7 per cent of firms that used agency labour reported difficulties in hiring sufficient numbers of workers on these terms. Of firms utilising self-employed workers, 37.8 per cent reported similar problems, although it should be recognised that less than half of the respondents reported making use of self-employed workers, and overall the amount of work being allocated to this form of labour was shrinking. It is notable that a high proportion of firms who had experienced problems hiring sufficient levels of subcontract labour (88.6 per cent), agency labour (95 per cent) and self-employed workers (80 per cent) had also experienced problems recruiting direct labour. This finding suggests that the same firms are experiencing problems hiring sufficient amounts of labour across the range of contract alternatives, implying a shortage of suitably skilled labour regardless of contract form.

The increased use of contingent labour raises the question of training for workers employed on such contracts. Results from the survey reported elsewhere (see Forde and MacKenzie, 2004) revealed a position towards the training of contingent labour that would arguably contradict the logic of externalisation often associated with the use of such contracts – essentially that training will be provided by someone else. Only one in ten respondents disagreed with the statement that training contingent labour was an investment in the future of the industry, while almost half (43.9 per cent) disagreed with the statement that it was not their responsibility to train contingent labour (Forde and MacKenzie, 2004: 83). These perceptions towards training for workers who are not tied to the firm by direct employment contracts fits with the tradition of the strong occupational labour market within the sector, in which it is assumed workers can easily carry their skills between employers. This tradition has certainly informed training policy within the sector and provides the rationale for the CITB levy-based redistribution of training costs, a practice that is extended to the training of contingent labour. Given the strong presence of contingent labour within the sector, and the proportion of the total available labour pool it represents, the training of such workers is important to the maintenance of skills and the future of the industry.

Looking more closely at the types of training offered, there is evidence to suggest that the use of contingent labour does not offer a viable long-term alternative to maintaining the sector's skills profile. The training offered to contingent labour is often more limited in scope than that given to direct labour. Training offered to contingent labour is most commonly 'on the job', rather than formal on-site or off-site training (Forde and MacKenzie, 2004). Caution should therefore be exercised when drawing conclusions regarding the *quality* of the training offered to contingent workers. Further, evidence from qualitative research suggests that much of the training offered to contingent workers is related more to health and safety compliance than to the transmission of production skills. It is important to recognise, therefore, that the provision of training to contingent workers may reflect self-interested, penalty-avoidance imperatives, in terms of ensuring all workers on-site meet the requirements of health and safety legislation. Furthermore, industry-based apprenticeships, seen by many as a long-term solution to skills shortages in the sector, are generally attached to direct employment contracts rather than contingent contract arrangements. According to one recent survey of construction workers, 92 per cent of workers within the industry who had received apprenticeships did so while directly employed, as did 86 per cent of apprentices who are now self-employed (CITB, 2003b).

2.5   Conclusion: A casual response to recruitment difficulties?

Recruitment difficulties among construction employers are widespread as evidenced not only by this research but also by other surveys, most recently a report from the Office of National Statistics (ONS) on labour shortages in skilled construction trades (see Ruiz, 2004). These difficulties are pronounced for the recruitment of both direct staff and contingent labour. Employers appear to be adopting a dual response to difficulties recruiting direct staff, with an internal response typically taking the form of upgrading the skills of existing employees and an external response being increased recourse to subcontract arrangements. Yet neither of these strategies is commensurate with attracting new workers into the industry, one of the key aims identified in recent Construction Workforce Development Strategies (see CITB, 2002).

The survey reported here points to the widespread use of contingent labour throughout the construction and civil engineering industry, with subcontracting being the most commonly utilised form. In many cases, the amount of work and, to a lesser extent, the range of tasks undertaken by these forms of labour have increased in recent years. Yet this is not to suggest that the move to contingent labour was necessarily a strategic one. The fact that many firms had made a conscious shift away from contingent to direct labour, but at the same time had increased the amount of work allocated to contingent labour, suggests that for many the use of contingent labour was a constrained choice: they are perhaps being compelled to use these labour forms as an alternative to direct labour. This observation suggests that while a return to direct labour may be desirable in order to obviate the problems associated with skills development and the use of contingent labour, such a course of action is by no means straightforward.

Within both the industry and government, concerns have been raised over the level of skills shortages in the construction industry and the low numbers of new workers being attracted into the sector (see, for example, DfEE, 2000; CITB, 2002). One strand of policy to address this problem has focused on increasing the skills of the existing workforce by encouraging workers to train for recognised qualifications, and also on increasing the number of multi-skilled workers (DfEE, 2000: 76). An alternative strategy has sought to attract workers into construction from non-traditional recruitment grounds beyond the customary core of 16–19-year-old males (ibid.). Yet the survey evidence presented in the current study underlines how the widespread use of contingent labour may impact upon initiatives that seek to address the complexities of skills reproduction.

According to some observers, the use of contingent labour has led to employers placing emphasis on providing workers with skills related to 'conceiving, scheduling and managing projects' rather than 'operational, site-based skills (which) are important, but do not vary much from company to company, and so are not the real drivers of competitive success or failure' (ibid.: 10). Production skills, it is argued, can be 'bought in' on a contingent basis as and when required. While this argument echoes the axiom found in the human resource management literature, which espouses the firm's blend of skills as a unique source of competitive advantage not easily copied by business rivals, this rhetoric is difficult to reconcile with an industry that remains so dependent on an occupational labour market for resourcing production. Downplaying the importance of production skills in this way is also at odds with the ongoing concerns over the general shortage of labour in the industry (CITB, 2002). Furthermore, the suggestion that such production skills can 'simply be bought in when needed' (DfEE, 2000: 10) may increasingly prove to be a flawed assumption with potentially deleterious consequences for the industry as a whole. It appears that problems with the availability of contingent labour are already emerging, which suggests that the logic of externalisation cannot ameliorate the longer-term problems of attracting new people into the industry. The survey results suggest that the same firms that have experienced problems recruiting direct labour have also experienced problems recruiting contingent labour. These results are consistent with a recent report from the Office of National Statistics, which highlighted employer difficulties in recruiting sufficient self-employed workers in skilled construction trades (Ruiz, 2004). The problem, it appears, is one of a shortage of suitably skilled labour regardless of contract form, with firms moving between contract types to meet their demands for labour. The result is further casualisation of employment practices within construction, and an increased reliance on contingent labour to address labour shortages.

Unions representing workers in the sector have voiced concerns regarding the casualisation of employment. The Union of Construction, Allied Trades and Technicians (UCATT), the largest union representing workers in the construction sector, has consistently highlighted the negative consequences of casualisation. The rise in self-employment – much of which is seen to be bogus – and the increase in subcontracting, is said to have directly undermined industry-wide initiatives, including the implementation of safety and training standards (see UCATT, 2005). The survey results presented in this chapter suggest that in terms of addressing the perceived shortage of labour within the UK construction industry, further casualisation of the employment relationship and the increased use of contingent labour offer no long-term solution to the problem.

References

Barker, K. and Christensen, K. (eds) (1998) *Contingent Work: American Employment Relations in Transition*, Ithaca: Cornell University Press.

Bresnan, M. and Fowler, C. (1994) 'The organisational correlates and consequences of subcontracting: Evidence from a survey of South Wales businesses', *Journal of Management Studies*, 31, 6: 847–864.

Clarke, L. and Wall, C. (2000) 'Craft versus industry: the division of labour in European housing construction', *Construction Management and Economics*, 18: 689–698.

Construction Industry Training Board (2001) *Construction Workforce Development Planning Brief 2001–2005*, Norfolk: CITB.

Construction Industry Training Board (2002) *Construction Workforce Development Plan 2002*, Norfolk: CITB.

Construction Industry Training Board (2003a) *Construction Skills Foresight Report 2003*, Norfolk: CITB.

Construction Industry Training Board (2003b) *The Effect of Employment Status on Investment in Training*, Norfolk: CITB.

Dainty, A., Briscoe, G. and Millett, S. (2001) 'Subcontractor perspectives on supply chain alliances', *Construction Management and Economics*, 19: 841–848.

Dainty, A., Ison, S. and Root, D. (2005) 'Averting the construction skills crisis: A regional approach', *Local Economy*, 20, 1: 79–89.

Department for Education and Employment (2000) *An Assessment of Skill Needs in Construction and Related Industries*, London: DfEE.

Druker, J. and White, G. (1997) 'Constructing a new reward strategy: Reward management in the British construction industry', *Employee Relations*, 19, 2: 128–146.

Forde, C. and MacKenzie, R. (2004) 'Cementing skills: Training and labour use in UK construction', *Human Resource Management Journal*, 14, 3: 74–88.

Gospel, H. and Fuller, A. (1998) 'The modern apprenticeship: New wine in old bottles?', *Human Resource Management Journal*, 8, 1: 5–22.

Green, F. (1999) 'Training the workers', in P. Gregg and J. Wadsworth (eds), *The State of Working Britain*, Manchester: Manchester University Press.

Green, S.D. and May, S.C. (2003) 'Re-engineering construction: Going against the grain', *Building Research & Information*, 31, 2: 97–106.

Harvey, M. (2001) *Undermining Construction*, London: Union of Construction, Allied Trades and Technicians.

Hillage, J., Regan, J., Dickson, D. and McLoughlin, K. (2002) *Employer Skills Survey 2002*, Department for Education and Skills Research Report 372, London: DfES.

MacKenzie, R., Forde, C. and Robinson, A. (2006) 'Changes in employers' use of contingent labour: Evidence from the construction sector'. Paper prepared for presentation at the IIRA XIV World Congress, Lima.

Royal Commission on Labour (1892) *Reports: Commissioners*, 18, 4, London: HMSO.

Ruiz, Y. (2004) 'Skills shortages in skilled construction and metal trade occupations', *Labour Market Trends*, March, 103–112.

Union of Construction Allied Trades and Technicians (2005) 'About us: UCATT's history'. Accessed online at http://www.ucatt.org.uk/about1.htm.

3 The nature of the employment relationship in the UK construction industry

A flexible construct?

*Ani Raidén, Michael Pye and
Joanna Cullinane*

Abstract

Recent developments in employment relations (ER) show that 'traditional' full-time, permanent employment is becoming less prominent in many industries. The construction industry presents a unique arena for discussion in this field due to its long history of atypical employment. In addition, the structure of the industry influences the nature of employment, containing as it does a large number of small- and medium-sized firms in coexistence with a few large organisations. The relationship between these is commonly a contracting or a partnering arrangement: a large contractor typically acts as a managing agent who employs the services of subcontractors to carry out 'bundles' of work.

Within the industry, a significant proportion of the subcontracted operatives are considered self-employed and it is generally understood that the term 'employee' in such organisations includes only the main contractor's directly employed staff. Yet by some definitions of the term, many self-employed staff might actually be 'employees'. This chapter therefore discusses the complex nature of ER in the UK construction industry with the aim of establishing whether the current norm is a multi-tier contracting/partnering arrangement or a misinterpreted contract of employment. An analysis of secondary data from national statistics and industry literature is used to construct the argument. The findings from the analysis that follows suggest that the notion of the employment relationship is used both flexibly and ambiguously but it is established that there is frequent misinterpretation of the definition of employee and the employment contract. An agenda for further discussion and research is developed on that basis.

Keywords: employment relations (ER), employment relationship, employee, worker and self-employment.

3.1   Introduction

The construction industry is increasingly being encouraged to foster better employment relations (ER) with its workforce and to engender greater employee participation in order to propagate improvements in its productivity and performance (Druker and White, 1995, 1996; Respect for People Working Group, 2000; DTI, 2005). However, questions remain as to whether the industry possesses a stable enough employment platform to underpin attempts to improve its performance.

The focus of such questions about the nature of ER in the industry arises from its tendency to utilise a core-periphery model of employment in which a minority comprising project and operational senior managers are in the 'core' and the bulk of the labour force are subcontracted, self-employed people in the first peripheral. The viability of improvements in the industry's performance hinges upon the experience of this vast group of workers, many of whom seem to be considered self-employed primarily to provide numerical flexibility for organisations in the industry (Langford *et al.*, 1995: 55). Indeed by some definitions, it becomes clear that much of the industry's labour force is too closely tied to one organisation to be truly defined as self-employed, they are in fact in 'bogus self-employed', much to the disadvantage of the tax authorities (Harvey, 2001, 2002). Consequently, the aim of this chapter is to establish, through critical discussion, whether the norm in relation to the nature of the employment relationship in the UK construction industry is actually a multi-tier contracting arrangement (self-employment) or in fact a misinterpreted contract of employment.

To foster this critical discussion, the chapter is constructed thematically. At first, the wider ER framework is established. Then follows a review of the characteristics of UK construction industry ER and some of the latest European influences. Analysis of pertinent construction industry statistics then precedes the discussion and conclusions, which suggest a response to the questions in the title and aim of this chapter.

3.2   Employment relations framework

Employment relations encompasses the employment policies and practices that deal with the 'dynamic, interlocking economic, legal, social and psychological relations that exist between individuals and their work organisations' (Bratton and Gold, 2003: 485). This definition implies that ER is a term used as a framework of analysis for matters related to work, employment and the employment relationship. It is most often conceived as the generic label which encompasses those matters that arise between parties in the context of the relationships that exist between workers/employees and employers. In that context, a key definitional distinction is made between the 'contract *of* service' (contract of employment), between a worker and an employer, and 'contract *for* service'. In the former, the employer takes the worker 'into service' in return for a wage or salary, while the latter is a commercial/economic relationship between two or more (nominally) equal parties in which a principal contracts with an agent for the supply of goods and services in return for consideration.

These different employment relationships are governed by quite different common law, statute and regulation. And although these bodies of law make 'clear' distinctions between the two types of employment (at least in normative terms), in practice there have always been 'grey' areas in definitions. The practical imprecision arises from the multiple jurisdictions and non-parallel developments in the common law of employment, taxation law, welfare law and other statutory instruments. In theory as well, recent developments in strategic supply chain management have resulted in re-conceptualisation of what constitutes an organisation. In particular, organisational structures and boundaries have broadened and it is increasingly becoming less clear who is inside the organisation and who is outside it.

Perhaps as a result of these developments, 'traditional' full-time permanent employment is becoming less prominent (Harrington, 1997). The decline in 'normal' employment is generally captured under the label 'atypical employment' or the rise in non-standard employment. Non-standard forms of employment describe employment relationships in which the worker is either a part-timer or on a temporary or fixed-term contract, and also increasingly include workers who are not legally 'employed' by an organisation, such as subcontractors, self-employed and agency workers (Winch, 1998; Ward *et al.*, 2001).

The usual way in which the growth in non-standard employment has been explored or explained has been through the lens of organisational motivations to achieve 'flexibility' (see, for example, Atkinson, 1981). However, another source of explanation seems to present itself from the pervasive adoption of strategic value chain management, after Porter (1985). This explanation arises from the conceptualisation of the value chain's inclusion of human resources as a secondary activity; that is, human resources are simply considered as a variable cost, which supports but does not primarily provide the creation of value. Similar normative rhetoric stresses the strategic imperative of 'sticking to core business' and of organisations divesting themselves of non-core functions including the labour components associated with these functions (Flynn, 1995; Allan, 2000).

At the most obvious level, these types of rhetoric permit the divestiture of staff who are not primary profit producers; at a deeper level, such conceptualisations have encouraged a reshaping of where the organisation's boundaries fall. In place of the traditional clear boundaries, in which the identification of those inside the organisation (employees) and those outside the organisation (suppliers, customers, etc.) was clear-cut, we increasingly have a situation in which complex chains of relationships exist not just with material suppliers but also with individual and commercial suppliers of labour (Flynn, 1995). Such complexity is ubiquitous in the construction industry and it could be argued that both peripheralisation and bogus self-employment in the industry arise from and are obscured from analysis by virtue of this complexity.

3.3 The characteristics of the UK construction industry employment relations

The structure of the construction industry features a particular set of characteristics that differentiate it from many other sectors of the economy. First, at the level of the industry it is fragmented and disparate: a large number of small- and medium-sized firms co-exist with a few large organisations. Equally, a large number of small- and medium-sized support/stakeholder organisations, other than employers, co-exist with a few central bodies (Dainty *et al.*, 2004: 16). Informal networks and regional construction forums form the main communication channels (ibid.: 17).

A second distinctive aspect to consider is the project-based nature of construction work. The relationship between companies is typically a contracting or a partnering arrangement on a project: a large contractor acts as a managing agent who employs the services of subcontractors to carry out 'bundles' of work. Significant proportions of the subcontracted operatives are considered self-employed and it is generally understood that employees within the project structure include only the main contractor's directly employed staff. The climate is highly competitively charged and yet there is a strong reliance on tried and tested ways of working to ensure coordination (ibid.: 20). Above all, this managing agent–subcontractor hierarchy means that relations between organisations are premised on project-by-project cost-effectiveness as the central motivation (Druker and White, 1995, 1996).

Partnering is one of the best-established 'new' management strategies in the construction industry. It was introduced in an attempt to initiate culture change in the industry by fostering relationships among the key stakeholders on a project: the owners, design professionals and contractors. It is expected that partnership relationships will be based on trust, dedication to common goals and an understanding of each party's expectations and values (Harback *et al.*, 1994; Slater, 1998). The partnership approach has gained increasing attention within the industry over the last decade; however, despite the clear normative emphasis on relationships and culture, the industry itself tends to presume or imply that implementing partnering is essentially a tactical problem of a technical-managerial nature (Bresnen and Marshall, 2002). Again, it could be argued that this rhetoric of partnership could (intentionally or otherwise) obscure from analysis the issues relating to the nature of employment relationships in the industry.

Defining the nature of the employment relationship

As mentioned above, organisations in the industry tend to classify only the main contractor's directly employed staff as employees. Other groups of workers in the industry are generally considered to be either self-employed or subcontractors. These divisions of workers into clear, defined groups reflect the (commercial) contractual clarity implied in the managing agent–subcontractor hierarchy but such clarity is overly simplistic with regard to the question of whether workers are self-employed, subcontractors and/or their sub-subcontracted operatives. As noted above (see Section 3.1), there is a certain degree of mislabelling of some workers as self-employed when they might be more accurately defined as employees. For instance, Druker and White (1996) observe that the Inland Revenue guidance identifies four main types of 'worker':

1   directly employed
2   in-house self-employed
3   casual self-employed
4   entrepreneurial self-employed.

Although in practice there is a general belief that the term 'employee' only corresponds with the directly employed, according to the Inland Revenue classification, both the in-house self-employed and casual self-employed categories of workers should often be considered to have an 'employment relationship' too.

Rainbird (1991: 210) offers further criteria for defining the nature of the employment relationship, including an assessment of the responsibilities for training and development provision: persons who have training and development provided for them might not actually be 'self-employed'. In addition, Winch (1998: 532) highlights two aspects which should determine whether a person is truly self-employed: first, the degree of continuity and intensity of the relationship and, second, the extent to which any capital is risked by the worker. On this basis, Winch makes a clear distinction between two groups of alleged self-employed workers: labour-only contractors and those who contract the supply of materials and plant as well as labour. He argues that since the latter risk capital, they are defined as small businesses (and therefore genuinely self-employed).

The implications of mislabelling in the employment relationship are significant. 'Self-employed' labour-only contractors are commonly said to enjoy the opportunities to seek well-paid, independent contract work that may appear higher in status than regular employment as a labourer or ground worker. But with this comes the drawback of limited or no rights to employee benefits or job security (Nisbet, 1997; Nisbet and Thomas, 2000; Leighton, 2002, 2005). Moreover, The Construction Employment Survey (Anders Glaser Wills and Contract Journal, 1998) highlights that many managerial personnel working for large companies on a contract basis report significantly poor ratings for four aspects central to ER: working conditions, salary levels, security of employment and provision of perks/fringe benefits. Only 3 per cent of the survey respondents rated their working conditions as being 'very good' (ibid.: 58). The remaining 97 per cent of the responses were divided equally between 'quite good' (47 per cent) and 'poor' (47 per cent) (with 1 per cent 'no response' rate). Similar responses were recorded in terms of salary levels: 'very good' 2 per cent, 'quite good' 37 per cent and 'poor' 57 per cent (with 4 per cent 'no response' rate). Most alarmingly, 72 per cent of the respondents rated security of employment as 'poor', with 22 per cent also rating it no more than 'quite good'. Only 3 per cent thought security of employment was 'very good' (with 1 per cent 'no response' rate). The provision of perks/fringe benefits received scores of 53 per cent 'poor' and 41 per cent 'quite good'. Again, only 2 per cent thought the provision was 'very good' (with 4 per cent 'no response' rate). The security of employment and provision of benefits are of particular interest here in relation to the respondents' perceptions of the nature of the employment relationship.

From the organisations' perspectives, the preference for applying the ambiguous definitions of 'self-employed' often arises from concerns to achieve labour flexibility as implied in both Atkinson's flexible firm model (see Section 3.1) and Porter's strategic value chain management model (see Section 3.2). Flexibility is imperative to construction organisations because of the risk and uncertainty that are an everyday feature of project and organisational management within the industry (Walker and Loosemore, 2003), so a capacity to respond to change is a prerequisite.

Unions and employment relations in the sector

Both unitarist and pluralist perspectives towards ER exist in the construction industry: managerial white-collar workers are bound by individual contracts, which commit them to working towards organisational goals, while blue-collar workers are served by collective bargaining (Loosemore *et al.*, 2003: 116). Croucher and Druker (2001: 56) highlight this as an unusual feature of the industry, since collective bargaining has long since ended in many other UK sectors.

Despite the existence of collective bargaining mechanisms for blue-collar workers, union membership in the industry has been in widespread decline since the 1990s. This decline mirrors both the general decline throughout the nation and the fact that trade union organisation has always been difficult to sustain within the construction industry because of the transient, mobile nature of the workforce (ibid.: 59). Other explanations for the decline in the construction industry focus on the independence of the large number of small firms (Tallard, 1991) and the generally inhospitable climate for construction unions (Evans, 1991). More recently, it has been recognised that the progressive decline in national union density may be due to increases in the proportion of groups traditionally found difficult to organise, such as white-collar and female staff, and also to the growth in self-employment, temporary and part-time work, and of course the move towards self-employment, outsourcing and subcontracting (Loosemore *et al.*, 2003: 136; Blackman, 2005).

The issue of union decline in the industry is not confined to the UK and such has been the extent of this problem that unions in many European countries have tried to combat the decline by beginning to accept the self-employed into the unions (EIRO, 1999, 2000, 2003). Such moves acknowledge that many of the self-employed are becoming a workforce controlled by large organisations (EIRO, 2000). As a result, they are just as likely to experience inequalities comparable to those present in the employee–employer power relations by becoming 'economically dependent' workers. By definition, they therefore move from the state of nominal equality with the principal to that of a tied dependent (Anders Glaser Wills and Contract Journal, 1998). Hence, Amicus, UCATT,

TGWU and GMB all also support any employers' efforts to drive the industry back towards direct employment (D'Arcy, 2001; Clarke, 2004; Jewson, 2004; Blackman, 2005; Contractjournal.com, 2005) (see Section 3.6).

3.4 Estimates on the extent of self-employment

In order to determine the extent and potential scope of the issue in the UK, national construction statistics sourced from the DTI (2003) and Labour Force Survey (LFS) (2002) were analysed. Initially, these revealed that in 2002 the total manpower level in the UK construction industry (averaged over the four quarters) was estimated to be 1,614,000 persons (range = 1,592,000–1,629,000) (DTI, 2003). Based on the LFS (2002), for the same period, an estimated 586,000 of these persons (an average with a range between 561,000 and 605,000) could be classified as being 'self-employed'. Therefore, it is possible to estimate that 36 per cent of the construction workforce in the UK is self-employed compared with 12 to 13 per cent of the total UK workforce (ibid.).

However, these statistics only make a distinction between those who are 'employees' and those who are 'self-employed'. Hence, they do not identify what portion of the workforce fall into the 'grey' area, somewhere between being an employee and being self-employed. More importantly, they do not estimate what proportion of the workforce might be falsely identified as self-employed when the persons are actually economically dependent (Burchell *et al.*, 1999; DTI, 2002). Moreover, in legal terms this dualistic definition ignores those members of the workforce who more accurately fall within the definition of a 'worker' than 'employee' under the terms of the Employment Rights Act 1996 and for whom it is not clear which way a tribunal might decide on the question of their status.

Indeed, Burchell *et al.* (1999) conclude that the LFS underestimates the number of people that could be covered by statutory employment rights by as much as 5 per cent. They estimated that of the total national workforce, 80 per cent are clearly employees or 'dependent workers' and 7 per cent are clearly self-employed. Removing from consideration the 1 to 2 per cent of persons who are unpaid family workers or in training schemes, this leaves approximately 12 per cent of the workforce whose employment status is unclear for a variety of reasons. On this basis, Burchell *et al.* (1999) argue that the percentage of the workforce that could be covered by the 'worker' definition could be as high as 92 per cent of the total workforce, or 5 per cent higher than the estimate in the LFS.

A comparison of the estimates based on the LFS for the number of persons who are 'self-employed' in the construction industry (36 per cent) with the figures for other sectors of the economy (12–13 per cent) reveals a significant difference of 23 per cent. There may be 'good' reasons for the differences in employment practices between the construction industry and other sectors of the economy (although these may also be a matter for debate). However, applying the conclusions of Burchell *et al.* (1999) to an analysis of the construction industry raises serious questions about the employment status of many persons in the industry. Accepting all the caveats about comparing like for like,[1] and accepting as true Burchell *et al.*'s estimate that 12 per cent of the UK workforce has an 'unclear' employment status, then on the basis of the 2002 construction industry figures an average of 194,000 construction workers fall into this category. Further, it could be argued that a minimum of 80,000–90,000 persons in this group could fall at least under the definition of a 'worker' if not an 'employee' as provided under the ERA 1996 and other legislation/regulation.[2] This figure of 80,000–90,000 is a conservative 5 per cent of the total construction workforce defined as 'self-employed' in the 2002 construction sector figures and 48 per cent of those with an 'unclear' employment status in the sector.

While accepting that the arguments made above are based on taking the conclusions of one more general study and applying them to the specifics of the construction sector, and that these arguments are clearly open to contestation, it is clear that this is potentially a substantial issue. Furthermore, what is certain is that (i) there are clearly gaps in the body of knowledge and understanding of employment relationship/status and consequently the nature of the ER in the UK construction industry and that (ii) these gaps warrant further research and clarification.

3.5   The impact of the tax system

In addition to causations from the changing structure of organisations and from the drive towards flexibility, Harvey (2001, 2002) and Briscoe *et al.* (2000) also note that levels of self-employment fluctuate in response to changes in construction taxation schemes. Harvey's (2002) discussion about the impact of measures such as the Finance Act 1970, the shifts in the application of the SC60/714 certification of self-employment during the 1980s, the 1997 issuance of IR148/CA69 leaflet relating to the tests for self-employment and the introduction of Construction Industry Scheme (CIS) in 1999 highlights the direct impact of taxation law and policy upon the prevalence of self-employment in the industry. For example, when the Finance Act 1970 was enacted, about 25 per cent of the private sector manual workforce was classified as self-employed (Harvey, 2002). Then, in 1980, with the shift in the application of the SC60/714 certificates (which effectively meant that employment status became a matter of self-declaration) the number of self-employed private sector manual workers rose to 67 per cent. But the 1997 IR148/CA69 reinstatement of the tests for self-employment reversed the trend again. This change in policy in 1997 caused 180,000–250,000 construction workers to change from self-employed to employed status and by 1999 the numbers of self-employed and employed workers in the industry equalised. The 1999 introduction of the CIS once more encouraged growth in self-employment, and statistics post-1999 show a 10 per cent increase in self-employment in 1999–2001 (Harvey, 2002) and a further 13 per cent in 2001–03 (Lindsay and Macaulay, 2004: 402). Taking into account that in the latter period overall employment increased by 177,000, of which 102,000 (58 per cent) was self-employment (LFS, 2003), this means that self-employment has increased more quickly than overall employment (ibid.), albeit not as quickly as in the peak of the late 1980s and early 1990s.

Harvey (2002) concludes that 'the overwhelming majority of those currently designated as self-employed who have tax deducted at source [the CIS4 card holders] would be designated as employees were their status to be tested in tribunals of courts of law'. Briscoe *et al.*'s (2000) conclusions suggest a similar direct correlation between the industry's taxation schemes and the level of self-employment. This is interesting since Lindsay and Macaulay (2004: 401) conclude that, in general, considering all industries in the UK, tax changes cannot explain the rise in self-employment. Instead, they argue that the increase is due to genuine industry demand and economic factors such as house prices and interest rates (ibid.: 404). In the light of this, it will be interesting to see how the New CIS 2005 regulations (planned for implementation in April 2006) will impact on levels of self-employment. The reform is directly aimed at reducing fraud (Roberts, 2004). So far, it is clear from Harvey (2001, 2002) and Briscoe *et al.* (2000) that 'the rise and fall of self-employment in the construction industry is clearly related to changes in taxation regimes, their implementation, and compliance enforcement. The mass shifts from one employment status to another – and in both directions – are best explained by these major changes in labour market institutions and people's reactions to them rather than infinite multitude of economic choices taken by individuals independent of institutional context, or by some other changes taking place independently in "the economy" ' (Harvey, 2001: 28). It is also clear that bogus self-employment currently results in large-scale tax avoidance and it is estimated that the exchequer's income is thereby reduced by £1.5–2 billion per annum (Harvey, 2001).

Accepting this, and assuming that the CIS 2005 reform is successful at achieving its aim, one might hypothesise that self-employment statistics should settle to reflect the true levels of independent-entrepreneurial self-employment in the construction industry.

3.6   The European influences

The observations about the developments in the nature and shape of the employment relationship in the construction industry are paralleled by other, wider developments. European Union (EU) Directives have acknowledged the misnomer in the use of the term 'self-employed' and have used a wider definition of the employment relationship based around 'workers', whereas the current legal definition in the UK centres around 'employees' (see S230(1) of the Employment Rights Act 1996). In most cases this difference between the EU and UK emphasis has been dealt with by reference in UK law to entitlements that flow out of S230(3) of the Employment Rights Act 1996, which provides a definition of 'worker'. On this basis, it could be argued that there is already some shift occurring in the definition of a 'worker' in the UK and this might eventually impact upon the nature of the employment relationship within the construction industry.

Of potential significance to the construction industry, given the discussions and developments in the EU, is the possibility of the extension of the definition of 'worker' to cover the self-employed groups defined as 'economically dependent workers'. This term refers to those workers who are deemed to be falsely identified as self-employed because they are dependent upon one principal (EIRO, 2002). However, while it is clear that the EU directives are ensuring that UK law and policy is beginning to recognise the grey areas in the employment relationship associated with atypical forms of employment (see, for instance, DTI, 2002), the scale or extent of the impact of these changes upon the construction industry is not yet fully known (see estimates on the impact in Section 3.4).

3.7   Organisational return to direct employment

Whether in response to the CIS 1999's reinstatement of the widespread existence of false self-employment, from a drive to combat the low-cost, low-skill and low-productivity route, or through the role of unions or the EU, some of the industry's larger employers have recently undertaken serious efforts to compel the industry back towards direct employment (Carillion, 2001; Building, 2002; Jewson, 2004). Contractors such as Laing O'Rourke (Jewson, 2004), and major projects such as the Heathrow Airport Terminal Five (D'Arcy, 2001), have been at the forefront of this movement by introducing direct employment contracts for all workers. Some other organisations in the industry are likely to follow suit: Carillion (2001), Costain, Kier and Taylor Woodrow (Contractjournal.com, 2005) are also believed to be considering their options over the status of sections of their workforces. There is also conjecture that projects for the 2012 London Olympic Games may well specify that all contractors use directly employed labour (ibid.).

It is important to note that the trade unions have been instrumental in supporting such changes to the industry ER (D'Arcy, 2001; Building, 2003; Clarke, 2004; Blackman, 2005; Contractjournal.com, 2005) and although such direct employment relationships might be in the unions' interests, there has been some resistance in the form of industrial (or perhaps more accurately commercial) unrest led by self-employed workers who believed they were disadvantaged by such changes (Contractjournal.com, 2005).

The return to direct employment clearly runs counter to the philosophy of strategic and tactical management which took its inspiration from Atkinson's (1981) flexible firm and Porter's (1985) strategic value chain; but it reflects, perhaps, an acceptance that low-cost, low-skill approaches utilising bogus self-employed in high volumes is actually counterproductive. As an alternative, the unions suggest a contract of service (contract of employment) as a mutually beneficial route to good working conditions, job security and employment benefits for the workers, and high-skill, high-quality and high-productivity outputs for the employers. Collectively, these would trigger the kind of culture change that partnering aims to achieve.

3.8 Discussion

The aim of this chapter is to establish, through critical discussion, whether the norm in relation to the nature of the employment relationship in the UK construction industry is a multi-tier contracting arrangement (self-employment) or a misinterpreted contract of employment. This question arose from the reported drawbacks of significantly high levels of false self-employment, which result in costs for the workers (little or no job security and/or employment benefits), the industry (low-cost, low-skill, low-productivity culture) and the exchequer.

This section seeks to answer that question. However, it is accepted that the industry's distinctive structure and the inherent difficulties in defining the nature of the employment relationships within it present a challenging context in which to clarify whether the predominant types of relationship are based on the notion of multi-level contracting/partnering agreements or on the misinterpretation, deliberate or otherwise, of the contract of service (contract of employment). The discussion is divided into two parts. The first part explains how the interpretation of the evidence led to the notion addressed in this chapter's title – 'the nature of the employment relationship in the UK construction industry: a flexible construct' – and explores some of the implications of the acceptance of the current norm. The second part develops an agenda for further research and discussion on this basis.

The nature of the employment relationship in the UK construction industry: A flexible construct?

In considering the nature of the employment relationship in the UK construction industry, attention must be drawn at first to the significant variations that exist in the literature, according to which variables are used or suggested as the basis for the definitions of 'employee' and 'self-employed'. Building on the generic ER context (Flynn, 1995; Allan, 2000), variables prominent in the construction-specific material include terminology in contract law (contract *of* services vs contract *for* services) (Druker and White, 1996), training provision (Rainbird, 1991), continuity and intensity of the relationship between an employer and a worker, and the extent to which capital is risked by the worker (Winch, 1998). The EU perspectives suggest consideration of definitions of 'worker', 'employee', 'self-employed' and 'economically dependent worker' (EIRO, 2002). Overall, the current literature on the contract of service in the UK construction industry has some lacunae and is disconnected from the practice. Tests for differentiating between the (directly) employed and those in commercial contracts for service are included but the application of these definitions is rarely scrutinised. Harvey (2001, 2002) provides a refreshing diversion albeit mostly from the point of view of tax regulations. The EU directives also have an impact on the recognition of the grey areas in the employment relationship in the UK, and developing policy discussions may extend the law in respect to this recognition, but the extent of the impact upon the construction industry is not yet fully known.

Although fluctuations in the levels of self-employment have traditionally been associated with changes in tax regulation (Briscoe *et al.*, 1999, 2000; Harvey, 2001, 2002), with the impetus of trade unions, some of the larger construction employers have driven an agenda towards direct employment (Building, 2002; Clarke, 2004; Contractjournal.com, 2005). This, together with the CIS 2005 reform, was an attempt to reduce the extensive bogus self-employment. This reduction is significant when it is acknowledged that according to the LFS (2002, 2003) self-employment currently stands at around 36 per cent of the total workforce. The analysis of construction statistics (Section 3.4) reveals that a further 100,000 people could fall at least into the definition of a worker and be entitled to some further protections.

Thus, while the notion of multi-tiered contracting may be portrayed as a norm in the sector, which explains the use of self-employment as an organising concept, it is a false norm based on either the ambiguities that exist between the respective relevant laws or inadvertent misinterpretation of the law and the nature of the relationships, deliberate or otherwise.

The implications of the acceptance of this false norm are considerable. While it may be the organisations' staffing strategies, which focus on achieving short-term benefits (as suggested above) that have led to this common use of subcontracting and/or bogus self-employment, the implications have long-term consequences. In addition to the immediate drawback of workers being deprived of employment benefits and job security, for the industry this mass false self-employment means limited training input and associated limited skilled output. This deficiency in turn feeds the low-cost, low-skill, low-productivity culture. Arguably, such a culture reduces UK organisations' competitiveness in EU markets and simultaneously opens up the domestic market for international competitors.

At an individual level, the unclear definitions about self-employment also result in lack of clarity about the 'rights' and 'obligations' that parties have towards each other. This perhaps explains survey results indicating poor ratings for working conditions, salary levels and, in particular, security of employment and provision of benefits (Anders Glaser Wills and Contract Journal, 1998).

An agenda for further discussion and research

The false norm of self-employment is clearly unsatisfactory for workers, the industry and the exchequer and possibly for those employers who are concerned with wider issues than cost savings. Therefore, the moves towards direct employment are to be welcomed.

However, further research into the particular perspectives of trade unions, major employers, small firms, individual employees and the self-employed is undoubtedly necessary, particularly before any fundamental initiatives are planned. An in-depth study is needed to reveal the underlying trends and issues; a questionnaire survey of the industry structure in terms of the patterns of employment would be especially useful in generating large quantities of data. Specific questions to be addressed might include the following: What are the perspectives of the trade union organisations on the nature of the employment relationship in the industry? How do the major employers, small firms, those in genuine self-employment and individual employees view the nature of the employment relationship? What are the perceived and 'real' benefits of a 'worker', 'employee' and 'self-employed' to the individual, organisation and the industry? What is 'entrepreneurial' self-employment in the context of the UK construction industry? What is the true extent of self-employment in the UK construction industry?

These questions highlight only a few of the issues that remain unclear within the current literature or industry statistics. Nevertheless, answers to these questions alone would add considerably to our body of knowledge and understanding of the current situation and help steer away from the extensive use of 'estimate' (see Section 3.4). There are clearly gaps in our knowledge and understanding of employment status and consequently the ER in the UK construction industry. Much of this deficiency seems to stem from ambiguous definitions applied to those claiming to be self-employed in the industry. Accordingly, this is suggested as another major area of future research and discussion. In particular, it would be interesting to explore the true extent and nature of self-employment through both empirical research and clear statistical evidence. The nature of the current statistics offered for the industry may necessitate large-scale data collection, as also alluded to above. The Inland Revenue classification highlighted by Druker and White (1996), for example, may provide a useful structure for collecting this data and help examine the applicability of the wider UK/EU definitions, such as 'worker' and 'economically dependent' worker.

3.9   Conclusions

The notion of the employment relationship is used very flexibly in the construction environment. Clearly, there are some in the sector who are employees/workers as defined by both current employment and taxation law. Members of this group are involved in an employment relationship under contracts of service. Equally, there are people who are genuinely self-employed using the same sets of definitions. These people are involved in commercial relations under contracts for the provision of goods and services. They are not involved in an employment relationship. Among this latter group there may well be examples of multi-tier contracting in a chain of principals and agents in which each link in the supply chain is either genuinely 'self-employed' or exists as a genuine 'body corporate'.

What is equally clear is that there are a significant number of people working in the construction sector who, while they may currently be defined as self-employed links in these chains of principals and agents, are in fact not. Their relationship with the 'principal' is more akin to that of an employee or worker, broadly defined. While the notion of multi-tiered contracting may be portrayed as a norm in the sector, it is a false norm based either on the ambiguities that exist between the respective relevant laws or on the misinterpretation of the law and the nature of the relationships, deliberate or otherwise.

A number of themes discussed in the chapter have led to this conclusion, including the flexible interpretation of the tax regulations, the industry's larger employers' low-cost and/or value chain management strategies, and the lack of data for establishing the number of workers that operate in the 'grey' area between an employee and a self-employed, and those who may be termed self-employed incorrectly. The statistical estimates for these figures suggested nearly 200,000 persons having 'unclear' employment status, of which almost half could fall at least within the definition of a 'worker' according to EU regulations. Even the much simpler analysis of the number of employees and self-employed, excluding those potentially in the grey areas in between the two, must currently rely on estimates. Thus, an agenda for further research and discussion was suggested. This research must begin with an extensive quantitative and qualitative 'mapping' of the nature of the employment relationship in the industry and broaden to embrace wider discussion on the ER framework. This in turn should provide a useful contribution to the development of the academic comprehension of the employment relationship overall.

Notes

1 This is a comparison of estimated figures from the same data source – the Labour Force Survey.
2 For example, the National Minimum Wages Act, the Working Time Regulations, the Trade Union and Labour Relations (Consolidation) Act and so on. For a summary of legislation and the rights conferred on 'employees' and on 'workers' by these and other legislation, see DTI, 2002: 12–16.

References

Allan, C. (2000) The hidden organisational costs of using non-standard employment, *Personnel Review*, 29(2): 188–206.
Anders Glaser Wills and Contract Journal (1998) *Construction Employment Survey 1998*, Sutton: Reed Business Information.
Atkinson, J. (1984) Manpower strategies for flexible organisations, *Personnel Management*, August: 28–31.
Blackman, B. (2005) Construction sector, http://www.tgwu.org.uk/Templates/Internal.asp?NodeID=91798&L1=-1&L2=91798.
Bratton, J. and Gold, J. (2003) *Human Resource Management Theory and Practice* (3rd edn), Basingstoke: Palgrave Macmillan.
Bresnen, M.J. and Marshall, N. (2002) The engineering or evolution of co-operation? A tale of two partnering projects, *International Journal of Project Management*, 20: 497–505.
Briscoe, G.H., Dainty, A.R.J. and Millett, S.J. (1999) The likely impact of the tax system on self-employment in the British construction industry, in W. Hughes (ed.), *Proceedings of the 15th Annual ARCOM Conference*, Liverpool John Moores University, UK, 15 September, 1: 211–220.
Briscoe, G.H., Dainty, A.R.J. and Millett, S. (2000) The impact of the tax system on self-employment in the British construction industry, *International Journal of Manpower*, 21(8): 596–613.

Building (2002) Jarvis gets tough on agency tax-avoidance schemes, *Building*, Issue 3, 2 August.

Building (2003) UCATT calls for direct employment rules, *Building*, Issue 9, 7 March.

Burchell, B.S., Deakin, S. and Honey, J. (1999) *The Employment Status of Individuals in Non-standard Employment*, London: Department of Trade and Industry.

Carillion (2001) Sustainability 2000, *Company Environment, Community and Social Report,* http://www.carillionplc.com/sustain/per_4c.htm (accessed 25 August 2005).

Clarke, R. (2004) *Amicus Industrial Sector Fact Sheet for Construction & Contracting*, London: Amicus.

Contractjournal.com (2005) Union warning of revolt over worker status, http://www. contractjournal.com/home/Default.asp?type=2&liArticleID=47596&liSectionID=11 (accessed 25 August 2005).

Croucher, R. and Druker, J. (2001) Decision-taking on human resource issues: Practices in building and civil engineering companies in Europe their industrial relations consequences, *Employee Relations*, 23(1): 55–74.

Dainty, A., Ison, S. and Raidén, A. (2004) *An Engagement Model for the East Midlands Construction Forum, Final Report and Recommendations*, Loughborough: Loughborough University.

D'Arcy, J. (2001) Direct employment set for Terminal Five, contractjournal.com (accessed 25 August 2005).

Druker, J. and White, G. (1995) Misunderstood and undervalued? Personnel management in construction, *Human Resource Management Journal*, 5(3): 77–91.

Druker, J. and White, G. (1996) *Managing People in Construction*, London: IPD.

DTI (2002) *Discussion Document on Employment Status in Relation to Statutory Employment Rights*, London: Department of Trade and Industry.

DTI (2003) *Construction Statistics Annual*, London: Department of Trade and Industry.

DTI (2005) The People Agenda: Respect for People, http://www.dti.gov.uk/construction/respect/peoplerespect.htm (accessed 26 September 2005).

EIRO (1999) Trade unions open their doors to the self-employed, *European industrial relations observatory on-line*, European Foundation for the Improvement of Living and Working Conditions, Dublin, http://www.eiro.eurofound.eu.int/print/1999/07/feature/se9907178f.html.

EIRO (2000) Unionisation of the self-employed, *European industrial relations observatory on-line*, European Foundation for the Improvement of Living and Working Conditions, Dublin, http://www.eiro.eurofound.eu.int/print/2000/02/feature/es0002277f.html.

EIRO (2002) Economically dependent workers, employment law and industrial relations, *European industrial relations observatory on-line*, European Foundation for the Improvement of Living and Working Conditions, Dublin, http://www.eiro.eurofound.eu.int/2002/05/study/TN0205101S.html.

EIRO (2003) Building workers' union to organise self-employed, *European industrial relations observatory on-line*, European Foundation for the Improvement of Living and Working Conditions, Dublin, http://www.eiro.eurofound.eu.int/print/2003/08/feature/dk0308102f.html.

Evans, S. (1991) The state and construction performance in Britain, in H. Rainbird and G. Syben (eds), *Restructuring a Traditional Industry: Construction Employment and Skills in Europe*, Oxford: Berg, pp. 25–41.

Flynn, G. (1995) Contingent staffing requires serious strategy, *Personnel Journal*, 74(4): 50–56.

Harback, H.F., Basham, D.L. and Buths, R.E. (1994) Partnering paradigm, *Journal of Management in Engineering*, 10(1): 23–27.

Harrington, A. (1997) Hard core viewing (core/peripheral workers), *Human Resources*, 33: 52–54.

Harvey, M. (2001) *Undermining Construction: The Corrosive Effects of False Self-employment*, London: Institute of Employment Rights.

Harvey, M. (2002) Taxation regulations and the entrenchment of false self-employment in the Uk construction industry, *The Chartered Institute of Taxation Technical Article*, http://www.tax.org.uk/showarticle.pl?n=&id=597&p=1 (accessed 25 August 2005).

Jewson (2004) Direct employment contracts for all workers, http://www.jewson.co.uk/en/templates/news/industryNewsArticle.jsp?year=104&archiveflag=false&itemId=5100016 (accessed 25 August 2005).

Labour Force Survey (2002) National Statistics, http://www.statistics.gov.uk/CCI/nscl.asp?*ID*=5006&x=6&y=10.

Labour Force Survey (2003) National Statistics, http://www.statistics.gov.uk/CCI/nscl.asp?*ID*=5006&x=6&y=10.

Langford, D., Hancock, M., Fellows, R. and Gale, A. (1995) *Human Resource Management in Construction*, Harlow: Longman.

Leighton, P. (2002) Defining the employee and genuinely self-employed: Are we ever going to get these definitions right? *Industrial Law Society Annual Conference*, Oxford, UK.

Leighton, P. (2005) Flexible labour markets, intermediation and the rise of the 'strategic individualist': Challenges for law, *Managing Social Risks through Transitional Labour Markets – The TLM. NET Final Seminar*, 19–21 May, Academy of Social Sciences centre, Budapest, Hungary.

Lindsay, C. and Macaulay, C. (2004) Growth in self-employment in the UK, *Office for National Statistics Analysis in Brief, Labour Market Trends*, October.

Loosemore, M., Dainty, A. and Lingard, H. (2003) *Human Resource Management in Construction Projects: Strategic and Operational Approaches*, London: Spon Press.

Nisbet, P. (1997) Dualism, flexibility and self-employment in the UK construction industry, *Work Employment and Society*, 11(3): 459–480.

Nisbet, P. and Thomas, W. (2000) Attitudes, expectations and labour market behaviour: The case of self-employment in the UK construction industry, *Work Employment and Society*, 14(2): 353–368.

Porter, M.E. (1985) *Competitive Advantage: Creating and Sustaining Superior Performance*, New York: Free Press.

Rainbird, H. (1991) The self-employed: Small entrepreneurs or disguised wage labourers? in A. Pollert (ed.), *Farewell to Flexibility*? Oxford: Blackwell.

Respect for People Working Group (2000) *A Commitment to People 'Our Biggest Asset'*, Report from the M4I working group on respect for people, UK.

Roberts, G. (2004) Reform of the Construction Industry Scheme, http://www.boyesturner.com/news.asp?step=12&id=391 (accessed 19 August 2005).

Slater, T.S. (1998) Partnering: agreeing to agree, *Journal of Management in Engineering*, 14(6): 48–50.

Tallard, M. (1991) The context and limits of policies of social innovation in small and medium sized construction firms in France, in H. Rainbird and G. Syben (eds), *Restructuring a Traditional Industry: Construction Employment and Skills in Europe*, Oxford: Berg: 43–66.

Walker, D.H.T. and Loosemore, M. (2003) Flexible problem solving in construction projects on the National Museum of Australia project, *Team Performance Management: An International Journal*, 9(1): 5–15.

Ward, K., Grimshaw, D., Rubery, J. and Beynon, H. (2001) Dilemmas in the management of temporary work agency staff, *Human Resource Management Journal*, 11(4): 3–21.

Winch, G. (1998) The growth of self-employment in British construction, *Construction Management and Economics*, 16: 531–542.

4 Self-employment

Legal distinctions and case-law precedents

J. Craig Barker

Abstract

Employment law has struggled for many years with the problem of employment status. Throughout this struggle, case law has developed a number of tests in this respect. These have been described as hazy and it is certainly the case that applying these tests to workers within the construction industry has not stopped the perceived scandal of mass self-employment. A particular problem in the industry in recent years is the overuse of the labour-only subcontract and over-reliance on agency workers: the problem of casualisation. Tests developed by the courts to deal with casual workers appear not to help the position of individuals working on the basis of such contracts. Nor does the development of the category of 'workers' by the European Union (EU) give much cause for comfort. On the other hand, it is not the role of the courts to create contracts where none exist. Rather, their role is to identify the nature of the relationships which do exist between the parties to a contractual arrangement. If the problems of false self-employment and casualisation within the construction industry are to be solved, changes must occur within the industry itself rather than through the manipulation of the legal process.

Keywords: employment status, tests, casualisation, mutuality of obligations and contractual intent.

4.1 Introduction

The legal status of construction workers has long been a matter of controversy. Over many years, the common law has developed a number of general tests of employment status. As will be shown in this chapter, many of the key cases have involved workers within the construction industry. The question of employment status generally has become increasingly important as successive governments have sought to introduce legislation providing not only for specific employment rights but also in order to ensure increased regulation in relation to, for example, matters of health and safety. In spite of these developments, the question of the employment status of many workers within the industry remains one of considerable complexity.

The construction industry is unique in the number of individuals within it who are considered to be self-employed. Figures of 2001 indicate a conservative figure of 361,000 (Harvey, 2001), although the true figure could be considerably higher. According to the 2001 Labour Force Survey, 35 per cent of the construction workforce is self-employed, a figure greater than in other sectors of the national workforce (Forde and MacKenzie, 2007). On the other hand, it is clear that very few individuals who work in the construction industry can be said to be 'in business on their own account'. Why is it that the 'true' status of such individuals cannot be revealed? In other words, why can the law not identify those relatively low-skilled, low-paid workers who should be protected and bring them within the umbrella of employment law while at the same time allowing those who are truly self-employed to benefit from that status? In order to answer this question, it is necessary first to examine the various general tests of employment status that have been laid down by the courts.

4.2   The development of tests of employment status

In the early cases, which examined more clearly the personal master/servant type relationship as opposed to the more multifaceted relationship with which we are more familiar today, the issue of control was central to the relationship. According to the Court of Appeal in *Yewens* v. *Noakes* (1880), this required examining whether the master/employer controlled or had the right to control not only what the servant/employee did but also the manner in which he did it. Clearly, this test can still be applied to many employment relationships today. Nevertheless, it was recognised early on that this test, when taken to its logical conclusion, could produce bizarre results. For example, in the case of *Hillyer* v. *Governors of St. Bartholomew's Hospital* (1909), it was held that nurses were not employed by a hospital when carrying out duties in the operating theatre because they were acting under the directions of the surgeon. As a result of decisions such as this, the test was expanded to consider a number of indicators of control. Thus, in *Short* v. *J & W Henderson Ltd* (1946), Lord Thankerton looked at whether the putative employer had power of selection of his employee, the right to control the method of doing the work and the right of suspension and dismissal. According to Lord Thankerton, if these factors were present and the worker received a wage or other remuneration, he was properly termed an 'employee'. Nevertheless, even with this more developed application of the control test, control alone is always going to be a problem in these days of advanced technology.

Lord Denning attempted to develop an alternative test in the case of *Stevenson, Jordan and Harrison Ltd.* v. *Macdonald and Evans* (1952). This test, which came to be known as the integration test, is almost the reverse of the control test. Thus, instead of examining the employers' rights over workers, the test looks to the workers to see how much, if any, independence they had. Was the person in question fully integrated into the employer's organisation? However, this test often takes us no further than the control test. Thus, determining whether the worker was under the control of his employer was often the only means of determining who was part and parcel of a business.

It quickly became apparent that there was no one single test which could deal with all instances, particularly difficult cases. The result was the development of the economic reality or multiple test. This is really a form of composite test, looking at the issue from all sides and considering a number of factors. The leading early case in the development of this test was *Ready Mixed Concrete (South East) Ltd.* v. *Minister of Pensions & National Insurance* (1968). In this case, the employing company sacked its drivers, sold its lorries to them and re-engaged them under a written contract. The lorries remained painted in the company colours and the contract specified that the drivers had to wear the company's uniform, use the lorries only for company business, place them at the company's disposal for a set number of hours and obey the foreman's orders. All of these factors pointed to there being a contract of service. On the other hand, the drivers had to maintain the lorries and pay running costs, they could hire substitute drivers, decide their own hours as long as they completed their allocated jobs within a specified time and they paid their own tax and national insurance. All of these factors pointed towards self-employment. On the facts of the case, it was held that the drivers were self-employed. According to Mr Justice McKenna, there were three requirements for a contract of service/employment to exist: (a) the employee agrees that in consideration of a wage or other remuneration that he or she will provide his or her own work and skill in performing some service for the employer; (b) the employee agrees, expressly or implicitly, that in performance of that service he or she will be subject to the employer's control to a sufficient degree and (c) that the other provisions of the contract are consistent with it being a contract of service.

The test was developed further in *Market Investigations Ltd.* v. *Minister of Social Security* (1969). The case concerned a market researcher who was engaged on a fixed remuneration for a particular research survey. The company specified the questions to be asked and the people who were to be approached. However, the worker was allowed to do the work when she chose within a specific time period. The contract contained no provision for time off or sick pay. Nevertheless, the court held that she was employed. Mr Justice Cooke suggested that the test is whether the extent and degree of control exercised by the master is consistent with a contract of service and there are not any signs of the person being in business on his own account, which would suggest the contrary. Accordingly, factors to be taken into account include the degree of control but also whether the worker provided her own equipment, whether she could hire helpers, whether she had some financial risk, whether she could profit from the job through 'sound management' and whether she had responsibility for investment and management.

One of the leading writers in the field of employment law, Norman Selwyn, has produced a very useful list of factors which should be taken into account by a tribunal in making its determination as to employment status as follows:

a the contractual provisions (*BSM Ltd* v. *Secretary of State for Social Services* (1978))

b the degree of control exercised by the employer (*Global Plant Ltd.* v. *Secretary of State for Social Services* (1972))

c the obligation of the employer to provide work (*Nethermere (St. Neots) Ltd.* v. *Taverna & Gardiner* (1984))

d the obligation on the employee to do the work (*Ahmet* v. *Trusthouse Forte Catering Ltd*)

e the duty of personal service (*Ready Mixed Concrete (South East) Ltd.* v. *Minister of Pensions & National Insurance* (1968))

f the provision of tools, equipment, instruments and so on (*Willy Scheiddegger Swiss Typewriting School Ltd.* v. *Ministry of Social Security* (1968))

g the arrangements made for tax, national insurance, VAT, statutory sick pay (*Davis* v. *New England College of Arundel* (1977))

h the opportunity to work for other employers (*WHPT Housing Association Ltd.* v. *Secretary of State for Social Services* (1981))

i other contractual provisions, including holiday pay, sick pay, notice, fees, expenses and so on (*Hamerton* v. *Tyne and Clyde Warehouses Ltd.* (1978))

j the degree of financial risk and the responsibility for investment and management (*Market Investigations Ltd.* v. *Minister of Social Security* (1969))

k whether the relationship of being self-employed is a genuine one, or whether there is an attempt to avoid modern protective legislation (*Young and Woods Ltd.* v. *West* (1980))

l the number of assignments, the duration of the engagement and the risk of running bad debt (*Hall* v. *Lorimer* (1994))

m the presence or absence of mutuality of the obligation to provide or do the work (*Carmichael* v. *National Power plc* (1998) CA, reversed (2000)) (Selwyn, 2004: 48).

Selwyn has noted that 'as long as the employment tribunals take these into account, their decision is a matter of fact, not law, and their findings (either way) cannot normally be challenged unless they took a view on the facts which would not reasonably be sustained' (ibid.). Nevertheless, it is not open to the tribunals simply to run a checklist of the various factors. They are required to look at the whole picture. As Lord Justice Nolan made clear in the case of *Hall* v. *Lorimer* (1994: 174),

In cases of this sort, there is no single path to a correct decision. An approach which suits the facts and arguments of one case may be unhelpful in another. I agree with the views expressed by Mummery J in the present case [that] 'it is not a mechanical exercise of running through items on a check-list to see whether they are present in, or absent from, a given situation. The object of the exercise is to paint a picture from the accumulation of detail. The overall effect can only be appreciated by standing back from the detailed picture which has been painted, by viewing it from a distance and by making an informed, considered, qualitative appreciation of the whole. It is a matter of evaluation of the overall effect of the detail, which is not necessarily the same as the sum total of the individual details. Not all details are of equal weight or importance in any given situation. The details may also vary from one situation to another.

4.3   The position of workers in the construction industry: Increasing casualisation?

It has been suggested that the construction industry has distinctive characteristics, such as its regular stops and starts, its mobility and its lack of standardisation, which require it to be treated exceptionally. However, Harvey suggests that 'the case for exceptionality is often over-stated, and even more so when it is used as an argument for the difficulty of maintaining continuing employment under a proper employment contract' (Harvey, 2001: 7). This argument is supported through a comparison of the UK industry with construction industries in the USA and in South Korea, examples of highly deregulated industries which, nevertheless, have far fewer self-employed workers than does the UK. Harvey concludes that 'the United Kingdom is completely out on a limb for its low levels of proper continuous employment and high levels of self-employment' (ibid.). The suggestion, therefore, is that high self-employment in the construction industry is not a necessity. Rather, it is a construct of the particular way in which the construction industry seeks to order its affairs. According to Harvey, the essence of the problem is increased casualisation which has become institutionalised within the industry through labour-only subcontracting supplemented by agency labour (ibid.: 22).

Labour-only subcontracting is not a new phenomenon. It was apparent in many of the traditional manufacturing industries from the 1880s until First World War, when the practice died out. However, while other industries maintained a more structured approach to industrial relations, the boom in construction after Second World War caused the re-emergence of this form of contractual relationship (Winch, 1998: 531). According to Winch, many reasons have been given for the re-emergence of the phenomenon. These include labour shortage, rising wage costs, rising state intervention, particularly in relation to taxation, the culture of the industry, and unemployment, as well as productivity and flexibility. Winch analyses each of these suggested reasons and concludes that

> as construction firms faced the falling demand and increasingly fragmented product mix of the 1970s, which were intensified in their effects at the level of the individual firm through the contracting system, flexibility became the dominant element in operating strategy. So far as human resources management is concerned, this has meant an increasing use of labour-only subcontracting rather than direct employment, because of the combination of productivity and numerical flexibility that it offers. Higher labour costs are offset by higher input and full variability as market conditions change. An added

bonus is that such contracts open up the possibility of self-employment and consequent savings in employment overheads. Workers have acquiesced in, and at times encouraged this strategy because of the fiscal advantages of self-employment and the culture of the industry.

(Winch, 1998: 537)

Employment law has never been able to come to grips with this form of employment relationship satisfactorily. The courts sought to deal head-on with the problem of labour-only subcontracting in the case of *Ferguson* v. *John Dawson & Partners (Contractors) Limited* (1976). In that case, the Court of Appeal accepted the decision of Mr Justice Boreham at first instance (unreported) in which he declared that

> There remains the question: what matters are alleged to be inconsistent with the relationship of master and servant? First, there is the fact that the plaintiff and the defendants both regarded or labelled him as a 'self-employed labour only contractor'. Secondly, there is the fact, and fact it is, that the plaintiff was supposed to pay his own insurance stamp as a self-employed person; that his wages were free of tax and that the obligation was on him as on a self-employed person to account to the Inland Revenue properly for tax. But I ask this question: when all the other indicia point to the relationship being that of master and servant, are these inconsistencies which should compel me to a contrary conclusion? . . . I think not (826).

It is clear from *Ferguson* that where there is merely an agreement between the parties that the worker should be regarded as self-employed and all of the other factors point towards employment status then the agreement should be ignored.

However, the industry is constructed in such a way that these other factors increasingly point away from employment status. Part of the problem comes from the use of terminology and 'conceptual difficulties' existing around the malleability of concepts of 'subcontracting', 'labour-only subcontracting' and 'self-employment' (Winch, 1998: 531–532). Further confusion is caused by the multiplicity of contractual relationships at work within the industry. In many cases, it is difficult to distinguish between contracts involving capital risk, which are not properly employment contracts at all, and labour-only contracts, which may or may not result in the finding of employment status. An additional difficulty arises from the increased use of agency workers. As Forde and MacKenzie (2007) make clear elsewhere in this work, non-standard labour includes a high percentage of subcontracting (79.7 per cent) and other self-employed labour (47.3 per cent). However, it also involves a considerable degree of agency labour (35.6 per cent).

4.4 The problem of casualisation in the UK courts

The courts of the United Kingdom have increasingly been asked to consider the question of casualisation. However, these developments have occurred not out of complaints from workers in the construction industry but instead as a result of cases brought by casual workers in other industries seeking the protection provided by employment status. On the whole, casual workers have not been successful in persuading the court as to their employment status. In such cases, the courts have focused on two of the factors highlighted by Selwyn in the list of factors referred to above, that is the obligation to provide work and the obligation to carry out work. The combination of these two factors is referred to as 'mutuality of obligations'.

The leading early case on this point, which did in fact find in favour of the casual workers, was *Nethermere (St. Neots) Ltd.* v. *Gardiner* (1984), which concerned the status of homeworkers. The Court of Appeal held that, although the plaintiffs were paid by the piece and there was no obligation to work set times, there was an expectation of being given work and a corresponding obligation to accept and perform some minimum amount of work and that therefore they were employees. On the other hand, the lack of mutuality of obligations was used in the case of *O'Kelly* v. *Trusthouse Forte plc* (1984) to deny employee status to a group of so-called 'regular casuals'. Similarly, in the 1998 case of *Clark* v. *Oxfordshire Health Authority*, it was held that an individual who worked for a nurse bank was not an employee of the bank because of a lack of mutuality. In terms of the argument referred to above, mutuality of obligations is, of course, only one factor to be taken into account in determining employee status. On the other hand, it appears to be a very important one in the case of casual workers.

An attempt was made in the late 1990s to deal with the problem of regular casuals by arguing that they were subject to a global contract of employment. In other words, the multiplicity of individual hirings should be combined so as to create an overriding contractual arrangement. This approach found some favour among the judiciary as is evidenced by the Court of Appeal decision in *Carmichael* v. *National Power plc* (1998) in which the court found by a majority that tour guides who were employed on a 'casual as required basis' were employees. In that case, the Court of Appeal appeared to play down the significance of mutuality, at least in the sense that it does not override other significant factors. It considered that

> since the [Employment] Tribunal's only reason for concluding that the applicants were not employees was the absence of mutuality, and no other reason was advanced why they should not have been held to be employees, on the facts as presented, the proper conclusion must be that the applicants were employed under contracts of employment.

It would appear, however, that the Court of Appeal found, as a matter of law, that the applicants were employees because there *was* sufficient mutuality of obligation in that case.

The decision was appealed to the House of Lords. However, before turning to consider that decision, it is necessary to consider another Court of Appeal decision in the case of *Express and Echo Publications Ltd* v. *Tanton* (1999). In *Tanton*, the Court of Appeal held that the absence of an 'irreducible minimum of obligation' at least on the part of the employee is an overriding factor. Thus, where an individual was allowed in practice to provide a substitute on days that he was unable or unwilling to work, the Court of Appeal held that that was 'inherently inconsistent' with employment status. Relying principally on the decision of McKenna J. in *Ready-Mixed Concrete*, Lord Justice Peter Gibson held that

> it is necessary for a contract of employment to contain an obligation on the part of the employee to provide his services personally. Without such an irreducible minimum of obligation, it cannot be said that the contract is one of service.

Two issues would appear to arise from these two Court of Appeal decisions. First, what in fact is the requirement of an irreducible minimum of obligations? Different standards appear to have been applied by the Court of Appeal in each of these cases. Thus, in *Tanton*, the test appears to have been very strictly applied so that occasional substitution was enough to remove employee status whereas in *Carmichael* the test appears to have been less strict allowing for casual workers to be considered employees. Secondly, as Rubenstein points out in his editorial on *Tanton*, this case is a dangerous precedent. Thus, he notes that the case 'opens the possibility for employers and their advisers to draft contracts which will negate employment status for certain workers by including a substitution clause in their contract' (Rubenstein, 1999: 337).

As has already been noted, leave was granted in *Carmichael* for the employers to appeal to the House of Lords. The legal fraternity and, one would imagine, many employers and employees were hoping for a definitive statement from the House on the legal status of casual workers. However, their hopes were misplaced. In a rather technical decision based on a narrow point of law, the House of Lords allowed the appeal holding that the workers in question were not employees of National Power plc. This decision has been criticised by Rubenstein, among others, who has noted that

> the House of Lords decision in *Carmichael* v. *National Power plc* (2000) has turned out not to be the definitive decision on the employment status of casual workers, but instead a decision which shows that there are limits to how far the existing statutory definition of employment can be pushed.
>
> (Rubenstein, 2000: 1)

Many agreed with Rubenstein's somewhat harsh criticism of the decision of the House of Lords preferring the more casual-friendly approach of the Court of Appeal.

However, the House of Lords was simply following an approach which should be consistently applied through all of the cases concerning employment status.

That is, to examine each case individually, paying particular regard to the intention of the parties and examining all of the available evidence, including not only the written contract of employment, if indeed there is one, but also the conduct of the parties. The House of Lords based its unanimous decision on whether an exchange of letters between the parties did or did not amount to the entirety of the legal relationship between them. The House found that it did not. According to Lord Irvine of Lairg,

> the industrial tribunal must be taken to have decided that the [exchange of letters was not intended to constitute an exclusive memorial of their relationship] but constituted one, albeit important relevant source of material from which they were entitled to infer the parties' true intention.

(p. 45)

Essentially, what the House of Lords was saying is that the issue of employment status is a matter of fact and a decision of an Employment Tribunal cannot be interfered with unless it could not reasonably be sustained.

It is worth noting that Lord Irvine did specifically add, albeit in obiter, that had the appeal turned exclusively on the true meaning and effect of the documentation, it would have been held, as a matter of construction, that there was no obligation on the company to provide casual work or on the applicants to take it and that there was therefore an absence of the irreducible minimum of mutual obligation necessary to create a contract of service. This decision is undoubtedly a harsh one from the perspective of casual workers who may work exclusively for a single employer, as was the case in *Carmichael*. Nevertheless, to try and create a global contract of employment from a situation where none existed would have done a considerable disservice to the employer.

It might be possible to make an argument in favour of the global contract approach within the construction industry. The situation in *Carmichael* appeared to be unique, which is perhaps one of the reasons why it was appealed all the way to the House of Lords. Nevertheless, this type of 'hiring as required' approach is common within the construction industry. However, the precedent of the *Carmichael* case would be a difficult one to overcome. As will be argued below, it is not so much a problem of the law undermining the global contract of employment approach, but rather an approach within some sections of the construction industry which works to avoid the finding of employment status by the courts.

4.5   The employee/worker distinction

The issue relating to mutuality of obligation has been complicated further by recent developments arising out of the UK's membership of the EU. Specifically, the UK has recently implemented the EU's Working Time Directive 1993 in the form of the Working Time Regulations (WTR) 1998. The regulations apply to 'workers' rather than 'employees' and include provisions on, among other things, working hour restrictions, night working, annual leave and rest breaks. Regulation 2(1) defines a worker as

> an individual who has entered into or works under (or where the employment has ceased, worked under) – (a) a contract of employment; or (b) any other contract… whereby the individual undertakes to do or perform personally any work or services for another party to the contract whose status is not by virtue of the contract that of a client or customer of any profession or business undertaking carried on by the individual.

Clearly in terms of paragraph (a) all employees are also workers for the purposes of the Regulations. Paragraph (b), however, extends the definition of worker to persons other than employees to include persons who 'undertake personally to perform any work or services'. While similar to the concept of mutuality, this requirement is not identical to mutuality, which requires obligations on both sides of the contractual arrangement. However, the requirement of personal service is one of the factors discussed above in relation to employee status and was, indeed, the first of the factors to have been identified by McKenna J. in the leading case of *Ready-Mixed Concrete* referred to earlier.

It is worth noting the 2001 decision of the Employment Appeal Tribunal (EAT) in the case of *Byrne Bros (Formwork) Ltd* v. *Baird* (2002). The case concerned a claim for holiday pay by Baird, who was one of a number of workers working directly for Byrne Brothers as carpenters or labourers at a particular construction site. Baird had signed a subcontractor's agreement, which specifically provided that he was not entitled to holiday or sick pay. The EAT found that there was no obligation on Byrne Brothers to offer work and no obligation on Baird to accept an offer of work. Accordingly, there was no contract of employment. The agreement further provided that where a worker was unable to provide personal service, he was entitled, at his own expense and with the permission of the employer, to provide an alternative. Byrne Brothers argued that the effect of the provision was to remove Baird from the remit of the Regulations. However, the EAT dismissed the appeal holding that 'the limited provision permitting the substitution of an alternate worker was not inconsistent with an obligation to provide services personally'. The EAT further noted that 'self-employed workers in the construction industry were exactly the sort of worker whom the Regulations were designed to protect'.

The development of new laws providing a level of protection to workers who are not otherwise entitled to the benefits of employment are, undoubtedly, to be welcomed. Nevertheless, the rights provided to workers by the Regulations are considerably fewer than those provided to employees by the relevant employment legislation, such as the right to claim unfair dismissal. Furthermore, it is still necessary for workers to prove their status in order to benefit from those limited rights. Accordingly, the courts are now faced with another series of difficult decisions on worker/non-worker status. What has become quickly apparent is the willingness of the courts to take a more purposive approach to such decisions than might be apparent in relation to the employee/self-employed distinction discussed above. The case of *Baird* referred to above is an excellent example of such an approach.

Nevertheless, another recent case arising out of the construction industry has again highlighted the fundamental requirement that the creation of this new category of 'workers' is not intended to be a catch-all provision. In the joined cases of *Redrow Homes (Yorkshire) Ltd* v. *Wright* and *Redrow Homes (North West) Ltd* v. *Roberts & Others* (2004), the applicants worked for Redrow as bricklayers. Redrow provided bricks, pre-mixed mortar, forklift trucks and drivers, scaffolding and one labourer per site. The applicants provided their own hand tools. They were given a set of drawings and were subject to a building programme. They could regulate their hours to suit themselves as long as they stayed within the time periods laid down in the building programme. The applicants submitted a claim for payment each week. Payments were made into each person's bank account. The applicants worked on the basis of an official order accompanied by a document headed 'Conditions and Acceptance of Order'. Clause 6 of the document provided that 'the contractor must at all times provide sufficient labour to maintain the progress laid down from time to time by the company, and shall supply such labour with all necessary tools and equipment'. The question before the tribunal was whether the applicants were workers for purpose of the Working Time Regulations and therefore entitled to paid annual leave.

Redrow argued that, as a result of Clause 6, the contract did not contain an obligation on any of the applicants to do the work personally. They simply had to ensure that there was sufficient labour provided. However, both Employment Tribunals and the EAT unanimously held that the applicants were workers within the meaning of the Regulations. The Court of Appeal upheld these decisions finding that the 'Conditions and Acceptance of Order' was a standard form contract governed by Clause 1 which provided that 'the contractor, having had an opportunity of inspecting [the] conditions of contract, shall be deemed to have noted its provisions, and hereby agrees to be bound by them *in so far as they are applicable to his sub-contract*' (emphasis added). Having noted that 'it is not suggested that the applicants worked under a contract of employment' (*supra*, 721), Lord Justice Pill went on to conclude that

> In my judgment, the intention of the parties when the contracts were made involved, in each case, an obligation on the applicants to do the work personally. That makes sense of Redrow's decision to contract with bricklayers individually. The scheme for payment points strongly in the direction of contracts with individual bricklayers to do the work personally . . . I agree with the decision of the EAT that it was the intention of the parties that personal services be provided.

> (p. 724)

Nevertheless, Lord Justice Pill reached his conclusion after sounding an important note of caution:

In my judgment there is force in the submission that employment tribunals should not be deflected from a consideration of the definition of 'worker' and from a consideration of the terms of the contract in that context by general policy considerations as to the nature of employment and self-employment. The reasoning of the tribunal in *Roberts*, with its long citation from *Byrne Brothers*, appears to come close to saying that, because the applicants ought to come within the definition of worker, it follows that they do. The Regulations leave parties free to enter contracts and, whether or not the contract includes an obligation to do the work personally, is a metre for construction.

(p. 723)

Essentially, the position of the Court of Appeal in relation to the interpretation of the worker/non-worker distinction differed very little from that of the House of Lords in *Carmichael* in relation to the employer/self-employed distinction. That is, that ultimately the courts can do no more than interpret the contractual arrangement between the parties.

Where does this leave us in relation to the employment status of construction workers? It is submitted that if the parties choose to enter into an agreement to provide casual work in a way that does not involve mutuality of obligations in order to secure employment status or provide for personal service in order to secure worker status, then the courts cannot conjure into existence a global contract which does not exist. Of course other factors come into play in determining employment status if not worker status. However, within the construction industry those other factors, including the issue of taxation, have historically pointed towards self-employment status.

It may be argued that individuals in the construction industry are not in a position to negotiate the terms of their contractual arrangement with their 'employer' in the light of cultural pressures within the industry. This may well be the case but what it highlights is not a failure in the law but rather a failure in the 'way in which [the industry] is institutionally organised' (Harvey, 2001: 18). As part of this process, it is certainly the case that the industry as a whole should seek to engage with the judiciary and the legal profession more generally in order to highlight the uniqueness of the industry and to seek greater understanding for its specific problems.

4.6 Conclusions

The employment status of workers in the construction industry has always been a matter of considerable controversy. Increased casualisation has complicated matters still further. Furthermore, while the introduction of the new category of 'workers' has ensured that those who cannot properly be regarded as employees are, nevertheless, entitled to a limited number of protections, the issue of who can properly be described as workers introduces another level of complication into the equation. Legal precedent in this area is undoubtedly complex and, in many cases, inconsistent. On the other hand, the decision of the House of Lords in *Carmichael* in relation to employment status and that of the Court of Appeal in *Redrow* in the case of workers make it clear that, in reviewing all of the evidence available to the relevant tribunal, the principal role of those tribunals is to identify, so far as possible, the intention of the parties to the relevant contract. It is up to the industry to decide if it is in its interest to continue to manifest an intention of casualisation rather than permanence.

References

Druker, J. (2007) 'Industrial relations and management risk in the construction industry', in this volume, in Dainty, A., Green, S., and Bagilhole, B. (eds), *People and Culture in Construction: A Reader*, Oxford: Taylor & Francis.

Forde, C. and MacKenzie, R. (2007) 'Concrete solutions? Recruitment difficulties and casualisation in the UK construction industry', in Dainty, A., Green, S. and Bagilhole, B.M. (eds), *People and Culture in Construction: A Reader*, Oxford: Taylor & Francis.

Harvey, M. (2001) *Undermining Construction: The Corrosive Effects of False Self-Employment*, London: Institute of Employment Rights.

Rubenstein, M. (1999) 'Editorial', *Industrial Relations Law Reports*, 337.

Rubenstein, M. (2000) 'Editorial', *Industrial Relations Law Reports*, 1.

Selwyn, N. (2004) *Selwyn's Law of Employment*, London: Butterworths.

Winch, G. (1998) 'The growth of self-employment in British construction', 16 *Construction Management and Economics*, 531.

Cases cited

Ahmet v. *Trusthouse Forte Catering Ltd* [1982] IDS Brief 250.

BSM Ltd v. *Secretary of State for Social Services* [1978] ICR 894.

Byrne Bros (Formwork) Ltd v. *Baird* [2002] IRLR 96.

Carmichael v. *National Power plc* [1998] IRLR 301, CA.

Carmichael v. *National Power plc* [2000] IRLR 43, HL.

Clark v. *Oxfordshire Health Authority* [1998] IRLR 125.

Davis v. *New England College of Arundel* [1977] ICR 6, 11 ITR 278.

Express and Echo Publications Ltd v. *Tanton* [1999] IRLR 367.

Ferguson v. *John Dawson & Partners (Contractors) Limited* [1976] 3 All ER 817, IRLR 346.

Global Plant Ltd. v. *Secretary of State for Social Services* [1972] QB 139.

Hall v. *Lorimer* [1994] IRLR 171.

Hamerton v. *Tyne and Clyde Warehouses Ltd.* [1978] ICR 661.

Hillyer v. *Governors of St. Bartholomews Hospital* [1909] 2 KB 820.

Market Investigations Ltd. v. *Minister of Social Security* [1969] 2 QB 173, [1968] 3 All ER 732.

Nethermere (St. Neots) Ltd. v. *Taverna & Gardiner* [1984] ICR 612, [1984] IRLR 240.

O'Kelly v. *Trusthouse Forte plc* [1984] QB 90, [1983] 3 All ER 456.

Ready Mixed Concrete (South East) Ltd. v. *Minister of Pensions & National Insurance* [1968] 2 QB 497, 1 All ER 433.

Redrow Homes (Yorkshire) Ltd v. *Wright* and *Redrow Homes (North West) Ltd* v. *Roberts & Others* [2004] IRLR 720.

Short v. *J & W Henderson Ltd* (1946) 62 TLR 427.
Stevenson, Jordan and Harrison Ltd. v. *Macdonald and Evans* [1952] 1 TLR 101.
WHPT Housing Association Ltd. v. *Secretary of State for Social Services* [1981] ICR 737.
Willy Scheiddegger Swiss Typewriting School Ltd. v. *Ministry of Social Security* (1968) 5 KIR 65.
Yewens v. *Noakes* (1880) 6 QBD 530.
Young and Woods Ltd. v. *West* [1980] IRLR 201.

Legislation cited

Working Time Directive (Directive 93/104/EC).
Working Time Regulations 1998 (1998/1833).

5 Industrial relations and the management of risk in the construction industry

Janet Druker

Abstract

This paper explores the tension between two distinctive – and apparently contradictory – perspectives on industrial relations in the UK construction industry. On the one hand the industry is marked by unilateral decision-making by employers and a vigorous assertion of the managerial prerogative. The majority of the workforce is self-employed and without a formal mechanism for a 'voice' within the industry, lacking both collective representation and support. On the other hand the industry is characterised, through its various sub-sectors, by institutional interest representation at national level and formal national-level industrial relations machinery providing for sector-wide, multi-employer collective bargaining. This provision has outlasted similar arrangements in other sectors which, typically, were dismantled in the last decades of the twentieth century.

It is argued that, from the employers' perspective, the framework for industrial relations is sustained as a form of risk management. Given the unpredictability of the human element within the construction process, national collective bargaining represents the machinery for damage limitation and for cost containment in managing people. The national machinery for bargaining is not seen by employers as a vehicle for enhancing and extending employee commitment or worker 'voice'. Nor does it serve significantly to extend participation in decision-making. Provision for employee involvement within the industry is absent, despite the formal recognition of trade unions within the sector. There is little evidence that employers have turned to concepts such as 'high commitment' or 'high performance management' – noted in some other sectors – as a means of implementing change. The various initiatives since the Latham Report ten years ago suggest a change in employer orientation, yet to date there is little evidence of substantive innovation in industrial relations management.

5.1 Background: Construction and risk management

The management of risk is central to business activity as decision takers are bound to weigh the significance of risks as they consider options and choices, setting directions for their future activities and commitments on the basis of imperfect knowledge and understanding. Risk management can be understood in a variety of different ways. It is commonly conceived in terms of the management of financial risk, although at project level the financial issue is bound up with questions of time as well as cost. However, the experience of major global disasters, whether artificial or natural in origin, has encouraged risk avoidance in the protection, insurance and back-up for physical and technological resources too. Within this context, questions of environmental management, safety management and issues of legal compliance are closely interrelated.

Within the construction sector, shaped by intense competition for contracts and associated financial pressures within contract management to deliver to price, to time and to specified outputs, the management of risk provides a reference point for critical decisions in pricing, planning and work organisation (see, for example, Hayes *et al.*, 1986).

Success in winning contracts is central to the survival of private contractors and subcontractors yet the significance and benefits will differ over time and between contracts. Not all contracts offer the prospect of immediate profitability. Some will depend upon variations that are anticipated and, with tight profit margins, variable interest rates and restrictions on credit, a contract may be important in sustaining income and managing cash-flow in a period of contract famine. In such difficult circumstances, whether we are discussing major projects or work undertaken for domestic clients by jobbing builders, decision takers within the industry are concerned with the creation of conditions in which the future can be planned and controlled and risks minimised or managed. Contract management, in this context, can be understood as a process of negotiation and balance between multiple stakeholders. Managing risk is central to this management process and necessitates a tension between the possibility of conflicting interests and demands relating to cost, time and quality.

Without contracts there is no employment, and strategic decisions concerned with contract formation and with risk management impact upon the creation of social relations – on the culture – experienced both within and between organisations. As contracts vary in location, size and type, so too does the demand for skills and the balance between different trades or professional activities. On each occasion a project is costed on the basis of a particular labour price, and there is a risk that labour may not be available or may not be available at that estimated price. There is a risk too that competitors may seek advantage by bidding on the basis of low labour costs, with the possibility of driving out higher priced competitors. From time to time prices might be raised, and the individual contractor is vulnerable to lack of prior knowledge and assurance about competitor behaviour in this situation.

The issue of risk in this situation derives from the variability – or the unpredictability – of human behaviour. Hence contractors have sought to create and sustain a framework of shared expectations and rules. It is important that this framework draws together different employer interests, going beyond the individual employment relationship and creating formal institutional interest representation for employers, through employers' associations. By this means, employers can combine to set standards – for example, in relation to pricing decisions – and to lobby governments on behalf of shared interests.

Trade unions face particular difficulties in working across organisational boundaries where organisations may work together through contractual or supply chain networks (Marchington *et al.*, 2004). In such cases, workers may be affected not only by the actions of their own managers but also by others. This is the problem that has, historically, faced unions in the construction sector and it goes some way to accounting for their emphasis on multi-employer collective bargaining over many decades. The major construction unions date their origins at least to the nineteenth century. The institutional framework of industrial relations in construction in the UK is, in a formal sense, remarkably continuous with the picture presented by Dunlop in his comparative study undertaken in the mid-twentieth century (1958, reprinted in 1993). Trade union activities are defined particularly by the complex nature of many projects, involving different trades brought together in a fixed work location for a finite period. High levels of subcontracting and the small size of many firms are constraints on union organisation, further compounded by the vulnerability to the weather on outside work and the hazardous nature of work itself.

The discussion of industrial relations in the construction sector must be located within a wider context, marked by waning trade union influence and the diminished influence and impact of collective bargaining – not only within Britain but also in other developed economies (Gospel and Wood, 2003: 2–8). In the private sector, union density was below 20 per cent in 2004 and although union membership has been stabilising in recent years there are questions about the future of worker representation in Britain (Heery *et al.*, 2004). The advent of new statutory employment rights in the 1990s – for example, rights to a National Minimum Wage or to paid holidays – may have encouraged recourse by workers to statutory redress through Employment Tribunals, as an alternative or an addition to work-based support through trades unions. Other forms of worker involvement – for example through Joint Consultation Committees – have also been noted, although again not necessarily to the exclusion of trade union representation (Gospel and Wood, 2003: 12).

This chapter highlights the tension that exists within the management of industrial relations in the UK construction industry. Contrasting perspectives are evident since, on the one hand, the industry continues to be marked by institutional interest representation at national level with formal industrial relations machinery providing for collective bargaining. On the other hand, though, the construction industry is often characterised by unilateral decision-making by employers and a vigorous assertion of the managerial prerogative. The chapter explores the underlying reasons for this tension.

It begins with a discussion of risk management as a framework for social and cultural exchange, within construction contracts. There follows an analysis of the national framework of industrial relations, outlining key reasons for the longevity of national collective bargaining in construction. The contrast to be found between the terms and conditions of employment set out in working rule agreements and the conditions applied in practice within the major sub-sectors of construction are discussed. Successive studies have commented on the pressures within the private sector of the industry to outsource or subcontract significant parts of the construction process. Subcontracting is endemic within the UK since it enables major contractors to minimise risk and to download the costs and the responsibilities of employment, drawing on skills as and when they are required (Beardsworth *et al.*, 1988). The use of self-employed labour remains important within the construction industry, despite forecasts some years ago that this form of labour engagement was to diminish or even to disappear (Macaulay, 2003).

Of course there is variation in practice between different trades, interests and collective agreements. The origins of the engineering construction industry agreement in 1981 differ from those much earlier in building and civil engineering (Korczynski, 1997). Engineering construction is distinctive because of its clients (including power, oil, chemicals and petrochemicals, nuclear waste reprocessing and food manufacture) and its working practices; it retains stronger traditions of direct labour than in, say, building or civil engineering. Electrical contracting also has its own distinctive history, although, as in the building industry, there are pressures for employers to subcontract, with evidence of a growing use of self-employed labour (Gospel and Druker, 1998). Yet there are sufficient similarities to justify discussion here of trends across the wider construction industry and the development of an analysis which is founded on identification of common influences.

During the last 20 years the language and practice of human resource management (HRM) has influenced the ways in which the employment relationship is managed in many sectors. Driven by initiatives from the US, the influence of HRM has spread throughout Europe and indeed across the globe as senior managers have espoused the importance of 'people' to business results. This is not the point at which to review the associated debates, but it is important to note that successive studies have shown that human resource initiatives will take diverse forms and have different significance depending upon context and upon the attitude of senior decision takers. The growing recognition of the importance of human effort and commitment to business success has inspired a plethora of initiatives, not all of them as thorough-going as the rhetoric that accompanies them. (For a review of some of these see, for example, Boxall and Purcell, 2003.) Given this emphasis on change, the continuity suggested by the survival of traditional industrial relations machinery may seem surprising. However, Hyman's analysis of wider trends in employee representation in the UK suggests that surviving industrial relations machinery constitutes a 'hollow shell', with an absence of independent representation of worker interests (Hyman, 1997). It is a point that is pertinent to the discussion that follows.

5.2   Risk as a reference point for industrial relations practice

Central to the management of risk is the intention of controlling as much as is practicable within the planned sphere of activity, reducing uncertainty to the very minimum (Bernstein, 1996). Not everything can be controlled, however. Human action deriving from the decisions of others represents a central source of unpredictability. Drawing on the work of Von Neumann, Bernstein points to game theory as an explanation for uncertainty, and for the ways in which it might be managed (Macrae, 1992, cited in Bernstein, 1996: 232). As Bernstein puts it, real life is 'a game of strategy, combined with contracts and handshakes to protect us from cheaters' (Bernstein, 1996: 232).

What game theory and an understanding of risk can offer us here is an appreciation of the factors that influence different parties to the construction process in formulating the framework for their relationships and interaction. Contractors or their agents – managers acting on their behalf – seek to reduce the uncertainty deriving both from the nature of commercial contracts and also from the behaviour of workers by setting in place a framework of rules through which behaviour can be regulated and controlled.

Managers and workers on a particular construction project operate within the parameters of strategic decisions already taken. The construction project is spatially and temporally framed and is of fixed duration. Contractors operate on the basis of a commercial contract – and of course such contracts are necessarily bounded by the requirement and processes that may be specified. The process of subcontracting – and the inherent organisational fragmentation – carries with it the further risks inherent in contractual complexity and human interaction. Workers engaged for the project are expected to perform according to pre-specified output criteria, according to a critical path that will lead to a timely completion. However, the worker is hired on the basis of an expectation or a promise to deliver a certain amount of work in a specified time period and the project is vulnerable to any slippage in delivery on that promise – or to any human intervention impeding rather than hastening progress.

Like other processes that are required within a fixed time-scale – in the printing industry or in engineering production, for example – the financial return on the construction project may be undermined by delays in completion.

The project is dependent upon human performance and requires from the entire workforce adequate skill, willingness to engage with the project at the specified price and a capacity and willingness to perform at the pace required.

5.3   The institutional framework of national multi-employer bargaining

The organisations associated with collective bargaining within the construction industry can be best understood as growing out of this need to manage risk and to reduce uncertainty. Employers historically sought to ensure a common industry-wide framework for training and for pay systems and structures. In the event of workers individually or collectively challenging such a framework, employers needed to be confident that they would be a part of a wider support network involving other employers – the 'contracts and handshakes' in Bernstein's words – that would represent their interests. Employers' associations and trades unions in the building industry can trace their histories back at least to the end of the nineteenth century. Although the names and structures of organisations have changed over time, the major associations can demonstrate remarkable continuity in their operations since that time.

Relations between employers and unions developed unevenly prior to 1914, but by the inter-war years associations of employers and trade unions were co-ordinated along national lines and agreements for national multi-employer collective bargaining were established in building, civil engineering, electrical contracting and plumbing.

Employer interest in collective bargaining derived from three key factors, all of them concerned essentially with the management of risk. First, rates of pay set by collective agreements were intended to ensure a level playing field in the competition for contracts, removing one area of variability in the contracting process (Gospel, 1985). Employers' associations endorsed collective agreements that regulated the price of labour. Secondly, closer relations with trades unions that derived from formal and ongoing collective bargaining structures provided employers with an 'insurance policy' in controlling and containing conflict, since the human factor in the construction process was one of the most unpredictable that they might have to deal with. National procedures meant that, in the event of a dispute, national conciliation procedures were expected to govern behaviour and to restrain militancy at site level. As one commentator put it, 'procedures are treaties of peace and devices for avoiding war' (Marsh, 1966: vii).

Thirdly – and paradoxically – the recognition of trade unions at national level managed the risk that employers might be vulnerable to a push for union recognition. National collective bargaining ensured that unions were a known and accepted phenomenon at national level. At the same time, there was no necessary or automatic implication that trades unions would be represented on site. Hence national collective bargaining was both the vehicle for familiarity and distance for employers in addressing issues of employee representation.

Despite the impact of significant disputes in the 1960s and a major national strike in 1972, national multi-employer, industry-wide collective agreements remain characteristic of the construction industry in the twenty-first century. This is not to suggest that there has been no change in practice. The political climate of the 1980s and 90s had a substantial impact on construction, encouraging self-employment. Yet, perhaps surprisingly, there has been significant continuity in the machinery for collective bargaining. Multi-employer agreements survive in electrical contracting, plumbing, heating and ventilating, and engineering construction. In 1980–81, a new agreement for the engineering construction industry was created, counter to the national trend, and reinforced the raft of national agreements which survived in construction (Korczynski, 1997). There was a review of the structure of national bargaining during the late 1990s but, significantly, national multi-employer bargaining was retained, although two separate agreements for building and civil engineering were brought together in 1997. The scope of surviving agreements including separate arrangements made for Scotland is illustrated in Table 5.1.

Table 5.1 National collective bargaining in the British construction industry

Construction Industry Joint Council	Construction Confederation	Union of Construction Allied Trades and Technicians (UCATT) T&GWU, GMB
Electrical Contracting Joint Industry Board	Electrical Contractors Association	Amicus
Scottish Joint Industry Board for the Electrical Contracting Industry	SELECT (Formerly the Electrical Contractors Association of Scotland)	Amicus
Plumbing Mechanical Engineering Services Joint Industry Board	National Association of Plumbing, Heating and Mechanical Services Contractors	Amicus
Heating Ventilating Air Conditioning, Refrigeration & Domestic Piping Industry	Heating and Ventilating Contractors' Association	Amicus
Scottish and Northern Ireland Plumbing Industry	Scottish and Northern Ireland Plumbing Employers' Federation	Amicus
Engineering Construction Industry National Agreement	Engineering Construction Industry Association, Thermal Insulation Conatruction Association (TICA) and SELECT	Amicus, GMB & TGWU

This continuity in structures is significant because of the overall decline – demonstrated in successive surveys of industrial relations – in multi-employer, industry-wide bargaining as a source of pay determination (Millward *et al.*, 2000: 184–199). In other sectors, there has been a shift away from national bargaining. Where collective bargaining has been retained, there has been a trend to decentralisation, with single-employer bargaining including workplace bargaining more common than multi-employer bargaining (ibid.: 186). Workplace union membership density declined overall between 1980 and 1998, and the major time series of changes in union representation and impact reported in 1998 suggested that collective bargaining arrangements were increasingly a 'hollow shell' (Hyman, 1997; Millward *et al.*, 2000: 183).

National agreements were designed to address grievances and disputes and to delay industrial action where it was threatened. Viewed in the long term, they were remarkably successful in meeting these objectives and there were few large strikes. There was conflict in the 1960s when major disputes occurred in the building industry at the Barbican and Horseferry Road sites in London. Both regional and national disputes panels were invoked to contain them, although on this occasion the disputes ultimately sparked intervention beyond the industry itself (Report of a Court of Inquiry into trade disputes, 1967). None the less, the significance of formal national collective disputes procedures was reflected in the fact that significant militant union organisation was challenged by senior union officials as well as by employers (ibid.). The conflict led to predictions that the negotiating machinery would collapse:

> My personal view, and I think this is the view of quite a number of the people on our side of the industry, is that the National Joint Council and the negotiating machinery in the industry is likely to collapse within the next five years if something is not done to stop this method of employing labour.
>
> (Harry Weaver, General Secretary of the National Federation of Building Trades Operatives, quoted in Marsh and McCarthy, 1968: 61)

The years that followed – until the national building industry strike of 1972 – marked a watershed, as from that time on the pay rates set by the National Joint Council for the Building industry (NJCBI) fell increasingly out of alignment with those being paid unofficially at site level. Cash-in-hand payments and the use of 'lump' labour proliferated and union campaigning to counter these trends was given additional impetus through the work of a Communist inspired rank-and-file group, the Building Workers Charter.

While predictions of the ending of national bargaining proved premature, the national strike of 1972 marked a shift to its lesser significance in application. Concerted and politically focused employer action at the time of the 1972 strike was followed by changing approaches to contracting with the more widespread use of management contracting and construction management, involving greater use of systematic subcontracting. Direct labour gave way to self-employment and, since that time, real wages and conditions at site level, in the building industry particularly, have departed from the formally negotiated rates set by the NJCBI and subsequently the Construction Industry Joint Council. Trade union membership has diminished and union density in the private sector of building and civil engineering is now very low. Of course, in some sub-sectors of the industry, union membership continues to be above average – in electrical contracting, for example – but even here density has declined (Gospel and Druker, 1998). Two of the major unions – the Union of Construction Allied Trades and Technicians (UCATT) and the Transport and General Workers Union (T&GWU) – have seen significant decline in private sector membership.

The question then is why, when unions have such low membership and so little influence in large parts of the construction sector, the formal machinery survives? One answer could relate to innate conservatism within the industry; yet this interpretation could be misconceived since in so many respects the industry is far from conservative, as it adapts and responds to changing demands from clients and from the wider environment. An alternative answer is that continuity in the formal institutional arrangements for national collective bargaining derives from continuity in the advantages that accrue from these long-established arrangements in terms of risk management. Fragmentation within the industry may accentuate rather than diminish the need for negotiation.

The industry continues to be made up of many small firms (most of them without specialist personnel departments) and a very few large organisations.

Employers' associations offering professional advice and support continue to be relevant – and perhaps even more important for small firms as they seek to offset the risks of increased regulation. Collective agreements still provide a benchmark for estimating labour costs within the industry. Negotiated rates (consolidated and improved annually) are a useful reference point and the annual uprating process is recognised industry-wide (even where the rates themselves are not applied in practice). Since negotiated rates are often lower than actual rates (which may fall as well as rise in response to local supply/demand factors), they do not impede variability in terms of site-level practice. Industrial action has been rare since the national strike in 1972 but it can never be ruled out. Indeed the Engineering Construction Industry Association (ECIA) suggested that 'relative employment relations stability has been the single greatest benefit that the NAECI has brought for clients and contractors' (ECIA, 2004). The relationship with trades unions at national level still offers formal procedural arrangements with some illusion of influence and control in the event of disputes and the significance of this has not been undermined by new statutory entitlements to union recognition. Unions are party to national negotiating machinery and are consulted pragmatically by employers where site circumstances or client preference may dictate – typically larger sites or contracts that may be most at risk from effective union organisation and industrial action. The Major Projects Agreement (MPA), recently adopted by BAA, the airport company, at Heathrow Terminal 5, offers one example.

The coexistence of short-term, one-off projects, together with high levels of self-employment and overtly anti-union attitudes amongst employers, means that, by contrast with other sectors, trade union density is lower, unions are less likely to be represented and trade union representatives are rarely found at site level. Whereas in other areas of employment, joint consultative committees or works' councils may offer a route for worker representation, they are less common in the construction industry. Extracts from the 1998 Workplace Employee Relations Survey (WERS) data are presented in Table 5.2 showing both the lower trade union density and the lesser presence of consultative committees. Since some of the public sector (known for higher levels of union

Table 5.2 Employee representation in construction in 1998

	Workplace average construction (%)	Average all workplaces (%)
Trade union density	17	27
Recognised union	37	45
No union present	55 (of all workplaces)	47 (of all workplaces)
Aggregate density (% of employees who are union members)	30	36
No joint consultative committee	73	47

Source: 1998 Workplace Employee Relations Survey, Cully *et al.* (1999: 88, 92 & 99).

organisation) will be included in the 'construction' heading here, employee representation in construction is overstated in this table.

Table 5.2 suggests that trade unions are scarcely visible at the level of the establishment or workplace and, of course, the existence of a national level agreement does not mean that the individual employer is bound to deal with a trade union. Trade union officers invest much of their time and energy in recruiting and retaining members and the establishment of an effective union presence in the construction sector requires a high ratio of officers to members. Union leaders are unlikely to abandon national collective bargaining since it is one of their strongest points of influence within the industry, but they are well aware of the limitations of union power.

It seems unlikely, on the basis of current information, that the construction sector will demonstrate an increase either in trade union organisation or in employee representation through consultative committees in construction in the next round of WERS data. If the scope or depth of union influence were assessed on issues such as physical working conditions, staffing levels or other factors identified as indicators of union influence (Millward *et al.*, 2000) they would clearly be found very weak. At the same time, there is little evidence of consultation, or initiatives intended to encourage employee participation.

It might be argued that the national machinery has in some ways outlived its usefulness but, from the employers' perspective, it continues to mitigate risk. The benefits of such arrangements are questioned as the shape of the industry changes, with increased inward investment and tougher financial and competitive markets. The arrival of contractors from outside the UK has caused further tensions as, rather than recruiting locally, they often bring with them their own existing workforce operating to different standards. On the one hand, the collective machinery does very little to constrain employer freedom to engage labour on whatever basis the employer may choose and so the incentive for change or innovation is missing. On the other hand, it might be argued that the negative orientation of national bargaining – to avert risk – does nothing to promote innovation by addressing problems of employee motivation and commitment at site level. However, in order to understand fully the dynamics of industrial relations within the sector, we need also to review the ways in which self-employment contributes to work organisation and social relations within the industry. The formal structures described above constitute only a part of the picture. A critical factor – and one that cannot be ignored – is the widespread use of self-employed labour. It is this issue that is considered in the next section.

5.4   Industrial relations on site

Despite the appearance of formal structures associated with traditional pluralism, industrial relations at site level seem to characterise the 'bleak house' or 'black hole' scenarios in terms of management practice (Guest, 2001: 98). The approach at workplace level can best be described as unitarist although it is a form of unitarism that is thinly overlaid by pluralistic rhetoric.

There are a number of contextual factors to be considered. First, industrial relations practice is influenced by the size of the enterprise. Formal employment practices – for example, on recruitment and selection or discipline and grievance management – had to be used in larger workplaces and the influence of trade unions in smaller enterprises is known to be very low (Matlay, 1999). Given the predominance of small- and medium-sized enterprises (SMEs) in construction, it is unsurprising that the industry is characterised by informality on many employment-related issues which elsewhere might be governed by more formal procedures. Secondly, construction projects are often awarded at short notice and are one-off, with complex and multi-disciplinary requirements (Cheng, Dainty and Moore, 2005). Contractors and subcontractors engage labour in many cases solely for the time that it is required, without guarantee of continuity in employment contract between projects (Loosemore *et al.*, 2003). The pressures to move swiftly mean that recruitment is short-term and often by word of mouth, with recruits often coming via subcontractors or gangers, or through network or family contacts.

Employment is highly fragmented, with most workers – employed and self-employed – working in small units of employment. Informality may be perpetuated by training arrangements where learning is often undertaken 'on the job' without the benefit of formal arrangements. Apprenticeship training has declined as the scope for long-term, firm-based training has diminished.

The contract management processes and the subcontracting practices of larger construction companies necessitate a core workforce with key management skills – in surveying, accounting, project or contract management, for example – typically characterised by a professional identity that serves to engage and regulate this section of the workforce. Central to management requirements is the need to coordinate subcontract activity, including subcontractors' workers. Larger projects may involve dozens of subcontractors, each of them bringing in teams of workers who may or may not be directly employed. Some manual workers with core skills may be retained on a permanent basis, and self-employed workers may be regularly engaged with the legal status of 'worker' – working primarily for one 'employer' – while retaining self-employed status for tax purposes. Within this framework subcontractors undertake the brunt of labour management, providing the route through which activities can be expanded or collapsed in scope according to the scale of demand, a process which bears most heavily on itinerant and casual workers. It should be noted, however, that for those with specialist skills, these arrangements may facilitate greater continuity in engagement across projects. None the less, the informality and the fluidity of network-based engagement must inevitably distance workers at site level from the overall process of project management.

There are few signs that this approach to labour management, consolidated as the predominant approach in the years following the 1972 strike, is changing. The self-employed workforce in the UK grew between 1977 and 1997 with negative implications for training and innovation (Winch, 1998). There was a fall in the number of the self-employed in the late 1990s as a consequence of changes in Inland Revenue policy (Consultative Committee on Construction Industry Statistics, 1998). This was a short-lived trend, however, and the number of self-employed workers in construction has been increasing since 2000, despite forecasts to the contrary. Although the fastest rate of increase in self-employment across the economy over this period was in banking, finance and insurance, construction remains the industry with the highest number of self-employed, topping 700,000 in 2003. There was a rise of 53,000 self-employed workers in the sector in that year alone, of whom the majority were male, full-time workers aged between 35 and 49. The skilled trades represent the largest self-employed group by occupation, including many occupations within the construction sector (Macaulay, 2003).

The widespread use of self-employed labour can be clearly located within a strategy of cost and risk management. From the employers' perspective, national insurance and employment overheads are minimised. It is true that some employment rights – notably the National Minimum Wage (NMW) and minimum holiday entitlement – have been extended to 'workers' and might therefore be deemed to include the self-employed. However, the self-employed worker finds it more difficult to exert a claim for employment rights than someone who is explicitly and directly employed and the numbers able to enforce a claim are still relatively restricted. Significantly, there is no continuity in commitment for the self-employed worker. Such a worker is engaged solely when required and retained only for so long as the job in question lasts. The self-employed are often paid by the task rather than by the hour, encouraging task completion. Unionisation is less common among the self-employed and collective identity in the form of unionisation is less likely to threaten work organisation with disputes. In assessing risks, it is essential to recognise that in an era of skills shortage it is retention that is at risk here, not industrial action. Strikes among self-employed operatives are almost unknown. Here then, from the employers' perspective, is a strategy for risk management that accommodates the pressures of project-based requirements, while minimising the potential for industrial action.

The informality that governs labour engagement in the industry has survived despite predictions to the contrary. While there was some speculation in the late 1990s that government policy and Inland Revenue regulation of self-employed labour would affect employment structures within the industry, it now seems clear that this is not to be the case. Whichever political party is in power, governments are unlikely to seek policies concerned with greater regulation of small firms and successive governments have been prepared to tolerate, and even to encourage, self-employment within the construction sector.

It could be argued that self-employment is the preferred form of labour engagement for the workforce since it provides for higher wages, supplemented by tax concessions and at least the appearance of independence and choice in work activities. Trades unions are less popular with the self-employed. However, employers create the framework for employment and it is clear that this form of labour engagement is rather exclusive since it relies particularly on men between the ages of 35 and 49. The effects on diversity and opportunity and on training and safety within the industry have been noted elsewhere, but with a focus on industrial relations it is important to note the implications for employee 'voice'. The self-employed have few routes to representation and are less likely than the directly employed to seek trade union membership or support. The predominance of self-employment within the industry might seem to reinforce the view that the workforce is expendable since there is no route within the industry through which individuals can in practice articulate grievances or contribute to change.

More significantly, perhaps, this model of industrial relations provides little scope for culture change, since its inherent flexibility constrains the scope for continuous or improved investment in human relations. It must be understood then as one of the ways in which labour is managed and controlled – but one which may inhibit the potential for more dynamic innovation in the human resource contribution within the industry.

5.5   Conclusion

It is argued above that, from the employers' perspective, the dual approach to industrial relations is sustained as a form of risk management. Given the unpredictability of the human element within the construction process, the formal industrial relations machinery is intended to provide the machinery for damage limitation and for cost containment at the same time retaining the maximum scope for employer choice in labour deployment. Voluntary regulation does not, for the most part, provide a vehicle for day-to-day management of the employment relationship, nor is it intended as a means of encouraging participation or involvement within the enterprise. Notions of employee commitment or involvement within the industry are not addressed, despite the formal recognition at national level of trade unions within the sector. There is little evidence that employers have turned to concepts such as 'high commitment' or 'high performance management' – evident in other sectors – as a means of implementing change.

Paradoxically, the arrangements for informal and self-employment in construction also provide for risk management. The human potential for resistance is given significant weight both in the formulation and maintenance of national machinery for collective bargaining and in the alternative routes to employment that govern day-to-day site practice. The dual track approach, with the ambiguities inherent within it, carries immediate advantages in terms of employer flexibility and choice. Employers are able to access the benefits of these two distinctive and apparently contradictory approaches. However, in defending their interests in the face of risk, employers cannot also achieve the objective of 'involving everyone in sustained improvement and learning, and a no-blame culture based on mutual interdependence and trust' (Rethinking Construction, 1998). An alternative approach involving a changed employment status for some sections of the workforce might provide greater scope for innovation and productivity enhancement if perceptions, values and potential of all sections were thereby to be opened up. This advance would require not only direct employment but also review of pay, status and reward for construction operatives: a fundamental questioning of status and of stake within the industry. In this way, the potential for employee 'voice' might be enhanced – and without it the industry may fail to harness skills, realise potential and raise productivity.

References

Beardsworth, A.D., Keil, E.T., Bresnan, A. and Bryman, A. (1988) Management transcience and sub-contracting: The case of the construction site. *Journal of Management Studies*, 25 (6) November, 603–625.

Bernstein, P. (1996) *Against the Gods: The Remarkable Story of Risk*. Wiley & Sons, New York.

Boxall, P. and Purcell, J. (2003) *Strategy and Human Resource Management*. Palgrave Macmillan, Basingstoke, Hampshire.

Cheng, Mei-I, Dainty, A. and Moore, D. (2005) What makes a good project manager? *Human Resource Management Journal*, 15 (1), 25–37.

Consultative Committee on Construction Industry Statistics. (1998) *The State of the Construction Industry Report*. Issue 9, September.

Cully, M., Woodland, S., O'Reilly, A. and Dix, G. (1999) *Britain at Work as Depicted by the 1998 Workplace Employee Relations Survey*. Routledge, London.

Dunlop, J. (1958, reprinted 1993) *Industrial Relations Systems*. Harvard Business School Press, Boston.

Engineering Construction Industry Association (ECIA) (2004) *Overseas Workers – The Perspective of Engineering Construction*. ECIA, London.

Gospel, H. (1985) 'The development of bargaining structure: The case of electrical contracting. In C. Wrigley (ed.), *A History of British Industrial Relations*, Vol. II. Harvester, Brighton.

Gospel, H. and Druker, J. (1998) The survival of national bargaining in the electrical contracting industry: A deviant case? *British Journal of Industrial Relations*, 36 (2) June, 249–267.

Gospel, H. and Wood, S. (2003) Representing workers in modern Britain. In H. Gospel and S. Wood (eds), *Representing Workers: Union Recognition and Membership in Britain*. Routledge, London.

Guest, D. (2001) Industrial relations and human resource management. In J. Storey (ed.), *Human Resource Management: A Critical Text*. Thomson Learning, London.

Hayes, R.W., Perry, J.G., Thompson, P.A. and Willmer, G. (1986) *Risk Management in Engineering Construction: Implications for Project Managers*. Thomas Telford, London.

Heery, E., Healy, G. and Taylor, P. (2004) Representation at work: Themes and issues. Chapter 1 in Healy *et al.* (eds), *The Future of Worker Representation*, pp. 1–36. Palgrave Macmillan, Basingstoke.

Hyman, R. (1997) The future of employee representation. *British Journal of Industrial Relations*, 35 (3), 309–336.

Korczynski, M. (1997) Centralisation of collective bargaining in a decade of decentralisation: The case of the engineering construction industry. *Industrial Relations Journal*, 28 (91), 14–26.

Loosemore, M., Dainty, A. and Lingard, H. (2003) *Managing People in Construction Projects: Strategic and Operational Approaches*. Spon, London.

Macaulay, C. (2003) Changes to self-employment in the UK: 2002 to 2003. *Labour Market Trends*, December, 623–628.

Macrae, N. (1992) *John Von Neumann*. Pantheon Books, New York.

Marchington, M., Rubery, J. and Cooke, Fang Lee (2004) Worker representation within and across organisational boundaries: A case study of worker voice in a multi-agency environment. In Healy *et al.* (eds), *The Future of Worker Representation*, pp. 82–102. Palgrave Macmillan, Basingstoke, Hampshire.

Marsh, A.I. (1966) *Dispute Procedures in British industry*. Royal Commission on Trade Unions and Employers' Associations. Research papers 2 (part 1). London, HMSO.

Marsh, A.I. and McCarthy, W.E.J. (1968) *Disputes Procedures in Britain*. Royal Commission on Trade Unions and Employers' Associations. Research papers 2 (part 2). HMSO, London.

Matlay, H. (1999) Employee relations in small firms – a micro-business perspective. *Employee Relations*, 21 (3), 285–295.

Millward, N., Bryson, A. and Forth, J. (2000) *All Change at Work: British Employment Relations 1980–98, as Portrayed by the Workplace Industrial Relations Survey Series*. Routledge, London and New York.

Report of a Court of Inquiry into trade disputes at the Barbican and Horseferry road construction sites in London (1967) Cmnd.3396. HMSO, London.

Rethinking Construction (1998). The report of the Construction Task Force to the Deputy Prime Minister, John Prescott, on the scope for improving the quality and efficiency of UK construction. Department of the Environment, Transport and Regions, London.

Winch, G. (1998) The growth of self-employment in British construction. *Construction Management & Economics*, 16 (5) September, 531–542.

6 Divergent divisions of construction labour

Britain and Germany

Linda Clarke and Georg Herrmann

Abstract

This chapter compares the divisions of labour in the construction industries of Britain and Germany. It shows how the nature of construction skills, in terms of vertical (hierarchical) and horizontal (functional) divisions in Germany, has generally become less manual and more abstract, as evident from the dramatic decrease in unskilled, labouring work and from the general increase in APTC (administrative, professional, technical and clerical) employment. By contrast, in Britain divisions appear to have remained relatively stable over the past thirty years. These differences are borne out in comparisons of actual projects and the employment profiles of firms. In the UK, the horizontal division of labour – both professional and operative – is much deeper and more fragmented than in Germany. At the same time, the vertical skill structure is far less permeable, so that career progression paths are unclear. These differences are shown to go together with the institutional frameworks governing each system and characterised in Britain by weak institutional linkages and individualisation of the employment relationship. In contrast, the German industry has a high degree of institutionalised linkages and collective processes, including collective bargaining of skilled rates covering not only all operative levels but almost all professional levels.

Keywords: labour, Germany, comparison, professional and housebuilding.

6.1 Introduction

This chapter is about the very deep difference between the division of labour in the British construction industry and that in a continental country of a comparable size in terms of population, such as (West) Germany. The differences are observable at all levels: at the macro level or what might be termed the level of the labour process; at firm level, the level of production where capital and labour combine; and at site level, the level of the work process. They are attributable both to the institutional framework governing the constitution of skill divisions and to the nature of the employment relation itself. The British or Anglo-Saxon model of weak institutional linkages, individualisation of the employment relationship and managerial prerogative is contrasted with the German model with its regulated labour market, high degree of institutionalised linkages and collective employment relations (Hall and Soskice, 2001). The chapter draws on research conducted to compare innovation and skills in social housebuilding projects and firms in Germany and Britain.[1] The result in the British case is much greater labour intensity on site and a lower level of productivity than found in the German model.

6.2   Divisions of the labour process

We would expect, given the development of the production process – notably greater mechanisation and use of prefabrication – to find heavy physical, labour-intensive labouring work gradually replaced by more logistical, planning and coordination activities so that the proportion of manual staff employed decreased at the same time as the proportion of non-manual staff increased with the more abstract nature of work. In this respect, a link between a relative increase in professional work and a relative decrease in unskilled labour should be observed. More advanced building techniques generally require a higher professional input, higher levels of skills overall on site – and hence greater training effort – and less labouring work.

This is certainly the case for Germany. As shown in Table 6.1, while their numbers in absolute terms have constantly and dramatically decreased, the proportion of skilled workers employed in West German construction rose from 48 per cent in 1950 to 61 per cent by 1990, only to fall subsequently to 56 per cent by 2003 (Die Deutsche Bauindustrie, 2005). Much more dramatic has been the simultaneous fall in the employment of unskilled workers, or rather those without a formal three-year training: by two-thirds in absolute terms between 1950 and 2003, and by half in proportional terms, from 36 per cent to 17 per cent of the workforce. Conversely and even more pronounced is the increase in non-manual labour, *Angestellte*, roughly the equivalent of Britain's Administrative, Clerical, Professional and Technical staff (APTC), by more than fourfold between 1950 and 2003, from 4.8 to 21.8 per cent of the total, and a more than threefold increase in absolute terms. Finally – and in accordance with the increasingly skilled nature of labour – the proportion of trainees in the total workforce increased from 3.9 per cent in 1960 (discounting the rather exceptional circumstances in 1950) to 6.1 per cent in 1998, though since declining slightly with the recession in German building activity, to 5.3 per cent in 2003. From the introduction of the dual system of training in the 1970s, the proportion of trainees has remained relatively steady – between 5 and 6 per cent of the overall West German workforce.

This example of the West German construction industry demonstrates the long-term structural change in the nature of work. Productivity gains were achieved through high levels of capital spending and the mechanisation of the

Table 6.1 Employment in the West German construction industry

	Total	Skilled		Non-skilled		APTC (Angestellte)		Trainees	
	Numbers in thousands	Numbers in thousands	% of total	Numbers in thousands	% of total	Numbers in thousands	% of total	Numbers in thousands	% of total
1950	834	401	48	299	36	40	4.8	94	11.3
1960	1338	717	54	485	36	84	6.2	52	3.9
1970	1495	874	58	448	30	145	9.7	28	1.9
1980	1203	678	56	297	25	159	13.2	70	5.8
1985	974	578	59	182	19	147	15.1	66	6.8
1990	982	604	61	183	19	157	16	39	3.9
1998	757	424	56	135	18	152	20.1	46	6.1
2003	564	315	56	96	17	123	21.8	30	5.3
% change 1970–2003	−62	−64		−79		−14		7	

Source: Die Deutsche Bauindustrie (2005) *Baustatistisches Jahrbuch 2004/05*, Frankfurt am Main.

industry in terms of tools and equipment, such as new lifting, (micro) electrical and electronic equipment. Building methods were industrialised with the use of concrete, related modern shuttering systems, advances made in the building products market and electronic data processing (Janssen, 1982–2001; Reus and Syben, 1991; Gluch *et al.*, 2001). Over the period from the 1970s to the 1990s, the dual system of vocational training went alongside steady increases in productivity to which collective bargaining agreements were linked.

Labour productivity in construction in Britain through to the 1990s was consistently lower than in Germany, especially in areas such as formwork (Edkins and Winch, 1999). And when we look behind this, at changes in the division of labour, a totally different picture emerges from the German situation (Table 6.2). The absolute numbers of APTC employed have, for instance, decreased, while they have remained remarkably steady as a proportion of total employment, now standing at 17 per cent of the workforce, almost as was the case in 1970 (DTI, 2004). These figures are in sharp contrast to Germany, where the proportion of *Angestellte* more than doubled in the same period (Table 6.1). There is no suggestion in Britain of any proportionate increase in total professional work itself (i.e. including those professionals not employed by contractors), which would indicate the more abstract nature of construction activities. Indeed, Labour Force Survey figures shows that between 1985 and 2000 professionals as a proportion of all employment in construction fell from 16.5 to 14.6 per cent (EC, 2002).

Other comparisons between the German and the British construction sectors are more difficult to draw due to lack of comparable statistics. Measuring the proportion of trainees to total employment is, for instance, extremely difficult given that in Britain training itself is so very varied in nature. Also, British statistics are on the basis of entrants rather than on the basis of trainee numbers, as in Germany. While in Germany the training for construction trades is at least three years – that is to an equivalent above National Vocational Qualification (NVQ) Level 3 – in Britain it very often stops when someone has achieved NVQ Level 2 or even below. In any attempted comparison, therefore, like is not being compared with like. An added difficulty is to measure actual employment in Britain, given that the proportion of self-employed in the industry rose dramatically from 22 per cent of the workforce in 1970 to 29 per cent in 1980, then to 42 per cent by 1990, peaked at 46 per cent in 1996 before falling again to 36 per cent in 2002 (DTI, 2004: Table 6.2). At the same time, those directly employed fell by half, from 65 per cent of the workforce in 1970 to only 39 per cent in 1990 and 37 per cent in 2002.

What we can discern from the statistics is that total employment in British construction has not fallen by anything like the same amount as in Germany. In 1970, it was approximately 1.8 million, falling to 1.7 million in 1980, where it remained in the 1990s, before falling to 1.6 million in 2002: an overall decline of 10 per cent. In the same time period in West Germany the numbers fell by over half from 1.5 million in 1970 to less than 700,000 by the beginning of the new millennium.

Table 6.2 Employment in the UK construction industry

	Total	Directly employed operatives		Self-employed		APTC		Trainees*	
	Numbers in thousands	Numbers in thousands	% of total	Numbers in thousands	% of total	Numbers in thousands	% of total	Numbers in thousands	% of total
1970	1802	1170	65	405	22	333	18	84	4.7
1980	1696	975	57	495	29	346	20	69	4.1
1985	1492	725	49	470	32	297	20	49	3.3
1990	1703	668	39	718	42	317	19	46	2.7
1995	1375	436	32	621	45	238	17		
2000	1508	578	38	545	36	248	16		
2002	1613	591	37	586	36	270	17	(34)	
% change 1970–2002	–10	–50		+45		–19			

Source: Department of the Environment (DoE), *Housing and Construction Statistics* and Department of Trade and Industry (DTI), *Construction Statistics Annual*, London: HMSO.

Note
* Figures for trainees were discontinued by DoE in 1989 and those for 1990 are therefore from 1989. The figures exclude those in public authority Direct Labour Organisations. The figures for 2002 shown in brackets are for Further Education First Year Entrants to NVQ 2 and 3 courses.

It is much more difficult to assess changes in the numbers of trainees. The Department of the Environment series for trainees, which discontinued in 1989, show a significant decline from 84,000 or 4.7 per cent of the workforce in 1970 to only 46,000 or 2.7 per cent of the workforce in 1989, compared with a rate almost double in Germany (DoE, 1981, 1991). Since then, trainee figures have related to first-year entrants – that is, to intake – and it is impossible to compare numbers in equivalent training. First-year entrants training to NVQ Levels 2 and 3 numbered 34,000 in 2002, that is 2.1 per cent of the total (though this figure does not include those in their second or even third years and does not reflect dropouts – estimated at about 40 per cent of the total) (CITB, 2004). What is apparent is that the decline in trainees from 1970 went together with the fall in direct employment by 50 per cent from 1.2 million in 1970 to less than 600,000 by 2002, reinforcing the now widely accepted association between the rise in self-employment and low levels of training (Winch, 1998).

Figures for the unskilled or untrained are rather more difficult to obtain. In 1975, 29 per cent of the UK construction workforce was categorised as 'labourers', a figure that rose to 31 per cent by 1980 and stayed at this level until 1989 when statistics of this kind, broken down by trade, discontinued (DoE, 1981, 1991). This situation contrasts strongly with the decline in non-skilled workers, or rather those not formally trained, in Germany in the same period from 30 per cent in 1970 – similar to the British level – to 19 per cent by 1990. By 1997, 78.5 per cent of the construction workforce in Germany held at least an apprentice qualification compared with only 41 per cent of the British construction workforce with its equivalent of an NVQ Level 3 – a figure which has since risen to an estimated 46 per cent (Richter, 1998; CITB, 2004).

The evidence therefore indicates stagnation in the British division of labour over the past 30 years, with the proportion of APTC and of professionals in the workforce remaining relatively constant or even declining, the size of the workforce itself remaining approximately the same, and the proportion of trainees declining at least until the 1990s. How do we explain this relative stagnation?

6.3   Institutional explanations

One important explanation is the institutional framework governing the construction of skills in Britain, so extremely different from that found in Germany (Clarke and Herrmann, 2004a). In focusing on this aspect, we draw on the work of Maurice *et al.* (1986) who similarly sought to show how the institutional framework governing social relations within firms is critical to explaining differences between countries. The principle on which this framework is based in Britain is one of governance by quangos, defined officially as 'a public body which has a role in the processes of national government but is not a government department or part of one, and which accordingly operates to a greater or lesser extent at arm's length from Ministers'. Through these quangos the state has handed over responsibility to employer-led Learning and Skills Councils responsible for vocational training in Further Education (FE) colleges and under which come the Sector Skills Councils, including ConstructionSkills, formerly the Construction

Industry Training Board (CITB), responsible for overseeing training in the industry in response to employer demand (Figure 6.1).

One important characteristic of the British system is the divide – the lack of direct link – between the construction industry, supposed to provide trainee work experience, and the FE colleges, although both refer to the Sector Skills Council, CITB-ConstructionSkills. Such a gap is critical given that 62 per cent of construction trainees are to be found in FE colleges, the majority of whom never enter the industry because they cannot obtain the necessary work experience (CITB, 2004). Of the 38 per cent of trainees who are modern apprentices based with an employer, the majority attend college on a day-release basis, a far cry from the block release system of training under the Standard Scheme of Training in the 1970s and 1980s. They therefore have a very minimal grounding in more abstract skills and underpinning knowledge – indeed far less than they received thirty years ago (Steedman, 1992).

A second characteristic of the British system is its employer-led and dependent basis. This means that any work-based training and, above all, the necessary work experience is dependent on individual employer goodwill and thus governed by short-term employer interests rather than the more long-term concerns of the employee. The state's role is also performed at arm's length, rather than through direct intervention. A third characteristic is the fact that skills as defined by NVQs are largely the result of individual trade lobbying and therefore numerous (over 50) and fragmented. They are neither subdivisions of industry activity nor directly related to the skill categories of collective

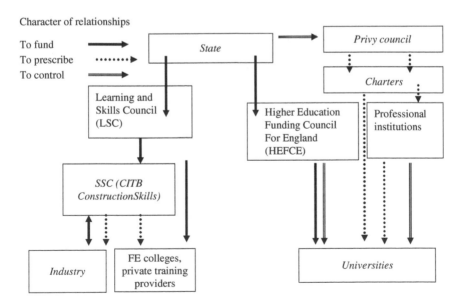

Figure 6.1 The structure of learning in the UK.

agreements. The vast majority of those NVQs taken (86 per cent) are concentrated in traditional trades. A further and critical aspect of the system is the constitutional divide between the formation and the regulation of operative and professional skills, making for a serious lack of permeability, in particular from operative to professional levels. Despite efforts by CITB-ConstructionSkills, career progression remains difficult, a factor that acts as a considerable obstacle to the development of intermediate and technical skills, so critical to the construction process in continental countries.

Unlike the system for operatives, accountable ultimately to the state or government, that for professionals is accountable to the Privy Council and governed by royal prerogative (Figure 6.1). The Privy Council, in exercising the prerogative of the crown, grants charters to each professional institution – defining its objectives and working structure to safeguard the public interest and above all the unique and exclusive skill area it covers. A charter is in effect a privilege granted in perpetuity, only revoked in the case of a serious breach. It allows each institution to control, on the one hand, entry into the profession via the categories of membership and, on the other, educational content via its validation of higher education courses. The institutions largely determine what their members learn, as in the traditional system of tutelage where the architect or surveyor was articled with a practice. When this system was replaced by academic learning, the controlling role of institutions continued, with the state playing no direct regulatory role except through the Higher Education Funding Council, which promotes, funds, monitors and assesses university teaching. The one exception is the statutory Architects Registration Board, which exists alongside the Royal Institute of British Architects (RIBA) and prescribes standards of architectural education and competence.

There are a number of reasons why this system acts as a barrier to the dynamic development of the division of labour and skills in the industry. One is that it is based on the practitioners of today determining the skill needs of tomorrow. Compared with countries such as Germany, the role of universities in Britain in the education of built environment professionals is very restricted because of concerns about accreditation and the importance of professional membership to student careers. At least 75 per cent of the content of accredited courses is defined by the professional institutions as core knowledge, leaving little room for educational or specialist content. Those who lay down guidelines, validate and accredit are professionals and businesses themselves, holding the danger of too close an adherence to practical prescriptive application and to existing divisions and practices. Basing the system of education and training on the immediate needs of employers and current practices means that outmoded practices are constantly perpetuated and reproduced and there is no clear means to enhance skill potential, to introduce new skills and to plan for needs at industry level. In addition, professional institutions are incorporated on the ability to demonstrate an exclusive area of knowledge or skill, which remains fixed. Any new or overlapping areas of activity/knowledge, or areas falling between institutions, become issues of conflict, rivalry, demarcation or exclusion. Examples include urban design and project management. The institutions have a vested interest in maintaining their monopoly and little incentive to cooperate or merge with other institutions. Ministerial accountability and the monitoring of the public interest are also weak and remote. Indeed, as the commercial interests of institutions have become increasingly important, so the aim of serving the public good is put into question. Finally, the professional skill structure is highly fragmented, with a myriad of institutions. The main institutions alone are numerous: Royal Institute of British Architects, Institution of Electrical Engineers, Institution of Civil Engineers, Institution of Mechanical Engineers, Institution of Structural Engineers, Royal Institution of Chartered Surveyors, Chartered Institute of Building, Chartered Institution of Building Service Engineers and Royal Town Planning Institute.

In contrast, in Germany the two main construction professions remain architects and engineers: a very much simpler division of labour. The state also plays an active role in the establishment, structuring, administration and training of professionals, and the university system is integrated in the state apparatus (Figure 6.2).

The construction skill structure is comprehensively mapped to incorporate professional, intermediate and operative skills so that the ability to move vertically and horizontally in the structure is transparent. At operative level in Germany the social partners – the employers and the trade unions – are involved at all levels in what is known as the 'dual' system of training. They participate in negotiating new skill areas, formulating curricula, exam boards and overseeing training in each firm. There is nothing like the degree of fragmentation of

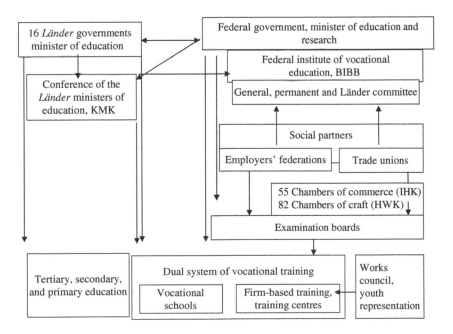

Figure 6.2 The institutional structure of learning in Germany.

occupations to be found in Britain, with a division into only 15 trade areas in which trainees are trained and qualified.

In summary, in Germany we find that skills are defined through a tripartite process between the state and the social partners. In the UK, by contrast and in neo-liberal tradition, the market or rather employer demand plays a prominent role in determining the supply and demand of operative skills. These skills have lost much of their general and transferable elements, though they retain at the same time their traditional character and divisions. As regards professional skills, the professional institutions have been granted skill monopolies and the powers to prescribe the content of learning so that divisions are maintained through institutional competition and rivalry. The divisions between professional institutions represent at the same time the divisions between professionals.

6.4   The level of the firm

At the level of the firm or of production, differences between Germany and Britain can be observed in the nature of both vertical and horizontal divisions. The vertical division concerns hierarchies within firms, whose structuring may be based on pay, status or qualification. The horizontal division, on the other hand, relates to different functions and activities within the construction labour process, whether between carpenters and bricklayers or between architects and engineers – based on historical tradition, training programmes and labour/trade/professional organisation.

Vertical divisions

In terms of vertical divisions, a key distinction can be drawn between, on the one hand, administrative and professional staff employed in the office and, on the other, management and operatives employed on sites. In Britain, the division between salary and wage earners, non-manual and manual workers, often still persists. Hierarchies for those in the office and those employed on sites in Britain are distinctly different, though two characteristics – pay and status – define their structuring and differentials are largely market-driven rather than regulated. As shown in Table 6.3, in one of the firms studied there is a very extensive pay differential both in the office and on site. In the office, in this case, the clerical account worker's pay was a quarter that of the contracts manager's. On site, the general operative or labourer's pay was half that of the site manager's.

The first characteristic of the British structuring is the fact that the pay is largely on an individual basis. Even when the collective agreement is applied, this refers only to those employed at skilled craft level and below. A second characteristic is that pay is not directly related to qualifications or even status: the skilled craftsperson may be paid the same or more than the foreperson on site or the quantity surveyor. Instead, pay level, in particular at site level, relates to experience and performance on the job in hand and on labour market conditions. The pay of a skilled carpenter, for instance, may therefore vary considerably both

Table 6.3 Hierarchical structuring of labour in the British construction industry, with example of one firm

Form of Pay	Education/training	Position in Firm/job	Professional/trade category	Basic gross pay	Bonus averages	Bonus averages	In proportion (%)
Individual pay		**Office Labour**					
	Academic/professional	Managing director		£42,000	£6,300	£48,300	191
		Technical production manager		£32,000	£4,800	£36,800	145
		Contracts manager		£25,000	£3,750	£28,750	114
		Project manager/estimator/Senior QS		£22,000	£3,300	£25,300	100
	Intermediate training/education	Site agent/QS		£19,000	£2,850	£21,850	86
		Buying		£14,000	£2,000	£16,000	63
		Assistant QS		£12,000	£1,800	£13,800	54
		Administration					
		Accounts					
		Trainee QS		£11,000	£1,650	£12,650	50
		Site Labour*					
	Intermediate training/education	Site manager		£28,000	£4,200	£32,200	122
		Foreperson		£22,000	£3,300	£25,300	95
	Vocational training	Skilled craft	Carpenter	£284.70	£300	£30,404	115
			Bricklayer	£284.70	£200	£25,204	95
			Plumber/electrician	£284.70	£175	£23,904	90
	Learning on the job	Skill level 4		£271.05	£150	£21,894	83
			Ground-worker	£214.11	£200	£21,534	81
		Unskilled/general operative		£214.11	£100	£16,334	62

Note
* 100 = £26,504 which is the average of the three skilled trades.

within a firm and across sites (Clarke and Harvey, 1996; Clarke and Wall, 1998). As a result there is no clear path of progression. The wedge driven between professional and operative levels through the institutional skill divide discussed above makes for a generally impermeable structure. At the same time, the lack of a direct relation between pay and qualification, and the individual and unregulated character of the pay structure, mean that career progression is not founded on moving up a clearly graded structure. Indeed, there is little incentive to progress from skilled operative to foreperson.

In Germany, hierarchical structuring is qualitatively different, being considerably more regulated, clearly defined and transparent (Table 6.4). After completing the three-year apprenticeship a worker has the options of different career paths including, the commonest, leading to site manager, *Polier* or *Meister*; another as building technician and, a third, following a degree course, as college trainer or building engineer. Large construction firms employ their own production personnel and promote a worker to foreperson once enough experience has been

Table 6.4 Hierarchical structuring of labour in the German construction industry

Form of pay	Education Training	Office labour			Site labour	
		General Category	Wage Category	Position in firm	General Category	Wage Category
Individual agreement	University	Managing director		General Director		
				Regional Director		
		Head of department	TH	Technical and commercial directors		
			T/K 7	Contracts manager		
Collective agreement	University of Applied Sciences	Project leader	T/K 6	e.g. site manager, estimator	Master	Foreperson
			T/K 5	e.g. buyer		
			T/K 4			
	Apprentice trade training	Technical and commercial experts	T/K 3	e.g. land surveyor	Skilled worker	I
						II
						III
	Limited formal training	Technical and clerical staff	TK 2	e.g. CAD technician	Semi-skilled	IV
						V
			T/K1		Labourer	VI
						VII

accumulated. Further progression depends on further training to become a *'geprüfter Polier'* (qualified foreperson, the equivalent of site manager). Unlike the British hierarchy, that in Germany is therefore structured through qualifications, built on formal programmes of training and regulated through the collective agreement, which covers all categories of labour up to senior management level. As a result, progression is far less dependent on the whims or policies of the individual firm. The majority of the workforce is covered by collective agreements and wage rates are broken down into six grades for office employees and eight for site staff, though there are four main categories: *Ungelernte* (untrained), *Angelernte* (semi-skilled), *Facharbeiter* (skilled), and skilled with further training.

In terms of wage differentials, therefore, the German system is much more transparent, though the actual differentials are similar to those found in the British firm studied. For office staff, for instance, the lowest paid position is almost half the rate of the technician and a quarter that of the project or contracts manager. At site level, too, the differentials found are similar to the British case, though far more regulated and less haphazard, making for a clear path of progression.

In summary, we can see that the hierarchical division of labour in the German construction industry, as demonstrated through detailed study at firm level, is structured through qualifications, themselves the result of a recognised scheme of training, and through the collective agreement, which applies throughout the state, covering 84 per cent of all employees in construction in the west and 51 per cent in the east (European Foundation, 2002). This means that progression and status are transparent: pay relates to the quality of labour itself and is outside the control of the individual firm. This is intended to ensure that competition is not on the basis of labour costs. In contrast, in Britain pay is individually determined on the basis of experience and performance within the individual firm rather than qualifications. This makes for a lack of both permeability and a clear structure of career progression, as well as a structure of pay that is individually determined and firm specific, so inviting competition on the basis of labour costs.

Horizontal divisions

The rationale underlying the horizontal division of labour in the British and German cases is also qualitatively different. In Germany, 343 occupational profiles are officially recognised by the state, maintained and upgraded, however, by tripartite negotiations between the state and its institutions and the social partners, that is the trade unions and employers' organisations. In construction, however, the vocational training system recognises 15 occupations, with the bricklayer as the dominant occupation. These include the newly introduced occupation of drylining, following a long process of negotiation. At a professional level, the main occupations are architect and building engineer, with the latter as the core professional occupation in construction firms. Building engineers complete a four- to five-year university or technical college education encompassing a wide body of knowledge and focusing not just on technology but also on construction law, management, finance and cost control. There are three times more engineers trained in Germany than Britain and they are found in all of the different functions, such as project, contract and general managers and quantity surveyor. Unlike Britain, there is no profession of surveying as the cost function is integrated into both the architect's and the engineer's role. The architect is traditionally the client's cost advisor and project manager.

At the level of the firm, as shown in Table 6.5, it is apparent that employment related to production predominates, with 85 per cent of all personnel in our example employed at site level, 15 per cent of who were trainees. The remaining 15 per cent of personnel are employed in the office, where the production department remains the firm's main skill base and technical knowledge predominates. In our example, only 9 of the 32 office personnel are employed in commercial functions. Indeed, for the successful operation of a firm, a combination of cost and production knowledge is essential. It is also critical to understanding the social organisation of the work process in German firms.

In sharp contrast is the organisation of the typical British construction firm, where site personnel of the main contractor represent a much smaller proportion of all personnel. In this respect, the firm given in our example (Table 6.6) is something of an exception in having as much as 76 per cent of its 478 employees employed on sites; in the other firms studied, site personnel represented less than half of all personnel. In the office of this firm, there is a very evident concern with costs rather than production compared with the German firm, with 51 per cent of office staff directly cost-related, whether in buying, estimating or surveying. In this the firm is more typical, similar proportions of cost-related staff being

Table 6.5 The German contractor: D1, a regional division

Office personnel										Site personnel							
Technical					Commercial						Operatives						
												Skilled					
Project management	Estimating	Technical office (head office)	Trainees (head office)	Others	Buying	Accounts	Controlling	Trainees	Others	Site managers	Ass.site manager	Bricklayers	Carpenters	Concreters	Un/semiskilled	Plant operators	Trainees
18	4	2	1	7	2	2	1	1	3	20	10	50	40	40	3	30	35
32					9 all in head office						228 = 85%						
41 = 15%																	
269																	

Table 6.6 UK contractor: UK1

Office personnel						Site personnel								
Cost						Production								
										Skilled				
Surveying	Estimating	Buying	Accounts	Admin	Others	Contracts manager	Site managers	Site management trainees/agents	Foremen	Bricklayer	Carpenter	Trainees	Unskilled personnel	Others
38	10	10	11	5	23	16	35		46	49	71	22	96	46
58 = 51%			55 = 49%				365 = 76%							
478														

found in the offices of our other case study firms. Unlike the German firm, too, there is no room for the engineer, an aspect that is typical of firms largely involved in housebuilding (Clarke and Herrmann, 2004b).

At a general level, there are four main professional occupations in the British case – architect, engineer, quantity surveyor and construction manager – each with its own professional institution. Each has a different function in the construction process, such as design, structures, cost and overall organisation of the process. The lack of engineering knowledge in the firm shown in Table 6.6 is symptomatic of the general situation of engineering in Britain (Roberts, 2002). In the housebuilding sector especially, technical engineering is incorporated in the one-off input of the structural engineer rather than in a firm's technical department. Together with this, the drive for cost reduction in large construction firms has seen production capacity largely banished to subcontracting firms.

The key difference between the British and the German horizontal divisions of construction labour at firm level is therefore in the emphasis in the German case on engineering and production knowledge and in the British on the cost function and with it the outsourcing of productive capacity to subcontractors.

6.5 Subcontracting

A heavy reliance on subcontracting is the logical concomitant of the British preoccupation with costs. Subcontractors come under the surveying as opposed to the production department. With respect to the division of labour, subcontractors tend to be structured according to traditional occupations, especially those that are 'labour-only', which usually relate to tried and tested areas of activity. Our survey of projects revealed just how important labour-only subcontracting is in the British case compared with the German, where such labour is employed directly rather than subcontracted. In the UK housing project shown in Table 6.7, for example, 42 per cent of subcontract values are labour-only, in particular bricklaying and groundworks. Labour-only subcontractors tend to rely on those classified as 'self-employed', that is in possession of a CIS card issued for tax purposes by the Inland Revenue. In this particular project, these labour-only subcontracters accounted for 57 per cent of operative input as recorded in the site diary. A key explanation for the traditional division of construction labour is therefore that such a large part of labour input is packaged into trade-based labour-only subcontracts, which in turn rely on 'self-employed' workers.

There are other aspects of subcontracting that impinge on the nature of the division of construction labour. As again illustrated in Table 6.7, sharp demarcations are drawn between the different trade areas represented in subcontract packages, whether between roofing, carpentry or brickwork. These demarcations result in particular problems at the interfaces between the activities of the different subcontract groups (Clarke and Wall, 1996). However, it also means that any changes proposed in the system, as for instance in the Latham report with its emphasis on partnering, tend to be directed at contract relations rather than at the employment relations embedded in the production process (Latham,

Table 6.7 Subcontracting and operative input on UK1

Subcontractor	Trade	Subcontract value £	As % of contract value	Operatives (peaks)	Total days	% Of operative input
1	Groundworks Plumbing and drainage External works	1,052,500	22.3	16 (30)	5301	23.9
	Brickwork	362,471		24 (30)	7295	33
2	Brickwork	507,000	8	2		
3	Specialist piling	220,000	3.5	2	257	1.2
4	Roofing	90,000	1.4	2	388	1.87
5	Plumbing/ heating	331,000	5.2	4	890	4
6	Electrical	167,000	2.6	4	719	3.2
7	Plastering/ screeding	216,000	3.4	(14)	1604	7.2
8	Carpentry	110,000	1.7	(15)	2127	9.6
9	Decorating	68,000	1.1	(9)	744	3.4
	Specialists (scaffolding etc.)	5,100–68,000	0.1–1.1	1–2	1181	4.9
	Subcontract values	3,371,571	53			
	Others				1555	7
	Trainees			1	132	0.6
	Total operative input				22,103	100

Table 6.8 Subcontract firms on D1

Subcontractor	Trade	Turnover 1999 £ m.	Total employees 1999	Subcontract value £	As % contract value	Operatives on site	Total days	% Operative input
1	Extern. insul./render	1.5	50	154,667	4.6	11 + 1 trainee	952	9.3
	Painting			30,933	0.9		304	3
2	Electrics	0.83	15	123,733	3.6	4 + 1 trainee	903	8.8
3	Plastering	0.4	8	NA		4	549	5.4
4	Ventilation/sanitary	NA	12	NA		3 + 1 trainee	503	4.9
5	Heating	0.5	8	NA		3 + 1 trainee	444	4.4
6	Screeding/floor layer	1.33	0	131,467	3.9	2 subsub-contractors	427	4.2
7	Window/door manu.	5.3	70			4	420	4.1
8	Tiling	0.83	15	135,333	4	2 + 1 trainee	325	3.2
9	Roofing, external cladding	2.8	40	94,733 50,267	2.8 1.5	3 + 1 trainee	314	3.1
10	Locksmith/ironworks	0.6	15	204,933	6	2 + 1 trainee	238	2.3
	Subcontract values			926,067	27.3			
11–13	Other Specialists						788	7.7
13 Specialists							6,167	60.4
Sub contractor input							6,682	65.4
Operative input							10,205	

1994). In addition, given the high degree of subcontracting – in this case 53 per cent of contract value – the structure of the industry, with its mass of small firms very often acting in the capacity of subcontractors, is reinforced. The reassertion of traditional trade divisions through subcontracting and self-employment and the restricted capacity of the subcontractors themselves also mean that similar levels of subcontracting – even higher in some cases – were found on our other sites, and have also been recorded at a macro level (Gruneberg and Ive, 2000).

In contrast, in Germany subcontracting is of a different order, above all because there is no labour-only subcontracting and no self-employment of the British kind. In the social housing project illustrated in Table 6.8, for example, 27 subcontracts were found but these represented only 27 per cent of the total contract sum, that is half the proportion of contract value recorded on our UK site. The spread of contract values is also much narrower and 13 of the subcontracting firms account for 60 per cent of operative input. The differences with the British case are that these are specialist firms, with many more in the building services area employing their own labour directly and almost entirely composed of forepersons/masters, skilled operatives and trainees. This specialisation is built on the specialist skills of the workforce, originally trained in the dual system, with its broad coverage of different trade activities in the first year and specialisation in one of the 15 construction occupations by the third.

In summary, in the British construction process, labour is locked into traditional trade divisions through the process of subcontracting to small firms, a high proportion of which provide (self-employed) labour only. In the German case, by contrast, subcontracting accounts for much less of contract value and subcontractors are specialist in nature, built on highly trained directly employed labour.

6.6   Deployment on site

The effects of this division of labour in terms of deployment on site can be seen when labour input according to different trade areas, as recorded in site diaries, is mapped throughout the course of projects. In Britain, as shown in Figure 6.3, what is immediately apparent is the front-loaded and highly labour-intensive nature of deployment, with the different trades coming in sequentially and overlapping. Much effort is expended in particular on groundworks, similar to the degree found in the 1970s (Lemassany and Clapp, 1978). Many of the operatives employed in this area are classified as labourers, with little or no formal training. Indeed, labourers accounted for 30 per cent of total operative time expended on site, indicative of much greater labour intensity than in Germany. The mountains of high labour input in particular areas on the UK site might, therefore, even be considered as representing a lack of training.

In contrast, in the German case illustrated in Figure 6.4, deployment tends to be end loaded, representing a gradual build-up of labour. There is also a high degree of direct employment – including bricklayers, carpenters and concreters – throughout the contract, accounting for 35 per cent of operative time.

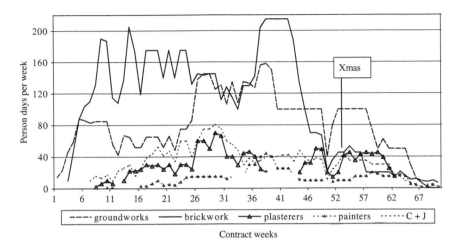

Figure 6.3 Main trades and finishing trades: the sequencing of work on UK1.

Figure 6.4 Labour deployment at D1.

If we compare the speed of production and the productivity of these two projects, what we find is that 20.8 square metres were produced per day on the German project and 28.4 square metres on the British. However, in terms of operative hours per square metre, the British project required 19.3 hours and the German only 13.9.

6.7   Conclusions

In conclusion, the fragmentation of the knowledge input into the construction process in Britain, ossified through the institutional framework regulating the divisions of labour, is observable at all levels. At a professional level, the separation of engineering, building technology, production and cost knowledge is reflected both in the organisation of firms and sites and in a preoccupation with cost rather than employment relations. And at operative level, the separation of different trades through subcontracting and self-employment matches contracting firms' emphasis on the cost function and means that productive capacity rests on a mass of small subcontractors. The cost and production functions are thereby separated in terms of different professions and different types of firms. At the same time the gulf between trade operatives and professions that exists in terms of institutional accountability is replicated.

No such gulf divides operative and professional in the German construction process, where the structure of occupations is permeable both at institutional and firm levels. The cost and production functions are integrated and contracting firms themselves employ production staff directly or subcontract specialist areas to firms which in turn employ their own skilled personnel directly. Skilled operatives are not separated contractually by trade and have themselves, through their training, considerable knowledge and experience of other areas. In this respect the division of labour in the German case is one that is integrated and at the same time collectively regulated and negotiated by the industrial social partners (unions and employers). This situation contrasts starkly with the British case where divisions of labour are maintained through the market and on the basis of employer demand.

Note

1 This was part of a larger study of social housing projects in Britain, Denmark, Germany and the Netherlands, supported by the Engineering and Physical Sciences Research Council and entitled 'Standardisation and Skills'.

References

Clarke L. and Harvey M. (1996) *Disparities in Wage Relations and Skill Reproduction in the Construction Industry*, Final Report to the Leverhulme Trust, University of Westminster.

Clarke L. and Herrmann G. (2004a) 'The institutionalisation of skill in Britain and Germany: Examples from the construction sector', in C. Warhurst, I. Grugulis and E. Keep (eds), *Skills that Matter*, Basingstoke: Palgrave Macmillan.

Clarke L. and Herrmann G. (2004b) 'Cost vs. production: Disparities in social housing construction in Britain and Germany', *Construction Management and Economics*, June, 22, 521–532.

Clarke L. and Wall C. (1996) *Skills and the Construction Process: A Comparative Study of Vocational Training and Quality in Social Housebuilding*, Bristol: Policy Press.

Clarke L. and Wall C. (1998) *A Blueprint for Change: Construction Skills Training in Britain*, Bristol: Policy Press.

Construction Industry Training Board (CITB) (2004) *Construction Skills Foresight Report 2003*, Bircham Newton.

Department of the Environment (1981) *Housing and Construction Statistics 1970–1980*, London: HMSO.

Department of the Environment (1991) *Housing and Construction Statistics 1980–1990*, London: HMSO.

Department of Trade and Industry (DTI) (2004) *Construction Statistics Annual 2003*, London.

Die Deutsche Bauindustrie (2005) *Baustatistisches Jahrbuch 2004/5*, Frankfurt am Main.

Edkins A. and Winch G. (1999) *The Performance of the UK Construction Industry: An International Comparison*, Bartlett Research, Paper Number 4, University College London.

European Commission (2002) *European Social Statistics: Labour Force Survey Results 2002*, Luxembourg: Office for Official Publications of the European Communities.

European Foundation for the Improvement of Living and Working Conditions (2002) *Collective Bargaining Coverage and Extension Procedures*, Dublin.

Gluch E., Behring K. and Russig V. (2001) *Baukosten und Bauhandwerk im internationalen Vergleich*, Institut für Wirtschaftsforschung, München.

Gruneberg S. and Ive G. (2000) *The Economics of the Modern Construction Firm*, Basingstoke: Macmillan.

Hall P. and Soskice D. (2001) *Varieties of Capitalism: The Institutional Foundations of Comparative Advantage*, Oxford: Oxford University Press.

Janssen J. (1982) 'Changes in Labour in the Post-War German Construction Industry' in *The Production of the Built Environment: Proceedings of the 3rd Bartlett International Summer School*, University College London.

Latham M. (1994) *Constructing the Team*, Final report of the joint governmental/industry review of procurement and contractual relations in the UK construction industry, London: HMSO.

Lemassany J. and Clapp M. (1978) *Resource Inputs to Construction: The Labour Requirements of Housebuilding*, BRE Current Paper 76/78, Garston: Building Research Establishment.

Maurice M., Sellier F. and Silvestre J. (1986) *The Social Foundations of Industrial Power: A Comparison of France and Germany*, Cambridge, Mass.: MIT Press.

Reus J. and Syben G. (1991) 'New technology in the West German construction industry and trade union policy', in H. Rainbird and G. Syben (eds), *Restructuring a Traditional Industry: Construction Employment and Skills in Europe*, Oxford: Berg.

Richter A. (1998) 'Qualifications in the German construction industy: Stock flows and comparisons with the British construction sector', *Construction Management and Economics*, 16 (5), 581–592.

Roberts, Sir Gareth (2002) *SET for Success: The Supply of People with Science, Technology, Engineering and Mathematical Skills*, London: HM Treasury.

Steedman H. (1992) 'Mathematics in vocational youth training for the building trades in Britain, France and Germany', *NIESR Discussion Paper No. 9*, London.

Winch G. (1998) 'The growth of self employment in British construction', *Construction Management and Economics*, 16 (5), 531–542.

7 Collaboration on industrial change in construction

On why Scandinavian union approaches are still modern

Christian Koch

Abstract

Technological, organisational and management changes put continual pressure on industrial relations (IR), people and cultures in construction. Scandinavian unions have developed responses to these changes in a characteristic three-level fashion. At the first level, the individual employee is the point of departure. Next comes formal cooperation at company level. And thirdly, cooperation at national level is carried out in a long-term collaborative three-party arrangement often labelled corporatism. The unions have maintained a high level of unionisation, and have been able both to defend co-determination in the workplace and to maintain influence on state policy and labour market issues.

In contemporary transformative capitalism, however, IR are understood as perpetually renegotiated. Two perspectives are described and discussed here. The first deals with the 'bottom-up' destabilising of IR. The second is the shaping of Lean Construction as an example of 'top-down' industrial change. At the bottom of the Danish construction labour market, labour-only contractors, casualisation and migrant workers are all growing trends, whereas non-declared work and do-it-yourself (DIY) have a more stable profile. From the top, the construction unions' participation in the shaping of Lean Construction over the last years is restating the long-term, three-level collaborative tradition. The cooperation around Lean Construction develops new, non-public, but cooperative institutions. This is innovative for the Scandinavian corporatism, because it implies a smaller role for the state. It occurs, however, as a part of a patterned picture of more and less state intervention and control.

The coexistence of relatively disconnected 'good', 'grey' and 'bad' company spheres is not qualitatively new, but has put certain companies under pressure. The leading large contractors have announced that they will start importing labour from Poland. And it may well be that medium-size contractors will increasingly employ a labour-only workforce.

Both construction unions' strategies and institutional development in Denmark are deeply embedded in a peculiarly longstanding modern culture, but elements of Danish practice can nevertheless be adopted in other national construction sector contexts, in balancing concerns for market, class and societal interests.

Keywords: Lean Construction, unions, corporatism and Scandinavia.

7.1   Introduction

Scholars with a social compromise collaborative agenda have often viewed Scandinavia as more or less the 'promised land'. Social welfare states, well-functioning economies and well-developed mechanisms of inclusion are some of the features often cited in this context (Mcloughlin and Clark, 1994; Jørgensen, 2002; Barbier, 2005).

From internal Scandinavian perspectives, however, industrial and societal developments seem less stable. At least since the Second World War, and increasingly so over the last ten years, developments can be perceived more as a constant renegotiation of the Scandinavian model. American business models and discourses are seen as particularly threatening to the Scandinavian model (Sandberg, 1995, thus focus on the threat of Lean Production in Sweden). Globalisation, mergers and acquisitions are increasingly prevalent, and changes in European law relating to labour market regulation and migrant workers, self-employment and precarious forms of work are further challenges. Scandinavian approaches therefore have to reform and reshape themselves to survive and develop.

This chapter takes a multidisciplinary, but predominantly Industrial Relations (IR), view of the construction industry's movement for change and innovation, which here is viewed as an example of transformative post-industrial capitalism. It focuses on the Danish Construction Union organisation BAT (the cartel of unions in the building, construction and woodwork sectors) and this union's role in the shaping of Lean Construction in Denmark. This focus enables the chapter to identify some general features of the role of the labour market parties and how they can contribute to change.

As a second aim, the chapter discusses how construction workers' unions strive at connecting the agendas of such industrial developments as Lean Construction with the 'bottom-up' developments of migrant workers, self-employment and precarious forms of work.

Given the corporatist, consensual setting, it is unsurprising that the Danish construction workers' unions engaged in Lean Construction when it was introduced by an alliance of consulting engineers and contractors at the end of the last millennium. The cooperation realised around Lean Construction involved building up new *private* institutions, which means that Lean Construction in Denmark is less of a state regulated initiative than many other industrial changes. In this sense, the development is less classical in its response to the need for industrial development and less typical of Scandinavian corporatism (Bang *et al.*, 2001, Jørgensen,

2002). It occurs in parallel with other 'top-down' initiatives of industrial development with union participation, such as the state-driven 'digital construction' programme and the quango of the Benchmarking Centre for Construction, which carries out large-scale benchmarking. These examples paint a patterned picture of greater or lesser degrees of corporatism. These developments all occur, however, against the background of an increasingly deregulated labour market; they therefore constitute the coexistence of an organised and a less organised sector, a duality which is not qualitatively new, but where volume and emphasis are shifting.

Finally, the chapter addresses the extent to which these experiences should be understood as (entirely) contextual and how other social compromises can or could be developed in other national and global contexts (Dawson and Koch, 1999). At least strategies of Scandinavian unions can be seen as an attempt to construe and reconstruct central values of modernity in a context where most deem modernity as a phase long gone, but where the conditions for collaborative agendas could still be present.

The chapter does not examine the content of Lean Construction. In its first phase, Lean Construction in Denmark was predominantly the implementation of Last Planner (1999–2003) (see, for example, Thomassen *et al.*, 2003). In its second phase, elements of Lean Design have been developed (Emmitt *et al.*, 2004, 2005). Finally it should be noted that the chapter's analysis was developed in 2004–05 and the developments have continued since then.

7.2   Performative capitalism and transformed corporatism

In recent years, several macro-society observers have argued for the existence of a profound transformation in a global society. A simple way to summarise (Bauman, 2000; Giddens, 1991; Burawoy *et al.*, 2000; Harvey, 1996; Castells, 1999) would be to use Thrift's notion of capitalism as performative, viewing capitalism as something always engaged in experiment (Thrift, 2005). What both Bauman and Giddens have pointed out, along with many others, is the gross tendency of decaying stability of contemporary modernity. Identity and institutions have to be performed and they become less stable than in previous modernity. Yet paradoxically, as noted by Bauman, modernity was also the time of heavy capitalism, thus having a characteristic set of opportunities and threats of its own. In a fluid or reflexive modernity, unionism and unions have to survive constant transformation in formulating responses to market, class and society issues (Hyman, 2001). In fact, both Castells and Bauman are quite pessimistic about the chances for organised labour to survive at all. Bauman thus celebrates the notion of fluidity and claims that the present stage of development celebrates nomadism above settlement and territorialism. Castells identifies an increase in concentration of capital along with disintegration of production and organisation units, creating new global common spheres of interest for employees, alien to traditional unionism (see also Koch, 2000a).

Political science scholars also note quite profound changes in the shape of the state and its practice of social partnership. Van Heffen *et al.* (2000) study new public management and argue for an understanding of diversity of governance forms and networking. Rothstein *et al.* (1999) argue that corporatism loses ground in a Scandinavian context, because society becomes complex and cannot be 'ruled' as the convening of a few blocks of interest, as early modernity could. One might add that, rather than the past being simpler than the present, corporatism ruled through a few interest blocks constitutes one way of simplifying complexity. As an important parallel to this argument, Marchington *et al.* (2005) study work fragmentation and increasingly ambiguous conditions of work. Yet, as this chapter will demonstrate, collectivity and individualism, local and central as well as local and global, are contradictions that can be tackled by a union institution.

7.3   Disconnected worlds? The top and the bottom of IR in construction in Denmark

At the beginning of April 2005, the trade union confederation of Denmark, LO, established and encouraged the blockade of a building site in central Copenhagen. The police had been advised four weeks earlier that the Latvian company operating at the site did not have the necessary papers. The client, a restaurant, had its representative on the street. He was furious and called the police. The police however declared that this was a legally warned labour market dispute, and they were thus prevented from further intervention.

This street-level action symbolises – paradoxically – the top-down view on how IR should function in Denmark. This view pictures a societal model of close collaboration in a long-term social partnership, solving conflicts in prescribed ways (Hyman, 2001). The society is dependent on a three-party corporatist welfare state, an institutional set-up at the macro level; a political structure within an advanced and regulated capitalism, which integrates organised socioeconomic producer groups through a system of representation and cooperative mutual interaction at leadership level and mobilisation and social control at mass level. (This definition of corporatism is derived from Panitch, 1980.)

Within this model there is a strong belief in negotiating agreements and contracts between the organised parties (unions and employers associations) and limiting public regulation. This division of turfs within corporatism is currently challenged by the labour market laws effected by the European Union (EU), which are moving the demarcation lines between labour market agreement and law to the benefit of the latter. This move reconstitutes central European models of labour market regulation but, in a Danish context, is viewed as a threat to models relying on negotiation and agreements between the labour market parties – what in Danish is called *samarbejdsmodellen*: the model of collaboration.

There is a tendency for the corporatist, consensual model to portray itself as all-encompassing. From the outside, IR seem extremely homogenous and consensual. Nevertheless, the presence of less-organised sectors of the economy is well known. In particular, non-declared and DIY activities have long been a grey area of the apparently well-organised model (Mogensen *et al.*, 1995).

Unemployment among immigrants and the other large social groups excluded from the labour market (Jørgensen, 2002) further challenges the image of broad inclusion. The recent restructuring of the labour market in general and the construction in particular augments and introduces further elements into this dual situation, including migrant workers, self-employment and increased project-based and other precarious, contingent forms of work (Table 7.1).

Since the Second World War, Scandinavian unions have developed particular responses to industrial developments: to new management, organisation and technological change. One of the central strengths of these responses is that they take as their point of departure the individual employee and his/her participation in and/or responses to change, but at the same time develop each into formal cooperation at workplace level as well as at regional and national levels (Koch, 2000a,b; Sandberg *et al.*, 1992). Unions participate in industrial development at all these levels and have established a range of institutions underpinning this participation, such as the four-year education systems for the crafts with formalised influence from the unions. It is thus part of the collaborative culture that employers take their counterparts, that is their employees into consideration even in the absence of organised representatives. And even small employers can see the advantage of collective agreements as a means of securing stability.

The coherence of union strategies and their strengths obviously vary. There is thus a whole spectrum from passive participation and cooperation to activity aimed at real influence (Clausen and Jensen, 1993). Some unions and

Table 7.1 Denmark at a glance (figures from 2004)

The geographical area of Denmark is 43,100 sq. km, with 5 million inhabitants: a size comparable to regions in the larger European countries	
Employment in Construction	147,000
Members of unions	80–85%
Members of BAT, Cartel of Unions in Building, Construction and Woodworks	138,000
Employers organised in 'Danish Contractors'	6,000
Employers organised in 'Tekniq', Danish Mechanical and Electrical Contractors	3,000
Employers organised in the Danish federation of small- and medium-sized enterprises	c.2,500
Single-man enterprises (estimated)	13,000
Migrant workers (estimated)	3,000
Non-declared work in percentage of legal work	25%
Unemployment among immigrants (estimated)	50%

scholars understand the most active union strategy to be when unions actively and independently build knowledge and action resources preparing them to interact with and co-shape new developments. They label this as the collective resource approach (Ehn and Kyng, 1987; Sandberg *et al.*, 1992; Clausen and Jensen, 1993; Kyng and Mathiassen, 1977).

Numerous general IR texts take a top-down view of the IR system (see, for example, Scheuer, 1998), thus highlighting the form of contract negotiation and cooperation between central parties as constitutive of IR and downplaying local processes. Local disputes are quantified and categorised as 'legal' and 'illegal' disputes according to the way they are organised; however, their content is rarely studied and the possibility of local developments entirely outside the IR system is disregarded.

Nevertheless, seen from a bottom-up view, the IR system tends to look like a closed shop, with its own circular problems and solutions. Conflicts and dilemmas stemming from everyday practice at workplace level tend to be 'negotiated' away, 'out' of the IR system. In the following sections, the two perspectives – top down and bottom up – are used to characterise IR in Danish construction, in order better to understand the contradictory context in which industrial development takes place.

The top-down view: BAT – the union institution

In the Danish construction sector, employers associations and unions cooperate closely at the central level. There are two main employers associations – Dansk Byggeri and Tekniq – and one major union player, BAT. They share a high degree of organisation and membership in the sector, giving industrial relationships an institutionally embanked, two-sided focus. Here the chosen focus is the union side.

BAT is a cartel of seven unions which together organise in the region of 138,000 members equating to 80–85 per cent of workers in the construction sector (Larsen *et al.*, 2003). BAT and other Danish unions were growing in membership until about 2000, but since 2001 BAT and its unions have been slowly losing members, mostly for demographic reasons: older members pensioned off outnumber younger members joining. In 2005 there was a 9 per cent fall in members from 1995 figures. In international terms, these figures are uniquely high (Ferner and Hyman, 1998). In a similar period, 40 unions were restructured into the present 19.

Danish construction workers are mostly loyal to one company. Figures from the labour market pension systems show that more than 70 per cent have stayed with the same company since 1993 (Odgaard, 2005). This figure is in stark contrast to the image in the Danish sector of the construction worker as a free agent.

As an organisation for building workers, BAT enjoys an unusually strong local presence in the sector, viewed in a European context. There are 6000–7000 elected shop stewards and safety representatives and a strong local union structure, albeit different from the various participating unions, for example 3F, the largest, has 50,000 members in construction, 77 local branches and 136 service offices. (3F translates to 'Common Trade Union' and resulted from the merger, in January 2005, of 'The Women Workers' Union in Denmark' and 'The General Workers' Union in Denmark'.) Contracts for the sector assure the employees the right to elect a shop steward and a safety representative in companies with just five employees. It should be noted that union merging is less of a survival game in Denmark than it is, for example, in the UK and Germany (see Waddington *et al.*, 2005).

As a mediator of collective bargaining between central and local levels, the BAT cartel operates as coordinator for the contracts for the area (7 main collective agreements). Most of them are negotiated centrally as minimum wage agreements supplemented with pension, maternity leave and other rights. In reality, an employee very rarely gets exactly the minimum wage, but is usually able to negotiate a higher pay.

At the central level, a secretariat was established along with BAT in 1990. Over the subsequent period more tasks were transferred to the cartel from the unions, to be handled by the secretariat and the political structure of the cartel. The establishment of a secretariat can be interpreted as professionalisation of the central work, since staff employed there are academics, a common development in Danish unions in this period. The main political work areas of the secretariat are work environment, business and sector development, co-determination and international issues.

The cartel has been able to gain institutionalised influence on legislation (e.g. health and safety), housing politics, public industrial development programmes and occasionally the four-year education systems for the crafts, where the formalised influence lies with the unions.

In recent years, this influence has included the establishment of a benchmarking evaluation centre, a series of experimental building projects, the development programme 'Digital Construction' and other developments.

The multilevel strength of the unions continues to be mutually reinforcing. The institutionalised central influence is thus based on a long-term development and embeddedness in IR and Danish construction cultures, and at the same time shaped and reshaped in an ongoing fluid and emergent process.

Bottom up: Fragmentation tendencies, 'arms and legs' enterprises, migrant workers, casualisation

Meanwhile in a remote city, and a remote quarter of residential villas, two carpenters were busy assembling a veranda for a client. It was Sunday and the client and the craftsmen had made a good deal . . .

In this section the focus is on the bottom of the labour market in construction. It covers some new forms of fragmentation such as labour-only contractors, casualisation and migrant workers. The longstanding and widespread practice of non-declared work is also part of this picture.

According to Danish statistics, the Danish building sector has around 147,000 employees (BAT, 2004c). About 30 per cent of those are employed with large contractors (110 companies employ more than 100 employees), whereas small single-craft-based companies account for about 88 per cent of the enterprises and around 50 per cent of the employees (less than ten employees per enterprise, Dansk Statistik, BAT, 2004a). Very many companies thus comprise fewer than 10 employees and act as subcontractors or small contractors at the lower end of the market (minor renovation, residential housing, etc.). Unemployment was 8.6 per cent as an overall average in 2004, but was expected to fall significantly in 2005. Some trades currently have full employment (BAT, 2004c).

Whereas the previous section took as its point of departure the well-organised upper part of the labour market, the attention here is on how a vast number of small enterprises operate at the lower end of the construction sector. Labour market institutions necessarily have little part in this discussion.

Arms and legs

In 2004 the number of single-man companies was in the region of 13,000, constituting 9 per cent of employees at that time. This was a remarkable increase compared to ten years previously. Between 1995 and 2001 the increase was 32 per cent (BAT, 2004b). BAT estimated that half of this group was legal and ordinary one-man enterprises (whereof only about 800 were members of an employers' association). The other half was labour-only craftsmen and workers, known as 'arms and legs' firms. There is a continuum between the legally operating one-man contractors and the 'arms and legs', which are defined by BAT as contractors, but which meet at the site without previous tendering or other formal papers and which do not supply their own materials nor manage themselves.

Medium-size contractors in particular might be tempted to use an increasing number of labour-only workers, since their contracts fit with subcontracting in this way (each contract of the trades within medium-size building contracts might more or less equal the work of say 1, 1.5 or 2 persons).

Casualisation

As noted above, Danish construction workers are mostly loyal to one company. For the remaining 30 per cent, employment at temporary projects with shifting employers is, to varying extents, common practice. The casualisation process also encompasses increased use of substitutes: there are no figures available, but there are at least 28 agencies offering this type of labour in Danish construction (BAT, 2004b).

Non-declared work

Non-declared work (i.e. in respect of tax declarations) in the Danish construction sector comprises 25 per cent of the official work performed (Mogensen *et al.*,

1995; Pedersen, 2003). In a comparative investigation, 20 per cent of the Danish workforce said they had done undeclared work, compared with 9.8 per cent of UK workers (Pedersen, 2003). In construction, plumbers and electricians are the most active in this respect, but carpenters are also active (Brodersen, 2003). However, much of this work is undertaken by students, apprentices and the unemployed (Pedersen, 2003). Undeclared work therefore appears to be more widespread in Denmark than in other European countries, but there are obvious problems in compiling the statistical evidence to substantiate this statement.

Do-it-yourself (DIY) is also playing quite a significant role in the residential construction market. Painting in particular, but also some carpentry, is frequently done by owners or tenants. A recent investigation found that DIY comprised 3 per cent of the Gross National Product (Brodersen, 2003). Research suggests that non-declaration diminished slightly over the late 1990s (Mogensen *et al.*, 1995).

Migrant workers

The construction sector is sometimes described as a magnet for migrant workers. Furthermore, with big variations of IR – with temporary and mobile characteristics, construction often goes to the extremes – employment relations can be both fully regulated and almost completely disorganised (Pedersen and Lubanski, 2004). It is assumed that the temporary and relatively labour-intensive character of building projects makes the market more open for migrants.

In May 2004 the EU was enlarged with ten new membership states, of which several (notably Poland and the Baltic states) are in the vicinity of Denmark. The enlargement opened new employment opportunities to East European workers. A legal path for this traffic was created in 2003 (the so-called East agreement – see p. 117–118), but it has not been much used. In May 2004 only around 20 workers from all of the new EU countries had applied. By March 2005, 150 workers were operating in construction on a legal basis, whereas 50 per cent of the total of 1600 legal workers were employed in agriculture and gardening (Gjesing, 2005).

In Denmark, most of the construction workers entering the country by autumn 2004 did so on an illegal basis. More or less dubious and temporal enterprises and illegal arrangements played an important role in enabling these entries. By December 2004, BAT had identified 109 cases of illegal entry involving 449 persons countrywide (Nissen and Odgaard, 2004); it further asserted that the known cases probably represented 20 per cent of the total, giving an estimate of 3000 persons entering illegally. Of the known cases, 60 per cent concerned Polish workers, 30 per cent Latvian workers and 10 per cent Estonian and Lithuanian workers. BAT estimated that 7000–8000 workers had entered Denmark illegally between May 2004 and the end of 2005. The figures for legal entry were substantially lower.

On a practical level, this influx of migrant workers led to language and technical problems. For example, rework was necessary on several building projects where detail elements did not comply with Danish norms.

The current labour shortage challenges the separation of a 'good' from a 'bad' sector of IR: three major contractors have announced that they will start using East European labour within the framework of the contracts with the unions (Danish Newspapers, December 2005). This move will allow contractors to pay a wage just above 100 Dkr (£10) per hour, whereas local wage negotiations would normally lead to a substantially higher wage: current prosperity levels mean that hiring a supplementary workforce might cost a contractor up to 400 Dkr (£40) per hour.

This section has focused on the bottom of the labour market in construction. Labour-only contractors, casualisation and migrant workers are becoming increasingly prevalent, whereas non-declared work and DIY activities have a more long-term, stable character. Nevertheless, the coexistence of a 'good company' sphere and a 'bad company', with a grey area in between, is not qualitatively new. There are trends strengthening the good company and the bad in parallel. However, as discussed above, large contractors have recently announced their intention to hire migrant workers and medium-size contracts might increasingly employ labour-only workforces.

It should be noted that several labour markets also coexist within the segmentation of the market for built services. Renovation and new construction of residential houses comprises one market; infrastructure, apartment and business buildings constitute another. Some construction work (notably DIY) never finds it way to a market and other work operates on the basis of unusually low pay (Pedersen, 2003). Care must be taken neither to overlook these segments nor to overestimate the barriers between them.

Finally, it should be remembered that the levels of self-employment and other casualisation, and of migrant workers, are considerably lower in Denmark than in many other countries. (For self-employment in the UK, see Chapter 2.)

7.4   Union responses to fragmentation

The Danish unions actively monitor several of the fragmentation tendencies described above. On the one hand, there are attempts, predominantly among union militants, to create a professional ethics within which non-declared work and DIY are deterred. On the other hand, there are active attempts to unionise both 'arms and legs' and migrant workers.

The handling of migrant workers is particularly interesting. All Danish labour market parties – state, employers and unions – wanted to address the EU expansion towards the East, and in December 2003 a so-called 'East agreement' was passed in parliament, regulating the admission of migrant workers (in parallel with several other countries putting transitional arrangements in place; Pedersen and Lubanski, 2004). The East agreement regulates admission of workforces from the new member states and demands sight of permits to work and stay in the country before employment can commence, thus departing from the EU ideology of the free movement of labour. As noted above, however, legal entry remains rare.

In 2004, the police, customs or tax authorities were notified of known cases of illegal entry, usually by local unions. The influx of illegal workers also led to a series of local union blockades, with demands that employers sign a Danish contract. It became clear during the autumn of 2004 that local police departments did not possess the resources, competence (and in some instances the will) to tackle all the cases. The parties behind the East agreement therefore agreed to develop a manual for professionals handling these issues. The ministry of employment launched this manual in December 2004, again with three-party support, thus furthering the new corporatist solution. However, the labour market conflict continued into the spring of 2005 and when the ministry issued an English translation of the manual, BAT asked for a more comprehensive effort from the authorities (Beskæftigelsesministeriet, 2005).

Fighting for contracts and organised conditions for migrant workers is arguably a 'good cause' for the unions: they obtain renewed support from local members and their efforts in this respect contribute to reshaping union strength (Odgaard, 2005). Local unions seem to have had some success in detecting a number of cases, leading to court convictions; in this way they create a climate in both the press and the sector that working conditions for migrant workers should follow the rules (law and contracts). The employers' official position, moreover, is that the contracts merely state a minimum wage, allowing local negotiations (Elgaard, 2004). The symbolic summit of the press campaign to date came in March 2004 when a union magazine revealed that the husband of a minister had hired migrant workers (on a farm) on an illegal basis; the result was several days of crisis for the government.

It is evident that whereas non-declared work and DIY have been coexisting with the formal labour market over a long period without much effective union response, the case is very different in respect of migrant workers. This labour market regulation is an interesting example of a renewed corporatist solution. Here the unions have successfully lobbied for state regulation and enforcement. And they have been able, simultaneously, to establish local activity, making the classic span between individual conditions and state regulation.

7.5   New industrial change: Lean Construction

The attention now shifts in this chapter from labour market developments to industrial developments. The key example in this regard is Lean Construction.

The Scandinavian unions have developed responses to new management, organisation and technological change over a decade, with the tackling of Taylorism in the 1950s as an important precedent. At that time, the characteristic Scandinavian response was to appoint a 'time study' shop steward, to help set the pace and the piece rate. Numerous later industrial developments have been tackled in a similar collaborative spirit.

Through the multi-level approach outlined above, the Scandinavian unions – unlike their British and American counterparts – have chosen actively to support development of the given enterprise (Boglind, 2003). In other words, they participate in organising the work in a way that allows change to be profitable to the individual, the enterprise and society.

Lean Construction

In Denmark, Lean Construction appeared as an agenda for change around 1999–2000. Initially, it was taken up predominantly by larger contractors such as MTH and NCC and was in this way an initiative for the 'good' companies.

In its first phase, roughly until 2003, Lean Construction in Denmark was predominantly the implementation of the planning and scheduling approach 'Last Planner' (Ballard, 1998; Thomassen *et al.*, 2003). In its second phase, a handful of companies have been carrying out experiments with Lean Design (Emmitt *et al.*, 2004, 2005). Larsen *et al.* (2003), representing BAT, notes that BAT soon started seeing Lean Construction as being in the interest of their members.

History

The story of Lean Construction in Denmark is one of several parallel and intertwining processes. At a broad media and public level, several reports and debates at the end of the 1990s paved the way for change (the Process and Product development Programme finalised in 2001, the report 'Construction in the 21st century' from the Academy of Technical Sciences (ATV, 1998) and 'The Future of Construction' from the task force of building policy (Bygge politisk task force, 2000). The political initiatives involved industrial players and partially provided a frame of reference for the alliance between NIRAS and MTH (a consulting engineering company and a contractor, respectively) on using Lean Construction. The alliance began experiments in 1998–2000 with building logistics and Lean Construction (in terms of Last Planner, Ballard, 1998). The basis of the work was, among other things, the NIRAS manager of development, Svend Bertelsen, which linked up with the IGLC international community during the nineties (Simonsen *et al.*, 2004).

MTH decided to create competitive advantage by implementing Lean on a broad scale by 2001. The company began an ambitious training effort to bring their building project managers up to speed. They also decided to try to keep information on implementation a business secret (ibid.). At the same time, BAT had requests for information about Lean Construction from members and shop stewards.

By the end of 2002, MTH had implemented Lean tools in about 30 projects (Thomassen *et al.*, 2003). Thomassen *et al.* (2003) provide an overview of 15 projects which implemented versions of Last Planner. These results were communicated to BAT, which saw how MTH evaluated that Lean Construction led to higher wages and fewer occupational accidents and did not affect employment.

In parallel to this development, other contractors, especially NCC, but also KPC, CEG and Hoffmann, began implementing Lean Construction ideas.

Bertelsen and Høgsted (2003) provide a selective account of six building projects with some of these contractors. In NCC an infrastructure was developed to support the implementation of Lean Construction in 2002 and the first project was carried out in the autumn of 2003 (Simonsen and Koch, 2004). This meant that two out of the three major contractors had embarked on Lean.

By the time the Danish branch of the Lean Construction Institute was established in 2002, it had become natural that the unions should be represented and the board of this private initiative included a member from the unions/BAT. In the summer of 2003, BAT representatives also participated in the International Group for Lean Construction 13th annual conference in Virginia, USA (IGLC 13) (Larsen *et al.*, 2003). Immediately afterwards, BAT was involved in organising IGLC 14 in Denmark. As a result, IGLC was held at a union school for shop stewards, another symbolic representation of the role BAT has obtained. BAT was also represented at IGLC 15 in Sydney, Australia (Buch and Sander, 2005).

It should be noted that some central players among contractors in Denmark have so far decided not to participate in the Lean movement, including Skanska Danmark, Pihl and Ove Arkil. In other companies, such as Hoffmann, implementation of some kind of Lean Construction is left to the initiative of the local project managers.

The role of BAT

BAT's engagement with Lean Construction began with the realisation that Last Planner, 'seven healthy streams' (Koskela, 2000) and similar initiatives were different from the contemporary public reports in their practical and decentral approach:

> We see it as the building worker put in the center. The 'seven streams', 'last planner'... it is about communication, cooperation and planning. It is about moving competencies downward... and the MTH was already engaged... and it was an answer to the critique from the public reports at the time.
>
> (Odgaard, 2005)

In 2002 BAT formulated a strategy for dealing with Lean (Larsen *et al.*, 2003). Understanding of Lean was to be integrated in vocational training and BAT aims to build a regionally based network of engaged unionists. BAT also used their membership of the newly established evaluation centre for benchmarking to redirect the Lean issue to another player (see p. 110, 121).

7.6   Discussion

In the development of regulatory efforts in the Danish labour market and IR, a series of shifts of 'weight' in the balance between state, unions and employers' organisations can be observed (see also Nielsen and Koch, 1997). Recent developments of the corporatist arrangement within construction do not differ from this general pattern. In 2000, when the state launched a massive critique of the building sector – identifying poor quality, high costs and so on – and proposed a series of reforms (Byggepolitisk Task Force, 2000), employers and unions responded by establishing the Benchmarking Centre for Construction. Such a centre was originally proposed in the state report, but was nevertheless set up as a private independent body with BAT as a board member. This centre launched 'round table' discussions of Lean Construction and employed the Lean Construction proponent Svend Bertelsen as research manager (Bertelsen and Høgsted, 2003).

In a similar vein, the climate for change created by the 2000 report meant that BAT saw the Lean Construction initiative as an answer to the report's criticisms. The initial development was not reliant on employers' associations or the state, but rather a small alliance of private enterprises. BAT later used its position on the board of the evaluation centre to participate in redirecting the Lean effort from this centre to the private Lean Construction Institute.

In this light, Danish construction unions' participation in the shaping of Lean Construction is less of a surprise. The unions have an ongoing cooperation with the employers' associations and the state regarding development agendas for the sector. The cooperation realised around Lean Construction encompasses building up new institutions and Lean Construction in Denmark, and is more a private than a state-regulated initiative. In this sense, the development is less classical in its response to the need for industrial development and slightly atypical of Scandinavian corporatism. On the other hand, the development of Lean Construction alliances and institutions occurs in a context with renewed corporatist initiatives and developments. There is thus a patterned picture of more and less corporatism and shifts of weight between the parties.

7.7   Conclusion

In the light of this narrative, two union officers on bicycles looking for illegal contractors with migrant workers in a summer-house quarter, or shop stewards participating in a coaching course within a Lean Construction frame, are two symbolic examples of modernity reconstituted.

The development demonstrates how Lean Construction becomes an agenda for the top level: 'good company' enterprises and corporatist arrangements prevail, and long-term cooperation patterns not only survive, but are also transformed as new compromises on development are shaped. In this sense, Lean is not an exception, but a scholarly example. Lean becomes part of the 'top' package: large, well-organised contractors, with contracts and cooperation with shop stewards; well-organised health and safety activities, with fewer accidents (Thomassen *et al.*, 2003); a transformed and modern corporatism realised.

This situation coexists, however, with that at the 'bottom' levels of the Danish construction labour market: economically unstable companies, poor health and safety, non-declared work and competition with migrant workers. Labour-only contractors, casualisation, and migrant workers are increasingly prevalent, and non-declared work and DIY persist as ever. Nevertheless, as has been argued above, there is nothing new about the coexistence of a 'good company' sphere and a 'bad company' sphere, with a grey area in between. Coexistent tendencies and trends strengthen the good company and the bad in parallel.

Union responses to non-declared work and DIY have not been very effective, but the case of migrant workers demonstrates another example of a renewed corporatist solution, as do union responses to Lean Construction. In the case of migrant workers, the unions have successfully lobbied for state regulation and enforcement. And they were simultaneously able to establish local activity, reconstituting the classic three-level span between individual conditions and state regulation.

The unions strive at separating the two agendas of industrial developments such as Lean Construction from the bottom-up developments of migrant workers, self-employment and precarious forms of work. But the two areas do occasionally mix, when contractors start hiring several layers of subcontractors. In such cases, a Last Planner activity is hampered by the impenetrable set of deep relations right down to the labour-only contractors, who will not participate in organised meetings, but nevertheless need management.

On the one hand, the Danish unions' experiences should be understood as strongly embedded in a set of contextual conditions. The union strength is a bundle of factors, but central to understanding why the politics regarding Lean can be successful in Denmark. On the other hand, the tensions between the worlds of 'top' and 'bottom' resemble patterns seen elsewhere in Europe and the USA.

Social compromises in other national and global contexts will need to be developed in different ways (Dawson and Koch, 1999; Koch, 2000a,b). Transformative capitalism will exhibit other types of traits than those dealt with in the Danish setting. But at least strategies of Scandinavian unions can be seen as an attempt to construe and reconstrue a modern society with basic values of egality, fraternity and solidarity, no matter how tarnished this trio might be. In other contexts, the balances between market, class and societal interests will be different. Other conditions would, however, usually leave space for parallel institutional reform, albeit with less scope in the labour market and more fragile implementation. This assumption rests in turn on the supposition that balancing concerns for market, class and society would be common to any construction sector.

References

ATV. (1998), Byggeriet i det 21. århundrede. Lyngby: Akademiet for de tekniske videnskaber (The Academy of Technical Sciences).

Ballard G. (2000), The Last Planner System of Production Control, Ph.D. Thesis, University of Birmingham.

Bang H.L., Bonke S. and Clausen L. (2001), Innovation in the Danish construction sector: The role of public policy instruments. In G. Seaden and A. Manseau (eds), *Innovation in Construction – An International Review of Public Policies*. Spon Press, London.

Barbier J.C. (2005), Apprendre Vraiment du Danemark? Research note. Accessed at www.bm.dk, September 2005.

BAT (2004a), *Aktivitetsrapport.* BAT København.

BAT (2004b), *Vikarer og Arme ben i bygge-og anlægsbranchen.* BAT København.

BAT (2004c), *Konjunkturanalyse af bygge-og anlægsbranchen* 2004. BAT København.

Bauman Z. (2000), *Liquid Modernity*, Polity Press, Oxford, England.

Bertelsen S. and Høgsted M. (2003), State-of-the-Art Rapport. Arbejdsgruppe Trimmet Udførelse Byggeriets Evaluerings Center. København.

Beskæftigelsesministeriet (2005), *Rules on Residence and Work in Denmark for Citizens from the New East European EU Member States.* Ministry of Employment, Copenhagen.

Boglind A. (2003), Facket och de magra åren. Sandberg Å (red) (2003), Ledning för Alla? Perspektivbrytning i arbetsliv och företagsledning. SNS förlag, Stockholm, pp. 119–163.

Brodersen S. (2003), *Do-it-yourself in North-western Europe. Maintenance and Improvement of Homes.* The Rockwool Foundation Research Unit, Copenhagen.

Buch S. and Sander D. (2005), From Hierarchy to Team-Barriers and Requirements in relation to a new Organisation of Building Sires. IGLC. Sydney. Accessed at www.iglc.net.

Burawoy M., Blum J.A., George S., Gille Z., Gowan T., Haney L., Klawiter M., Lopez S.H., Riain S. and Thayer M. (2000), *Global Ethnography: Forces, Connections, and Imaginations in a Postmodern World.* University of California Press, Berkeley.

Byggepolitisk Task Force (2000), *Byggeriets fremtid- Fra tradition til Innovation-Redegørelse fra Byggepolitisk Task Force.* By og Boligministeriet Erhvervsministeriet. København.

Castells M. (1999), *The Rise of the Network Society.* 2nd edn, Blackwell, Cambridge, Massachusetts.

Clausen C. and Jensen P.L. (1993), Action oriented approaches to technology assessment and working life in Scandinavia. *Technology Analysis and Strategic Management.* Vol. 5, no. 2.

Dawson P. and Koch C. (1999), Unions on the edge: Between inertia and agility. *Proceedings, the 13th ANZAM International Conference*, Tasmania, ANZAM, Auckland.

Ehn P. and Kyng M. (1987), The collective resource approach to system design. In P. Ehn and M. Kyng (eds), *Computers and Democracy.* Aldershot, Avebury.

Elgaard B. (2004), Press release from Dansk Byggeri. (Employers association).

Emmitt S., Sander D. and Christoffersen A.K. (2004), Implementing value through lean design, *Proceedings of 12th Annual Conference on Lean Construction* (IGLC), Helsingør.

Emmitt S., Sander D. and Christoffersen A.K. (2005), The Value Universe: Defining a value based approach to Lean Construction. *Proceedings of 13th Annual Conference on Lean Construction (IGLC).* Sydney.

Ferner A. and Hyman R. (eds) (1998), *Changing Industrial Relations in the New Europe.* Blackwell, London.

Forde C. and MacKenzie R. (2006), Concrete solutions? Recruitment difficulties and casualisation in the UK construction industry. In A. Dainty, S. Green and B. Bagilhole (eds), *People and Culture in Construction.* Taylor & Francis, London.

Giddens A. (1991), *Modernity and Self Identity.* Cambridge, Polity Press.

Gjesing F. (2005), Press communication from Dansk Byggeri (Employers association).

Harvey D. (1996), *Justice, Nature and the Geography of Difference.* Blackwell, Cambridge, Massachusetts.

Hyman R. (2001), *Understanding European Trade Unionism: Between Market, Class and Society*. Sage, London.

Jørgensen H. (2002), *Consensus, Cooperation and Conflict, the Policy Making Process in Denmark*, Edward Elgar, Cheltenham.

Koch C. (2000a), Building coalitions in an era of technological change: Virtual manufacturing and the role of the unions, employees and management. *Journal of Organisational Change Management*. Vol. 13, no. 3, pp. 275–288.

Koch C. (2000b), Collective influence on information technology in virtual organisations – Emancipatory management of technology? *Technology Analysis and Strategic Management*. Vol. 12, no. 5. Special issue on the intersection of Innovation Studies and Critical Management Studies, pp. 357–368.

Koskela L. (2000), An Exploration Towards a Production Theory and its Application to Construction. VTT Publications 408, Espoo.

Kyng M. and Mathiassen L. (1997), *Computers and Design in Context*. MIT Press, Cambridge, Massachusetts.

Larsen J., Odgaard G. and Buch S. (2003), A trade union's view of the building process. *Proceedings 11th Annual Conference in International Group of Lean Construction*, Virginia Tech. Blacksburg, Virginia.

Marchington M., Grimshaw D., Rubery J. and Willmott H. (2005), *Fragmenting Work: Blurring Organizational Boundaries and Disordering Hierarchies*. Oxford University Press, Oxford.

Mcloughlin I.P. and Clark J. (1994), *Technological Change at Work*. Open University Press, London.

Mogensen G.V., Kvist H.K., Körmendi E. and Pedersen S. (1995), *The Shadow Economy in Denmark 1994. Measurement and Results*. Spektrum, København.

Nielsen K.T. and Koch C. (1997), *Danish Working Environment Regulation. How Reflexive – How Political? – A Scandinavian Case*. AMOTEK work paper no. 4. København. 43pp.

Nissen E. and Odgaard G. (2004), Udenlandsk arbejdskraft og omgåelsesmuligheder i Danmark. In BAT (2004a), Aktivitetsrapport. BAT København.

Odgaard G. (2005), Interview with BAT representative (manager of secretariat).

Panitch L. (1980), Recent theorganization of corporatism reflections on a growth industry. *British Journal of Political Science*. Vol. 31.

Pedersen E.F. and Lubanski N. (2004), *Free Mobility and EU-s Enlargement-Migration of the Construction Workers May 2004*. Construction Labour Research, Copenhagen.

Pedersen S. (2003), *Sort arbejde i Skandinavien, Storbritannien og Tyskland*. Rockwool fondens nyhedsbrev. Juni (http://www.rff.dk/nyhedsb/jun2003.pdf).

Rothstein B. (1996), *The Social Democratic State. The Swedish Model and the Bureaucrastic Problems of Social Reforms*. University of Pittsburg Press, Pittsburg.

Rothstein B. and Bergsström J. (1999), *Korporativismens fall och den svenske models krise*. SNS Förlag, Stockholm.

Sandberg B.Å. (1995), *Enriching Production*. Aldershot, Avebury.

Sandberg B.Å., Broms G., Grip A., Lundstrom L., Steen J. and Ullmark P. (1992), *Technological Change and Co-determination in Sweden*. Temple University, Philadelphia.

Scheuer S. (1998), *Denmark: A Less Regulated Model*. In A. Ferner and R. Hyman (eds), *Changing Industrial Relations in the New Europe*. Blackwell, London, pp. 146–170.

Simonsen R. and Koch C. (2004), Shaping Lean Construction in project based organisations. *Proceedings International Group of Lean Construction Conference*, Helsingør, pp. 872–884.

Simonsen R., Bonke S. and Walløe P. (2004), Management Innovation Brokers – the Story of Lean Construction Entering Denmark. *Proceedings International Group of Lean Construction Conference*, Helsingør, pp. 859–871.

Thomassen M.A, Sander D., Barnes K.A. and Nielsen A. (2003), Experience and results from implementing Lean Construction in a large Danish contractor. *Proceedings International Group of Lean Construction Conference*, VirginiaTech.

Thrift N. (2005), *Knowing Capitalism*. Sage, London.

Torben Tranæs and Klaus Zimmermann (eds) (2004), *Migrants, Work and the Welfare State*. Rockwool Foundation, Copenhagen.

Van Heffen O., Kickert W. and Thomassen J. (2000), *Governance in Modern Society – Effects, Change and Formation of Government Institutions*. Kluwer, Dordrecht.

Waddington J. (ed.) (2005), *Restructuring Representation. The Merger Process and Trade Union Structural Development in Ten Countries*. Peter Lang, Bruxelles.

Waddington J., Kahmann M. and Hoffmann J. (2005), *A Comparison of the Trade Union Merger Process in Britain and Germany. Joining Forces?* Routledge, London, New York.

Part II

Implications for people management practices and culture

8 Warning

Working in construction may be
harmful to your psychological
well-being!

*Katherine Sang, Andrew Dainty and
Stephen Ison*

Abstract

Psychological well-being is an emerging topic within construction management literature. This chapter reviews existing construction management literature, identifying stressors experienced by those working in the construction sector and their consequences. Stressors identified include the culture of long working hours, high workload, time pressures, project-based working and poor work/life balance, leading to poor psychological well-being among construction professionals. Given the project-based structure and the prevalence of the competitive tendering procurement process within construction, there may be a link between the underlying characteristics of the sector and employee well-being, through the stressors those characteristics place on the workforce. The consequences of poor well-being are varied, but include absenteeism, high staff turnover and other performance concerns. The chapter then proceeds to summarise the business and ethical cases for improving the well-being of the construction workforce. It further offers practical suggestions for ways in which the sector, employers and individuals could improve employee well-being. Strategies might include less reliance on subcontracting, the appointment of a well-being champion within an organisation and stress management techniques. The chapter concludes by providing directions for future research. For example, the scope of existing research could be broadened to investigate well-being along occupational and gender lines, to determine if specific groups are at particular risk. Further research must also aim to explore the link between the structure and culture of the sector, working practices and employee well-being.

Keywords: construction industry, employees, psychological well-being, project culture and competitive tendering.

8.1 Introduction

Construction is an important contributor to the world's economy, both in terms of its proportion of gross domestic product (GDP) and its role as an employer.

Despite the sector's labour intensive production processes, little attention has been paid to the relationship between how the sector operates and the well-being of employees. Through a review of existing literature this chapter aims to explore the relationship between the structure and culture of the sector and workforce psychological well-being. It begins by reviewing the current profile of the construction industry, discussing the sector's structural and cultural characteristics, and highlighting areas which are of particular interest when considering psychological well-being. Characteristics include, among other things, competitive tendering, project-based structure/culture, long working hours, relatively low pay, the fragmented delivery of products and services, high workloads, time pressures, and a macho male-oriented culture. In combination, such factors are likely to be detrimental to the psychological well-being of those who work in the sector. The chapter concludes by providing a way forward, suggesting measures which could be undertaken by the construction industry, legislative bodies, organisations and the individual to improve well-being.

8.2   The current profile of the construction industry

The UK construction industry incorporates a wide range of occupations involved in the construction and maintenance of the built environment (CITB, 2002) employing approximately 10 per cent of the working population and accounting for around 10 per cent of the country's GDP (DTI, 2004). These statistics indicate the sector's national importance, in terms of both its economic contribution and its role as an employer. Any advances in the performance of the sector would in turn have a significant impact on the productivity of the country as a whole.

8.3   The structure and culture of the construction industry

The UK construction industry has a number of characteristics which shape the culture of the sector, which in turn affects how employees are treated. The sector is characterised by a project-based structure (Loosemore *et al.*, 2003), fragmentation (DTI, 2004), workforce homogeneity (Langford *et al.*, 1995) and a procurement system which hampers innovation (Blayse and Manley, 2004). All of these factors combine to form the structure and culture of the sector.

The construction industry is a labour intensive and low tech sector; as a result people are the most important and often the most expensive resource deployed within it (Loosemore *et al.*, 2003). Nevertheless, care for the workforce is often considered to be a secondary aspect of the construction process. In order to determine why this should be, it is necessary to examine how construction takes place, identifying those characteristics of the sector which impact how workers are treated.

Procurement

Procurement systems in the construction industry are an integral part of the sector's structure and culture. The sector largely relies on a price-based environment, with a competitive tendering procurement system (ibid.). Focusing on cost rather than quality is damaging not only to the end product, but also to the workforce. There are a number of problems associated with the tendering process. An emphasis on the lowest cost can result in working to the minimum required standard (Kashiwagi *et al.*, 2004) with significant implications for the quality of the end product and the health and safety of the workforce (Hide, 2003). The low profit margins associated with the tendering process can also mean that there is a strong desire to finish projects on time, avoiding additional labour costs. This desire, combined with increasing client pressure to produce a better quality end product, underlies the need to work long hours (Loosemore *et al.*, 2003). These time pressures and long working hours are associated with stress, anxiety, poor non-job-related well-being and burnout among construction professionals (Sutherland and Davidson, 1993; Lingard and Sublet, 2002; Lingard, 2003; Haynes and Love, 2004; Love and Edwards, 2005).

Project-based culture

Construction is a project-based culture, and it is generally believed that each construction project is unique, with its own problems and issues (Loosemore *et al.*, 2003). This culture has resulted in a fragmented workforce, with project, teams forming for only fixed periods before disbanding to begin new projects, valuable knowledge is thereby lost (Emmitt, 1999), The transient nature of the workforce also has repercussions for the health of employees. Construction workers need to be able to commute long distances to reach project locations, resulting in a longer working day (Loosemore *et al.*, 2003). There may be serious implications for families and communities (Giddens, 2001) and for employee well-being (Love and Edwards, 2005). Furthermore, a project-based culture does not foster a sense of job security as it often relies on the use of subcontracting, which further fragments the sector.

Subcontracting and self-employment

The majority of construction companies are small- and medium-sized enterprises (SMEs) (DTI, 2004), meaning that the majority of construction workers are either self-employed or work within small organisations. In 2003, 70 per cent of all private contractors employed three people or less (ibid.). Such reliance on subcontracting results in a casual labour force which is difficult to monitor, with a variety of consequences. For example, the process of competitive tendering can be undermined through cash-only casual workers undercutting legitimate contractors (Langford *et al.*, 1995). Any attempts to implement a human resource policy will be severely hampered not only by a casual workforce but also by subcontracting. Moreover, smaller organisations may lack the resources necessary to undertake strategies necessary to improve well-being (Rethinking Construction, 2003).

Skills shortage

The sector is currently facing a skills shortage in terms of both the numbers of people and the skills which they possess (CITB, 2004). Evidence suggests that if the sector's period of growth continues in parallel with declining numbers of students undertaking building-related degrees, the construction industry could have difficulty coping with its workload (Dainty and Edwards, 2003). The ageing population of the UK, increased participation in higher education and the poor image of the industry are seen as key factors in the shortage of new recruits (DfEE, 2000; MacKenzie *et al.*, 2000; Fielden *et al.*, 2001). The skills shortage will make it difficult for organisations to produce high-quality products within desired budgetary and time constraints, exacerbating problems associated with competitive tendering. The industry must work to improve its employment practices, if it is to improve retention of existing employees and recruitment into the sector.

Staff turnover, absenteeism and performance levels are indicators of employee well-being (Arnold *et al.*, 1991; Newell, 2002); they are also identified as drivers for change (Rethinking Construction, 2003). The skills shortage, health and safety concerns, staff retention, absenteeism and performance are drivers for change in employment practices in the construction industry (ibid.). As such they also act as drivers for an analysis of how existing employment practices affect those working in the sector. The following section defines well-being and discusses the concept within the construction management literature.

8.4   Well-being in the construction industry

There is no single definition of well-being, rather it is used to cover a variety of phenomena such as happiness, stress, coping and burnout. This chapter focuses on work- or job-related well-being, although the relationship between job- and non-job-related well-being is bi-directional: they have an effect on each other (Warr, 1996).

A number of themes emerge from the existing construction literature, namely that those working in the construction sector experience a number of stressors including long working hours, workload, role conflict and insecurity, relationships with other construction employees, job insecurity, low job satisfaction, time pressures, cost pressures, control and responsibility. It is important to note that it is not always possible to isolate a stressor from the resulting stress; for example, poor job satisfaction can be the result of stressors, but it can also act as a stressor. This section examines these stressors in relation to existing work, identifying how they are related to the structure and culture of the sector.

Long working hours

As has already been discussed, the traditional competitive tender approach to construction procurement has led to pressures on those working on projects.

These pressures include a need to complete the project within a given time and budget. As a consequence, workers face long working hours to meet a high work-load and strict budgetary constraints. This situation can be worsened if employees need to travel long distances to reach a site, thus lengthening the working day (Loosemore *et al.*, 2003). Many researchers have argued that time pressures and long working hours are damaging to those working within the sector (Sutherland and Davidson, 1993; Lingard and Sublet, 2002; Lingard, 2003; Haynes and Love, 2004; Love and Edwards, 2005).

Long working hours can also affect an individual's ability to spend sufficient time with their family (Haynes and Love, 2004) and can lead to conflict within relationships outside of work (Lingard and Sublet, 2002). Social support outside of work is an important mediator for job-related stress (Love and Edwards, 2005). If those working within the sector are unable to spend sufficient time with their family and friends then the effects of the long working hours and other stressors may be increased. In addition, there may be consequences for the well-being of those associated with the employee, such as partners and children. As already noted, job-related and non-job-related well-being are interlinked; any deterioration in one will therefore affect the other.

Workload

The need to meet the tight time and financial constraints resulting from compet-itive tendering means that construction employees can face a damagingly high workload (Sutherland and Davidson, 1993; Lingard and Sublet, 2002; Lingard, 2003; Haynes and Love, 2004). It is not only the amount of time that is spent working but also the workload itself that is important, although it would be reasonable to assume that the two are linked.

Like long working hours, workload is associated with relationship conflict (Lingard and Sublet, 2002), poor mental health (Sutherland and Davidson, 1993) and other measures of well-being (Haynes and Love, 2004). Lingard (2003) identified a relationship between high workload and burnout, a condition whereby the individual experiences emotional exhaustion, depersonalisation and reduced personal accomplishment (MacLeod, 1998). Burnout can have serious consequences for an organisation, as it has been associated with high staff turnover (Lingard, 2003), a matter which needs to be taken seriously in an industry which is experiencing a skills shortage and possibly a recruitment crisis (Dainty and Edwards, 2003).

Fragmentation and job insecurity

One of the key characteristics of the construction industry is that it is project based, which, as discussed earlier, results in a number of problems, such as frag-mentation, temporary teams and job insecurity (Loosemore *et al.*, 2003). The industry's fragmentation is demonstrated by the high number of small firms carrying out work (DTI, 2004). Haynes and Love (2004) have demonstrated that those working on small projects and presumably working for small companies experience poorer levels of well-being. They argue that this may be due to the extra responsibility placed on one person, as the company's future may rest on the success of the current project. This situation highlights the issue of job security, which is necessary for healthy well-being (Warr, 1996). If organisations are surviving from one project to the next then those working within such organisations are unlikely to experience job security. In addition, reliance on subcontracting (Langford *et al.*, 1995) further diminishes opportunities for job security as people may only be employed for as long as they are needed on one project. While job security has been alluded to in reference to job-related stress (Ng *et al.*, 2005), in previous construction literature there appears to be a need for further investigation.

Fragmentation and professional worth

Fragmentation is also the result of the increasing number of occupational groups within the sector. For example, project managers have taken over some of the roles traditionally occupied by architects and it has been argued that the architect's role is being eroded (Emmitt, 1999). This erosion may lead to a reduction in professional worth, which has been associated with well-being among civil engineers (Lingard, 2003). Further work is needed to understand how the social status of construction workers affects their well-being. If such work demonstrates that low professional worth is associated with poor well-being then the industry must work to improve the sector's image if it is to have a healthy workforce.

Temporary teams

A further consequence of the industry's project-based nature is the forming of temporary teams, brought together to carry out one project and then separating to work elsewhere (Loosemore *et al.*, 2003). The dynamics of the relationships between team members on construction projects have been shown to affect the well-being of the individual (Sutherland and Davidson, 1993; Sommerville and Langford, 1994; Lingard, 2003; Love and Edwards, 2005). For example, a lack of support from line managers can be a significant stressor (Sutherland and Davidson, 1993). In addition, if construction workers are working on temporary teams then they may not be able to form the relationships which are necessary to help cope with difficult situations. The relationships between team members can also affect the flow of communication (Emmitt, 1999), which is a mediator of stress and well-being (Sutherland and Davidson, 1993), and may have an impact on the final product.

Work/life balance

Working on a project-by-project basis may make it difficult for employees to take time off. This difficulty can affect their well-being (Lingard, 2003) and thereby exacerbate the work/life balance issues created by a culture of long working hours. A recurrent theme in the literature is the relationship between poor work/life balance and well-being within construction (Sutherland and Davidson, 1993; Lingard, 2003; Haynes and Love, 2004). Like job satisfaction, poor work/life balance can be seen as both a stressor and the result of other stressors, such as long work hours and heavy workload.

As has already been stated, the relationship between job-related well-being and non-job-related well-being is bi-directional. If, as has been suggested, working in construction means long work hours, high workload, resulting in poor work/life balance, insufficient time spent with the family and poor relationship satisfaction, then it may mean that the job-related well-being will be further affected and people may find themselves in a vicious circle. However, this question needs further research, as does the associated impact on the well-being of partners and children. There is also a need to understand how working in the sector affects the work/life balance of those without spouses and children.

Figure 8.1 places the factors discussed so far into diagrammatical form. It helps to demonstrate the basic relationships between variables, but should not be seen as a definitive explanation of relationships, as it is neither exhaustive in

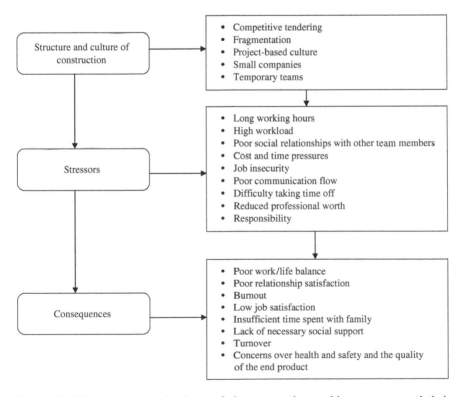

Figure 8.1 The structure and culture of the sector, the resulting stressors and their consequences.

terms of describing the structure and culture of construction nor a comprehensive statement of stressors experienced by individuals, or their consequences.

Although Figure 8.1 is a simplified representation of the factors involved in the poor well-being of construction workers, it helps to summarise the findings from existing literature. It is important to remember that factors such as low job satisfaction and poor work/life balance are in themselves stressors as well as consequences of stressors. Just as it can be difficult to separate the stress from the stressor, it can also be hard to differentiate between the structure and the culture of the sector, as the two are interlinked and mutually reinforcing. For the sake of clarity, they have not been separated in this diagram. However, there are factors categorised as stressors which also form part of the culture of the sector, such as long working hours. Further research is needed to develop a model of job-related well-being within the construction sector.

So far it is clear that at the most basic level construction operates in a way which is damaging to its employees. If work continues to be procured on a lowest-bid basis then it is clear that those working on a project will suffer through demands for long working hours, high workload and poor work/life balance. These problems are exacerbated by the fragmentation of the industry and a project-by-project approach to recruitment. The demands that tight financial and time constraints imposed by the lowest bid place on workers may be mediated by social support in the workplace, organisational support and a feeling of high professional worth. However, the forming of temporary teams may make it difficult to find the necessary support. Moreover, reliance on subcontracting and small businesses will often mean that organisations lack the financial and time resources needed to implement strategies to care for their employees' psychological well-being. In addition to these pressures, the project-based culture does not foster the sense of job security which is needed for healthy levels of well-being. The introduction of reports and initiatives such as Respect for People (2000) implies that those leading the sector are concerned about the welfare of those working within it. The evidence presented here suggests that while this may be so, little is being done to improve the lot of construction workers. In the early 1990s researchers highlighted the stress and associated problems experienced by those working within management positions within the sector (Sutherland and Davidson, 1993; Sommerville and Langford, 1994). In 2005 similar problems and stressors were still being reported (Love and Edwards, 2005) indicating that in over a decade the industry has failed to implement change. Suggestions as to how it might do so are made after the following section, which presents the business and ethical cases for improving employee well-being.

8.5 Improving the well-being of construction employees

The previous section demonstrated that there are aspects of working in the construction industry which are potentially damaging to the well-being of the workforce. This section demonstrates that poor well-being should be of concern to the sector, by presenting the business and ethical cases for improving employee well-being.

The business case

There is a multifaceted business case for the construction industry becoming more aware of the physical and psychological health of its employees. As discussed above, the skills shortage is an often-used argument for the industry to take more seriously its duty of care towards its employees (Constructing Excellence, 2004). In an industry which is having difficulty recruiting people at all levels, it is essential that steps are taken to encourage those already employed in the sector to stay. One way to achieve this would be for employers to determine the reasons for high levels of staff turnover. Initial evidence from the construction management literature suggests there may be a link between well-being and turnover. For example, the RIBA (2003) suggested that aspects of the architectural profession – such as long working hours and poor pay – lead female architects to seek alternative employment. Lingard (2003) highlighted similar concerns among civil engineers working in Australia.

Poor levels of well-being are also linked to absenteeism (Sutherland and Davidson, 1993). Statistics released by the HSE in 2004 suggest that approximately 13 million working days are lost each year due to stress, and between 8,000 and 18,000 construction workers reported stress, depression or anxiety (HSE, 2004). These figures indicate that occupational ill health and poor well-being are problems within the sector. Given the financial cost of absenteeism, these figures provide a convincing business case for the sector to take employee well-being seriously.

The impetus for improving the well-being of employees should not be purely financial. Employers have a duty of care towards the physical health of their employees through adherence to health and safety legislation, as illustrated by the Health and Safety Act 1974. Similar care should be taken towards the mental health of employees.

The ethical case

The bi-directional relationship between work- and non-work-related well-being (Warr, 1996) has resulted in an interest in the effects of work on marital and other relationship quality. Lingard and Sublet (2002) discovered that long working hours were significant predictors of relationship quality and satisfaction. Cano and Dina (2003) found evidence linking occupational stress (in a non-construction-based sample) among men with domestic violence, demonstrating that the relationship between work and non-work well-being can be very damaging to families. However, the impact of working in the construction industry on the well-being of family members is a significant gap in the research and needs further investigation.

Developmental psychologists have found significant evidence that poor parental well-being can have detrimental effects on child development, particularly if there is a lack of involvement between parent and child due to parental rejection, depression or stress (Shaffer, 1996). Subsequent damage to child development can be severe. Effects include aggression, difficulty interacting with others, poorer academic performance and antisocial behaviour (ibid.). Such evidence suggests that the impact of stress in the workplace can be far-reaching and have a long-term impact on communities and society.

Giddens (2001) explores the sociological impact of poor job-related well-being, emphasising the links between changing work roles, working hours, child development and the community. As well as being a source of stress and poor well-being, decreased job security in recent years may also drive people to work longer hours. This in turn is a predictor of burnout (Lingard, 2003) and stress, and may be particularly relevant in construction where employment status may change from one contract to the next. It can also have wider consequences, including an impact on a worker's ability to engage with their local community, for example involvement with school committees (Giddens, 2001). Considering the importance of the construction industry to the UK economy in terms of its contribution to GDP and its role as an employer, the sector must begin to consider the effect that it has on society in general and on local communities.

Poor employee well-being is harmful to the organisation as it has implications for retention, absenteeism, performance, and health and safety. This is a compelling business case for the sector and employers to monitor and improve employee well-being. However, poor job-related well-being has wider consequences to families, communities and society. The following section presents ideas as to how the sector could improve well-being, drawing on literature from both construction and non-construction sources.

8.6   Improving well-being: Practical suggestions

When a decision is made to improve employee well-being, a number of steps must be taken. The first is to decide who is responsible for implementing change. As well-being affects the individual, it may be necessary to focus efforts on an individual level. Alternatively, given the damaging effects of poor well-being to the organisation, those in charge of companies or departments should take responsibility for improving well-being. In a sector facing a potentially severe workforce shortage, efforts should be industry wide, undertaken by professional bodies, trade unions or organisations responsible for training. It has already been demonstrated that poor well-being can have an effect at the societal level; as such responsibility could fall to public bodies that can implement legislation aimed at maintaining the health of the workforce. A summary is provided in Table 8.1.

Policy level actions

The Health and Safety Executive (HSE) is a public sector body responsible for ensuring safety in all working activities (HSE, 2004). More recently the HSE has

Table 8.1 Summarising steps which can be taken to improve employee well-being

Policy	Sector	Organisation	Managerial	Individual
Enforce relevant legislation.	Broaden the scope of initiatives such as the Considerate Constructors Scheme to encompass employee well-being.	Integrate well-being management into HRM policy.	Audit well-being.	Be aware of well-being issues.
Use its role as a client to help raise awareness of employee well-being and measures which can improve it.	Professional organisations, trade unions and the Sector Skills Councils should help to support their members in auditing well-being and taking subsequent actions.	Appoint specialised managers or project teams.	Demonstrate effective and trustworthy leadership.	Be willing to improve their own well-being through measures such as relaxation.
		Provide necessary staff development.	Undertake well-being awareness and stress-management training.	Undertake necessary staff development, such as assertiveness training.

begun to focus attention on stress within the workplace, through commissioning research and publishing guidance to employers and employees. Legislation has also been introduced to restrict the number of hours worked per week. As of 2005, the Working Hours Directive has determined that the average working week must not exceed 48 hours (Constructing Excellence, 2004). However, this is an average to be taken over a period of four months, and does not prohibit a working week exceeding 48 hours. Current evidence suggests that a significant source of stress, poor work/life balance and poor well-being in the construction industry is the culture of long work hours. In order for the Working Hours Directive to be effective within construction, it must be adopted by organisations and employees. To achieve this, a major shift will need to occur in the culture of the sector, enabling employees to feel that taking advantage of the directive will not be detrimental to their career.

Sector level actions

The existing literature has demonstrated that the structure and culture of the sector has resulted in stressors which are potentially harmful to those employed within construction. If those leading the sector – for example, professional bodies, the sector skills council, trade unions and the government – are serious about respecting its workforce, then they must begin to change the industry at its most fundamental level. The way in which projects are procured must move away from a desire to seek the lowest tender. Those putting forward tenders have the responsibility to ensure that they allocate adequate financial, time and human resources in a way which is feasible to achieve without demanding long work hours and high workloads. This measure may mean that tender prices increase, in which case clients must be willing to increase their budgets. Although this may be difficult for many clients, professional clients such as the government are in a position to effect change.

It is not yet apparent if alternative procurement approaches such as PFI and PPP can improve the well-being of construction workers and this is an area in need of further investigation. However, if such approaches can result in a budget which accurately reflects the work necessary to complete the project on time, budget and specifications then the stressors experienced by those undertaking the task may be decreased. Such measures may help a move away from a project-based culture, which will in turn help to increase job security and form permanent teams, which can provide necessary social support.

Organisational level actions

If the construction industry is to improve employee retention, performance and recruitment, then organisations must seriously consider their role in monitoring and improving employee well-being. Due to the lack of construction-specific research in this area, the suggestions below have been taken from other sectors.

A well-being champion

Large organisations should appoint an influential person who is responsible for monitoring the well-being of employees, who can carry out well-being audits among employees, and implement any necessary measures to improve well-being (Schabracq, 2002). This responsibility could be part of a line manager's overall remit. Alternatively, if appropriate, the organisation could appoint a specialised well-being manager. Jordan *et al.* (2003) argue that it is important that there is top-level commitment within an organisation, thereby ensuring that stress-prevention measures are incorporated into the day-to-day running or culture of the organisation.

Leadership

Effective and sensitive leadership can have a positive effect on employee well-being. Important leadership qualities identified by Schabracq (2002) include respect for all employees, effective and honest communication with employees, awareness of the signs of stress and the ability to reward good progress, rather than focusing on negative issues. Sang *et al.* (2003) reported a lack of social skills among those working within the UK construction industry. If construction employees, particularly those with management responsibilities, do lack the social skills outlined by Schabracq (2002), then it may be possible for organisations to undertake training programmes to develop them.

Stress-prevention programmes

Stress prevention and well-being improvement programmes need to be integrated into Human Resource (HR) policy if they are to be truly effective. Not only can HR departments encourage staff development programmes (Jordan *et al.*, 2003) to develop social skills, there is also scope for them to undertake alternative routes. These include appropriate selection procedures to ensure that only those correctly qualified to carry out duties and tasks are employed (Schabracq, 2002). If particular positions are associated with stress and poor well-being, it may be possible to undertake job redesign, in order to remove or minimise stressful aspects of the job, including role ambiguity (Jordan *et al.*, 2003). Larger construction companies may be able to include well-being monitoring as a part of their HR department's remit. Such a measure should be able to integrate well-being into the HR strategy of the organisation. Large companies are able to take a leading role in workforce issues, and are in a position to increase awareness of well-being issues across the supply chain. SMEs, on the other hand, may lack the resources necessary to take up many of these suggestions. Thus, while the construction industry relies on subcontracting and SMEs, there is little hope of change. Instead, there needs to be a fundamental evaluation of the use of subcontracting, as it allows risk to be delegated onto smaller companies which may be unable to shoulder such a burden. Larger companies can set an example for smaller companies in terms of looking after their workforce, but they can also help by developing partnerships with small firms. By this means, the future of these organisations would be more secure, enabling them to plan workforce development.

Individual actions

There are actions each person can take to improve their own well-being. The first step is to identify symptoms of stress, which include excessive use of alcohol and cigarettes, juggling several jobs at once, poor eating habits and insufficient time to relax (HSE, 2004). Such stress can affect an individual's health, resulting in increased blood pressure, sleep disturbances and sick leave (Shirom, 2002). Once stress is recognised, the individual must take steps to improve their health. Possible means include relaxation, meditation, cognitive-behavioural therapy (changing thought processes, in order to reappraise a stressful situation and view it in a more rational manner), exercise, time management and assertiveness training (e.g. learning to say 'no' to a client) (Schabracq, 2002; Jordan *et al.*, 2003).

While construction companies may lack the resources to appoint a well-being specialist, or to develop a comprehensive HR strategy in this respect, they are able to encourage employees to take steps to improve their own well-being.

Many government and charitable organisations provide advice for individuals; for example, the HSE offers guidance on its website. Employers can make sure that employees are aware of these advice documents and support individuals who choose to try the suggested measures. It may be that the construction industry's macho culture (Fielden *et al.*, 2001) will militate against workers undertaking measures such as meditation or therapy. In such circumstances, employers must take the lead in encouraging the uptake of stress-prevention measures and breaking any associated taboos.

Small companies in the construction industry will need support to maintain a healthy workforce. This support can come from government organisations, such as the HSE. Other potential sources of support include the professional bodies and Construction Skills, the new Sector Skills Council (SSC) for construction. Such organisations are responsible for establishing the professional standards and training necessary to work within their fields. They are funded by their members and are in a position to provide advice to organisations and individuals about monitoring and maintaining healthy levels of well-being.

8.7   Directions for further research

While there is considerable scope for the sector to implement policies and practices which can improve the well-being of the construction workforce, research needs to be undertaken to determine the extent of the problem. An industry-wide study could assess well-being through the use of well established diagnostic tools, helping to determine if those working in the construction industry experience poor well-being. Such a study would help to broaden the scope beyond site managers, construction managers and civil engineers.

The scope of existing work could be broadened in a number of other ways. At present there is no empirical work indicating whether particular occupational groups are at increased risk of poor well-being. Such work would help direct measures to improve well-being towards those groups in particular need, and to tailor well-being programmes so that they meet the groups' specific needs. For example, those working in the trades and crafts will face different pressures than those working as architects. Consequently, sources of stress and poor well-being will be different and wellness programmes would need to ensure the needs of each group were met.

It is interesting to note that existing work has focused on men. While men and women working in the industry face similar pressures, it seems likely that women may face additional sources of difficulty. The difficulties experienced by women working in the construction industry are well documented (see, for example, Dainty *et al.*, 2000). Considering these difficulties, research must be undertaken to investigate their impact on the psychological health of women. RIBA (2003) linked aspects of the architectural profession with the exodus of women from it. Further work needs to be carried out to determine if women in other construction professions and occupations are affected in a similar way.

Any work which is undertaken to investigate the well-being of those working within the construction industry must consider the sector's structure and culture. The fragmentation of the workforce, the project-based culture and competitive tendering procurement all work together to form a culture which is probably unique to construction. The sector's culture necessarily impacts on the workforce, and any attempts to improve employee well-being must take this into account. Longitudinal research would help to investigate the nature of the relationship between factors. Study of teams working within alternative procurement approaches (e.g. Design and Build and partnering) would help determine if procurement methods have any impact on employee well-being.

8.8   Conclusions

This chapter has demonstrated that working in the construction industry is potentially damaging to well-being. Research conducted in the UK and Australia has highlighted that particular characteristics of the construction industry lead to poor well-being. These include long working hours, work overload and a lack of work-based social support. The findings suggest that the fundamental structure of the sector (for example, a proliferation of very small organisations and the use of competitive tendering) results in a number of cultural factors which are damaging to psychological well-being. The drivers for change identified in this chapter demonstrate the need for the sector to consider carefully the well-being of its workforce. There is a need for further research which can examine the dynamics of the relationship between employee well-being and the structure and culture of the construction industry. However, the existing research suggests that the construction industry must begin seriously to rethink the way in which it operates. The use of competitive tendering, the fragmentation of the industry and its project culture result in an environment of time and cost pressures, long working hours, high workload, and poor job security, work/life balance, relationships both at work and at home and job satisfaction. All of these stressors or variables place construction workers at risk of poor well-being. Those working in SMEs may be at greater risk of poor well-being as these organisations may lack the resources to implement appropriate well-being policies. The sector must determine a way forward which does not rely on the 'lowest bid' system, subcontracting and working from one project to the next. If changes can be made in these directions then those working in the sector may find it a healthier place to be.

References

Arnold, A., Robertson, I.T. and Cooper, C.L., 1991. *Work Psychology. Understanding Human Behaviour in the Workplace*. London: Longman Group UK Ltd.

Blayse, A.M. and Manley, K., 2004. Key influences on construction innovation, *Construction Innovation*, **4**(3), pp. 143–154.

Cano, A. and Dina, D., 2003. Are life stressors associated with marital violence? *Journal of Family Psychology*, **17**(3), pp. 302–314.

CITB, 2002. *CITB Skills Foresight Report.* CITB. Available at http://www.citb-constructionskills.co.uk/.

CITB, 2004-last update, national construction week. Available at: http://www.ncw.org.uk/ [3 February 2004].

Constructing Excellence, 23 November 2004, 2004-last update, working time directive [Homepage of Constructing Excellence] [Online]. Available: http://www.constructing excellence.org.uk/resourcecentre/publications/az.jsp?azID=17244&level=0 [3 February, 2005].

Dainty, A.R.J. and Edwards, D.J., 2003. The UK building education recruitment crisis: A call for action. *Construction Management and Economics,* **21**(7), pp. 767–775.

Dainty, A.R.J., Bagilhole, B.M. and Neale, R.H., 2000. A grounded theory of women's underachievement in large construction companies. *Construction Management and Economics,* **18**(2), pp. 239–250.

DfEE, 2000. *An Assessment of the Skill Needs in Construction and Related Industries.* Nottingham: DfEE Publications.

DTI, 2004. *Construction Statistics Annual 2003.* Norwich, UK: The Stationary Office.

Emmitt, S., 1999. *Architectural Management in Practice: A Competitive Approach.* Harlow, UK: Addison Wesley Longman.

Fielden, S.L., Davidson, M.J., Gale, A.W. and Davey, C.L., 2001. Women, equality and construction. *Journal of Management Development,* **20**(4), pp. 293–304.

Giddens, A., 2001. *Sociology.* Cambridge, UK: Blackwell Publishing.

Haynes, N.S. and Love, P.E.D., 2004. Psychological adjustment and coping among construction project managers. *Construction Management and Economics,* **22**(2), pp. 129–140.

Hide, S.A., 2003. *Exploring Accident Causation in the Construction Industry.* Unpublished PhD Thesis. Loughborough University.

HSE, 2004. *The Health and Safety Executive: About us,* http://www.hse.gov.uk/aboutus/ index.htm (accessed December 2004).

Jordan, J., Gurr, E., Tinline, G., Giga, S., Faragher, B. and Cooper, G., 2003. *Beacons of Excellence in Stress Prevention.* Sudbury: HSE Books.

Kashiwagi, D., Chong, N., Costilla, M., Mcmenimen, F. and Egbu, C., 2004. *Impact of six sigma on construction performance. 20th annual conference of the Association of Researchers in Construction Management (ARCOM),* K. Khosrowshahi (ed.). Heriot Watt University, September 1–3, pp. 13–23.

Langford, D., Hancock, M.R., Fellows, R. and Gale, A.W., 1995. *Human Resources Management in Construction.* England: Longman Group Ltd.

Lingard, H., 2003. The impact of individual and job characteristics on 'burnout' among civil engineers in Australia and the implications for employee turnover. *Construction Management and Economics,* **21**, pp. 69–80.

Lingard, H. and Sublet, A., 2002. The impact of job and organizational demands on marital or relationship satisfaction and conflict among Australian civil engineers. *Construction Management and Economics,* **20**(6), pp. 501–521.

Loosemore, M., Dainty, A.R.J. and Lingard, H., 2003. *HRM in Construction Projects: Strategic and Operational Approaches.* London: Spon.

Love, P.E.D. and Edwards, D.J., 2005. Taking the pulse of UK construction project managers' health: Influence of job demands, job control and social support on psychological wellbeing. *Engineering Construction and Architectural Management,* **21**(1), pp. 88–101.

MacKenzie, S., Kikpatrick, A.R. and Akintoye, A., 2000. UK construction skills shortage response strategies and an analysis of the industry perception. *Construction Management and Economics*, **18**(7), pp. 853–862.

Macleod, J., 1998. *An Introduction to Counselling*, 2nd edn, Buckingham, UK: OUP.

Newell, S., 2002. *Creating the Health Organization. Well-being, Diversity and Ethics at Work*. London: Thomson Learning.

Ng, S.T., Skitmore, R.M., and Leung, T.K.C., 2005. Manageability of stress among construction participants. *Engineering, Construction and Architectural Management*, **12**(2), p. 264.

Respect for People, 2000. *A Commitment to People: Our Biggest Asset, www.constructing excellence.org.uk*.

Rethinking Construction, 2003. *Respect for People Toolbox*. London: Rethinking Construction.

RIBA, 2003. *Why Do Women Leave Architecture? www.riba.org.uk*, London: RIBA.

Sang, K.J.C., Dainty, A.R.J., Ison, S.G. and Arnott, M., 2003. Investigating the Impact of Skills and Workforce Composition on the Performance of the UK Construction Industry, *1st Scottish Conference for Postgraduate Researchers of the Built and Natural Environment*, C. Egbu and M. Tong (eds), Glasgow Caledonian University, November, pp. 135–145.

Schabracq, M.J., 2002. What an organization can do about its employees' well-being and health: An overview. In M.J. Schabracq, J.A.M. Winnubst, C.L. Cooper (eds), *The Handbook of Work & Health Psychology*, 2nd edn, England: John Wiley & Sons, pp. 585–600.

Shaffer, D.R., 1996. *Developmental Psychology*, London: Brooks Cole.

Shirom, A., 2002. The effects of work stress on health. In M.J. Schabracq, J.A.M. Winnubst and C.L. Cooper (eds), *The Handbook of Work & Health Psychology*, 2nd edn, England: John Wiley & Sons, pp. 63–82.

Sommerville, J. and Langford, V., 1994. Multivariate influences on the people side of projects: Stress and conflict. *International Journal of Project Management*, **12**(4), pp. 234–243.

Sutherland, V. and Davidson, M.J., 1993. Using a stress audit: The construction site manager experience in the UK. *Work and Stress*, **7**(3), pp. 273–286.

Warr, P., 1996. Employee well-being. In P. Warr (ed.), *Psychology at Work*, 4th edn, London: Penguin Books, pp. 224–253.

9 Access and inclusivity of minority ethnic people in the construction industry

Andrew Caplan

Abstract

Despite no lack of interest in work in the construction industry, minority ethnic people are significantly under-represented as employees and particularly in senior posts. In part this is reflective of industry as a whole and certain 'traditional' industries in particular, and not only in this country. This chapter, based on four research projects conducted over the last five years, examines the structural reasons for this under-representation, barriers within education and training, and discriminatory practices which reduce opportunities for work. It suggests that the traditional culture and fragmented nature of the industry has been largely responsible for the industry escaping the kind of scrutiny to which other sectors have been subjected. However, there are encouraging signs that the industry is responding to societal pressures for inclusivity as well as addressing the shortfalls in essential industrial skills, which should mean an increasing utilisation of the skills that the minority ethnic population can bring to construction.

Keywords: diversity, racism, discrimination, construction, and under-representation.

9.1 Introduction

In recent years there has been an increasing emphasis within the construction industry on 'Respect for People', with a view to recruiting and retaining a skilled and effective workforce (Egan, 1998: para. 17). Issues of equality, diversity, education and the promotion of good practice have also been given greater importance in public statements and publications.[1] Although under-represented in the industry as a whole, minority ethnic people are employed in a range of construction trades and professional roles, continue to enrol on college and university construction-related courses in healthy numbers, and are progressing in several of the professions. However, there remain barriers to entry and progression, which mean that the industry continues to lose the opportunity to recruit skilled people from this section of the population. This is particularly problematic in the context of an overall shortage of trained staff.

Concerns about equal opportunities and the stark under-representation of women in both the trades and the professions within the British construction industry in the 1990s (Latham, 1994; Committee on Women in Science, Engineering and Technology, 1995) prompted thought about the participation and experiences of black and minority ethnic (BME) people (CIB, 1996: 2, 63). The Construction Industry Board (CIB) report stressed that if the industry as a whole continued to fail to attract more women, older workers and BME groups then construction companies would become 'less representative of the customers and clients who keep them in business'. Action with regard to BME people within the industry was hampered due to a lack of consistent data, monitoring and research.

The limited research undertaken pointed to the apparent numerical under-representation of BME people within the built environment sector (Grant *et al.*, 1996; Owen *et al.*, 2000; Twomey, 2001) despite the sector's expansion over the period 1996–98. Anecdotal evidence portrayed the industry as hostile to BME inclusion at all levels. However, it was unclear whether this hostility was the result of working-class xenophobic and racist opposition at site level, 'new racist' views among management (Barker, 1981) and the persistence of traditional practices, or a pre-emptive rejection of the industry by the BME community. Traditional sectors with a hard centre of so-called 'masculine' values similar to that of the construction industry – the police and the armed forces, for example – had been objects of scrutiny for implicit and explicit racist practices, but construction had escaped notice. This situation was to change with the publication of the Macpherson Report in 1999, its formulation of the concept of 'institutional racism', and its strong recommendations for the public sector and their implications for private sector compliance. The 'business case' for greater equality of opportunity was given a high profile by the Movement for Innovation (M4I) Working Group on Respect for People (2000: 5), which sought to respond to the challenge to find practical and effective ways for the construction industry to improve radically its performance on 'people issues' with a view to recruiting and retaining the best talent and business partners. The issue of diversity in the workplace became central to industry concerns (Commission for Racial Equality, 1995), and a 'Respect for People: equality and diversity in the workplace toolkit' was subsequently produced to encourage the construction industry to become 'more representative of the total labour available, to eliminate poor stereotypes and recruit and develop the best, most talented people'.

Attention was also drawn to the fact that while BME people were reasonably well represented on construction-related craft/trade training programmes and degree courses, they were significantly under-represented within the industry,[2] particularly at middle and senior management levels (CEMS, 1999: 6; Greed, 1999: 29). Reasons cited for this under-representation included a failure on the part of the industry and companies to develop awareness among BME communities of the wide range of opportunities in construction; the predominance of word-of-mouth recruitment practices and information about contracts which, it was found, tended to exclude BME people; a fear of discrimination and racism; the perception that the industry was a white, male-dominated environment where it was difficult for BME people to be accepted in an equal manner; and less-favourable treatment, widespread name-calling, harassment, bullying, and intimidation of BME people (CEMS, 1999).

Over the last five years the Centre for Ethnic Minority Studies, at Royal Holloway University of London, has been researching the nature of the relationship between minority ethnic individuals and communities and the construction industry. In particular, it has looked at their perceptions of and experiences in the industry: how they view it and why; and how they fare in it and why. To be clear, although this chapter's title refers to 'minority ethnic people', its main concern is visible minorities: black and Asian people. In fact, the underlying research has incorporated other ethnicities and other issues such as class, gender, and disability running parallel or intersecting with that of race and ethnicity, but this chapter will not directly address those issues.

The impetus for this work came from the Construction Industry Training Board (CITB) through its role in meeting the construction industry's need for a skilled workforce, and its perception that the industry and its training providers had failed to attract and retain a reasonable representation of minority ethnic people (Construction Industry Training Board).[3] The CITB commissioned a series of studies examining the reasons for the under-representation of minority ethnic people in construction. This body of work, carried out between 1999 and 2003, forms a coherent whole with regard to perceptions of, barriers to, and experiences in the construction and related industries by minority ethnic tradespeople and professionals, and it is upon these studies that this chapter is based.

9.2   Methodological approaches

The research was carried out largely through direct contact with individual minority ethnic people who were, had been, or aspired to be employed in the industry, and with smaller control groups of white people in similar circumstances, in addition to a sample of community and representative groups, and a spread of potential and actual employer organisations. The interviews, supplemented by questionnaires, were conducted through face-to-face and focus-group interviews held all over Britain, in mainly white areas as well as in the chief areas of minority ethnic settlement. The samples ranged from over 800 in the *Under-representation* study (1999) to 28 in the *People Profile* study (2003). In all, the experiences and views of approximately 1100 individuals were examined for these four studies – from secondary school students to graduate professionals; from unemployed tradespeople to senior managers in construction firms; and from ex-employees in the industry to members of professional associations. The research samples were collected from colleges, training agencies and providers, employers, job centres, and community advice centres. The *Under-representation* study also involved consultation with a sample of organisations – employer representation bodies, minority ethnic community and business associations, and the CRE – who would be key in the implementation of any changes in the industry. The study was benchmarked against three groups – the British Army, the Netherlands Construction Industry Training Board (SVB), and a number of UK-based construction firms – in order to identify elements of good practice that could be transferable to the UK construction industry.

Due to the concern about the low incidence of minority ethnic students entering the construction industry, one of the recommendations of the *Under-representation* study (no. 12: 71) was to chart career paths from college through employment into work (or otherwise). This study focused on General National Vocational Qualification students at Foundation and Intermediate Levels in two geographical areas – Yorkshire and the Humber, and the Midlands. It evaluated the experiences of minority ethnic and female students, and included control groups of white male students in each of the areas studied. The key objectives of the study were to

- investigate the perceptions and experiences of GNVQ Foundation and Intermediate students
- identify 'push/pull' factors which impact on further training and educational opportunities for GNVQ students
- identify the factors affecting retention and progression in the industry
- identify the degree of support such students receive in sustaining themselves through their course of study.

In addition to college and school students, the research team interviewed tutoring, lecturing, and managerial staff in the institutions, and consulted education advisers and key staff from the CITB and awarding bodies. In total, 94 students were interviewed for the study.

The *Retention and Career Progression* study (2000a), like the *GNVQ* project, concentrated on entry into the construction industry, but had a wider focus, looking in more detail at the barriers and incentives – push-pull factors – for minority ethnic people entering and staying in the industry. Its primary aim was to identify the reasons for losses to the industry of trained minority ethnic employees. To this end, interviews were conducted with 112 trade and college trainees, recent graduates, and employees of up to ten years' experience. The researchers also interviewed individuals who had been unable to find a job in the industry, and those who had left it for other work. The research design followed a 'career history' approach and sought to explore career influences and dynamics from the individual's frame of reference.

The fourth project for the CITB – *People Profiles* (2003) – aimed at increasing the amount of information on minority ethnic people working and training in the industry and disseminating good practice by producing a range of individual and company case studies illustrating the business benefits of diversity. The last project for CABE (2005) examined the representation of BME people within the built environment professions, barriers to entry into the professions, and progression in the sector. Researchers interviewed 37 BME students and 50 BME professionals (including 5 people who had left the sector or moved out of their original profession), 12 employer organisations who were from a cross-section of the built environment sector, lecturers and tutors from 6 universities across England, and 6 'key informants' representing different industry bodies and experience.

9.3 Findings

The construction industry cannot be considered in isolation from the rest of society where practices of institutional discrimination affect opportunities to succeed. In addition, there is an established set of customs and practices within the construction sector which can lead to institutionalised discrimination. This discrimination is not solely the result of action or inaction by particular individuals, but rather of an employment environment that operates to the detriment of minority ethnic people compared with their white counterparts.

Although the construction industry now recognises that in order to remain competitive and to address skills shortages, it has to recruit the best quality people, regardless of their race or colour, the study found (CEMS, 1999) that the industry:

- has a strikingly low representation of minority ethnic employees
- has not been successful in developing awareness amongst minority ethnic people of the wide range of opportunities in construction
- excludes minority ethnic people through the use of traditional, informal, word-of-mouth recruitment practice and information about contracts
- is perceived as a white, male-dominated environment in which it would be difficult for minority ethnic people to secure jobs or contracts
- could show few examples of positive initiatives to tackle the above issues.

Their attendance at college and university courses demonstrates that minority ethnic people are interested in construction-related study, but they remain grossly under-represented at all levels within the industry. This phenomenon has several causes.

- In part it is an information problem, that is a lack of information about the types of work available in construction, and the inaccessibility of existing channels of information (particularly family members working in the industry) to minority ethnic people.
- In part it is a visibility problem, with an overwhelming absence of minority ethnic role models at all levels, but particularly in senior positions throughout the industry.
- Lastly it is a 'race' problem, that is it concerns perceptions and experiences of direct and indirect racial discrimination in the industry.[4]

Perceptions

The construction industry is obviously concerned about the widespread lack of knowledge and understanding amongst the general public of the range of careers it comprises (Latham, 1994). In the main, the potential workforce still gets its image of the industry through walking past building sites which project an image of manual labour, mostly conducted outdoors. Although this obviously overstates the case, awareness (by younger people) of the possibilities of employment in the construction industry is strongest in the manual trades rather than in the professional roles (CEMS, 2000a: 8–10). Although levels of awareness differ between minority ethnic groups, the common external perception is dominated by a white male stereotype involved in hard, dirty, low status, unskilled or semi-skilled labour – supervised and directed by a white manager. This combination projects a negative image to minority ethnic people, particularly the young, in comparison to other career options.

A commonly held perception among employers and staff teaching on construction courses is that minority ethnic people (particularly 'Asians') are not interested in work in the industry because of its low status, but this perception was not borne out in these studies. Indeed, potential entrants (particularly young people) from all minority ethnic groups were interested in more information about jobs in the industry. GNVQ Construction and the Built Environment courses, although recruiting poorly compared with other GNVQs at Intermediate and Advanced levels, are in fact popular at Foundation level, but they recruit students who have very varied, or as yet unformed, career aims.

Familiarity with the industry at an early age remains a strong influencing factor for all students eventually going on to train for a job. White students had the advantages of concrete knowledge of the industry, greater awareness of the possibilities of work, and greater likelihood of relevant work experience and of positive encouragement from family members. The family or friends network, in addition to public sources such as daily newspapers and school careers staff, continues to be important in obtaining that crucial first job or work placement. The relatively high proportion of construction staff who were enabled to join through personal contacts and word-of-mouth recommendations reflects wider construction industry practice, and provides a commonsense explanation for the persistence of minority ethnic under-representation and the structural barriers to their entrance into the industry. For minority ethnic entrants, the wider influences – such as schools, colleges, careers advisers – are critical because their 'family and friends' networks are underdeveloped. Given the variability of schools' career advice, crucially important career decisions appear to be left more to chance, and accurate data of the wide range of trades and professions open to students may not be presented early enough.

Education

The research found that students chose the GNVQ Construction courses for a variety of reasons, but only a minority of all students had a firm intention of ever getting a permanent job in construction. Some students felt they had been forced onto the course by their school (possibly because their schools did not want low-achieving students to figure in their 'league table' scores). Others had been misinformed about the true nature of their course.

Although the form and content of GNVQ is outside the lecturers' control, there is considerable leeway for them to interpret the content and contexts of the course and deliver it in their own manner – by the attention they give to students, and the way they reward achievement and encourage further involvement in the industry. Although students were generally positive about these courses, all negative responses came from minority ethnic participants. Course teaching was largely unmonitored beyond official inspections. College lecturers were not always prepared by their training or the college support system to teach minority ethnic students, students who speak English as a second language, female students, or school pupils under the age of 16. Tutors exhibited aspects of cross-cultural confusion, misunderstanding, and lack of sympathy for the problems faced by their minority ethnic students – and assigned blame for problems in communication with them to the over-sensitivity of the students themselves. Some felt that female and minority ethnic students should be prepared to endure discrimination or harassment in the workplace without complaint, and refused to recognise that discrimination could impact on study or performance at work. Lecturers were often found to be ineffective in dealing with racial or sexual harassment, or in leading discussions about equal opportunities. Students were offered differential levels of academic and careers advice and encouragement along lines demarcated by their ethnicity or sex. On a national level, the management of GNVQ construction courses and the implementation of CITB policies were hindered by a lack of:

- management information, such as records on courses and statistical profiles of students (e.g. ethnicity, gender, prior achievement)
- adequate monitoring of the quality of delivery
- co-ordination of work between national, area, and curriculum centre staff.

Despite these criticisms, students were broadly happy with the teaching and relationships fostered on their course. Most tutors had attempted to provide an informal atmosphere in the classroom, deliberately emphasising the more popular, 'practical' aspects of the course, sometimes at the expense of the theoretical and academic core aspects – apparently in order to capture the attention and commitment of white male students who had been classified by the educational system as unsuitable for academic work (while at the same time disadvantaging the generally better performing female students). Thus, the cost of making the course responsive to the needs of some of the white male students was to alienate minority ethnic and the few committed female students who were disadvantaged by the 'matey' atmosphere. Given that it is the culture of the work site that is so often blamed for the failure of the construction industry to attract minority ethnic and female applicants, it would seem that the intrusion of this informal atmosphere into the classroom might be counterproductive. In effect, the education environment was reinforcing the pejorative aspects of the industry by de-emphasising the academic aspects of the course. Thus, the image of the construction industry as a bastion of 'laddish', white privilege was left unchallenged at the point at which the courses were taught. Instead, it was found that the workplace was entering the school or college, bringing with it an emphasis on a seductive set of apparently neutral values – friendship, respect, getting on – that actually operate, in an atmosphere dominated by white males, to disadvantage female and minority ethnic students.

Although, in general, the experience of being on the course reinforced students' interest in construction as a career,[5] in fact they got little information about working in the industry. Moreover, minority ethnic students received less attention from college staff, and were less encouraged than white peers to progress to a job. So, while it is true that most students enjoyed the teaching and their relationship with the teachers, it is also true that this positive relationship was not uniformly experienced.

Higher education

In higher education, too, there appeared to be a lack of attention to the particular needs and interests of minority ethnic students. This deficiency was illustrated by an inability to cope with an ethnically diverse classroom, blind spots in the curriculum relating to non-Western aspects of construction or architecture, and a lack of awareness or empathy on the part of lecturing staff to the academic and pastoral support needs of minority ethnic students. This exclusion from some of the key benefits of the educational experience – say, a work placement with a reputable company – represents an additional burden on minority ethnic students in progressing towards graduation and employment.

Recruitment

In addition to the negative perceptions and images of the construction industry, minority ethnic people were also discouraged by experiences of exclusionary and discriminatory practices. The combination of the very small numbers of minority ethnic people in the industry, coupled with the widely acknowledged word-of-mouth recruitment practices, means that the existing predominantly white workforce 'replaces itself' rather than incorporates new staff cohorts. This process starts early on the construction career path, where the lack of family and friends in the industry begins to have a disproportionate impact on minority ethnic people, who have to rely on 'cold calling' for work. For those who succeed at the early stages, not being part of the wider social network of white people means being excluded later from work or developmental opportunities. This experience is compounded by direct experiences of discrimination, creating a vicious circle of under-representation and exclusion, from which it is difficult to break free.

Networking and 'fitting in' to the construction culture were widely recognised by both tradespeople and professionals as important in getting and keeping employment. On the whole, minority ethnic people tended to be uncomfortable in the informal white male environment – for example, in interviews which concentrated on their social skills, pastimes, and interests. Some did 'fit in' by joining in such activities, more as a conscious career strategy than from a feeling of belonging. Minority ethnic people tended to depend on the more formal recruitment methods – agencies, job centres, newspapers – whereas white people tended more to informal networking – at the pub on a Friday night, face-to-face enquiries at building sites, and using the network of contacts developed through family, friends, and college. Minority ethnic people tended to rely more on their qualifications, and to delay face-to-face contact – understandable, given the experiences of racial discrimination that were cited to the research team, but also problematic in an industry where personal interaction is given such a high priority.

Racism

The minority ethnic respondents anticipated that the process of securing jobs, contracts, advancement, and development opportunities would be compounded by racism. Once any family pressures against work in construction were overcome, societal and institutional racism made it difficult to get into the construction industry and, if successful, hampered progression within it. Some of these inhibitors to progression were racist jokes, name-calling, harassment, bullying, and intimidation. Less overt, though hardly less subtle, than this direct racial prejudice was the stereotyping of abilities which resulted in discrimination in relation to types of jobs allocated[6] and to full access to training, development, and opportunities to broaden skills. Site agents and managers are crucial in controlling the interactions on site and the roles and responsibilities allocated to subcontractors, but the researchers found that agents and managers were insufficiently alert to forms of discrimination, such as exclusion, inappropriate allocation of tasks, and unacceptable banter. Although some white site managers had dealt very competently with one-off situations of racial discrimination, they appeared to act as individuals, with no support or guidance from their senior managers, raising doubts as to whether other managers had the competence and confidence to deal with racism in this way.

Of course, minority ethnic people have suffered from similar social behaviours in their past and have developed coping strategies, including the following:

- surface acceptance of racial remarks as an integral part of industry culture
- challenging racism as an individual, and accepting the consequences
- avoiding potentially confrontational situations with the client or employer professionals until absolutely necessary (e.g. using the telephone instead of face-to-face contact).

Work in a black-owned company can be an effective way for the individual to get necessary experience in the industry, although it does little to address the overall problem of the skewed ethnic profile of the industry. Both minority ethnic trainees and those employed in construction cited the value of role models when making a career decision, but the predominant image of those who are taking a lead in changing the face of the industry remains white, with a lack of 'black' faces in senior positions. Those aspiring to more senior roles felt strongly that they would need to get wider experience and even a higher degree, in order even to consider moving up the management ladder.

9.4   Profile of success

Doubtless, the industry has made progress in diversifying, and it would be remiss not to acknowledge the successes represented by the numbers of minority ethnic people at all levels within it. In the absence of concerted, targeted action to bring about a greater ethnic mix at all levels, and with only limited organisational cultural change, individuals have worked hard, braved discrimination, and made the most of their opportunities to win their place within the industry. It is possible, therefore, to suggest a career profile of success from these particular experiences.

There is no evidence that the construction and built environment trades and professions are despised by minority ethnic communities. Engineering, for example, was reported to be highly valued in Africa, Asia, and the Caribbean and within the respective diasporic communities in Britain. Once enrolled on a construction-related course, the more fortunate students were encouraged by their tutors to draw on their own cultural background for their work. The opportunity to combine employment with study (day-release, course work-placement, part-time job) was critical for giving students a realistic perspective about the industry and its demands. Some former students mentioned the value of working for a small firm in this context, where they had the chance to experience a range of work; others stressed the importance of combining study with practical experience on site, especially learning from more experienced and skilled operatives.

It is clearly important to build up as wide a range of skills as possible, to be good at what you are doing, and to accumulate contacts within the industry and professional associations. A British-African engineer summarised the professional skills required to work in the construction industry as follows:

- pride – always take that extra step to ensure that your job is done to the best of your ability
- honesty – take responsibility for your actions, whether good or bad
- integrity – always make decisions that are in the best interest of your employer, never on a whim or personal preference
- dedication – whatever it takes in time and effort, always see each of your projects through to completion and on time
- reliability – keep your management informed of every step that you take; do not rely on anyone else to ensure the job is well done
- analytical skills – always weigh the pros and cons; avoid jumping at the first solution to a problem that presents itself
- listening skills – when working in a team, which is usually the case, it is effective to listen and understand as opposed to merely waiting for your turn to speak.

None of this is surprising or indeed, one might think, particular either to minority ethnic communities or to the construction industry; these are what one might call the generic skills of success. Furthermore, there is a strong tendency among the successful at all levels to play down the debilitating aspects of racism in favour of stressing – first, the importance of personal and mental development and, secondly, the harmony of the site, working with colleagues from different nationalities and ethnic backgrounds. However, even the most sanguine respondents were aware of a level of subtle, unconscious assumptions and ethnic stereotyping, leading to a sense of profound discouragement, and eventually to a loss to the industry of good tradespeople and professionals – people who already have practical experience of moving between, negotiating with, and assimilating different cultures – because they 'are not being given the chance to show their value' (CEMS, 2000b).

But to identify the problem is not necessarily to identify a solution applicable to many different situations, trades, and professions. One successful minority ethnic manager saw his role as an *educator*: 'Use your diplomacy to influence people and get what you want, and endeavour to create a standard that almost exceeds all expectations' (CEMS, 2000a). Within such a strategy, his tactical approach to racism in the industry is to address it on its own ground: 'I am quite happy to have a laugh with comments made about my ethnic background as long as the person can take some "verbal" in return.'

Others find they need to accommodate racism within the industry culture by making adjustments to their behaviour. For some Muslims, for example, the pub culture is not an option, so they will need to find other ways to socialise if they are to fit in with their colleagues. The impression received is of minority ethnic people playing a waiting game while the industry becomes ethnically diverse by increments. In the meantime, they take some small comfort in the perception that 'not everybody is against you', and tolerate the fact that, as one project officer said:

> potentially there could be someone **judging** whatever you say because of your colour. I think that you have a job to do, and if you do that job properly, you tend to start breaking down those barriers. As well as getting all the right qualifications and experience, it is also important to get the little things right, such as being punctual, being polite and always delivering a smart professional account of oneself.

Success, therefore, in his view, depends on not letting racism upset your career path, taking advantage of every opportunity, and choosing your battlegrounds carefully because, as the researchers were told, 'at the end of the day, [if you are successful] you are going to be an asset to the company, and that is the way they see it'.

9.5   The industry agenda

Although it is now routine for most medium and large organisations to have an equal opportunities policy, it is rarely seen as having an impact on their core business. The low level of organisational awareness of the kind and types of discrimination which minority ethnic people experience calls into question the value of such policies. They can create a 'comfort zone', where an organisation and its leaders can feel and claim that at least a stance on equality has been taken, but they do not replicate the drive to tackle discrimination proactively that is sometimes seen within the public sector. Consequently, few construction companies – unlike, for example, educational institutions or government departments – actively set action plans, conduct ethnic monitoring, and assign staff with specific responsibility for organisational equal opportunities/diversity policies. Entrants to the industry – educated and trained in institutions which (in theory at least) publicly state their commitment to equality and diversity, and in some cases are able to support this with positive action, upon gaining a place in a construction company – find that these values are unknown, under-valued, or not prioritised.

However, in practical terms the key relationship for these minority ethnic employees is not with 'policy' or management but with the white colleagues they work alongside every day. They do not want to be associated with 'reverse' discrimination or even 'positive action' measures.[7] They consider these to be implicitly unfair and capable of creating a backlash which would ultimately do more harm to potential opportunities than they would ever be able to offset. However, action is clearly necessary to open opportunities to all, to provide a greater mix of staff, and to implant measures to ensure that previous discriminatory practices are not simply repeated. Such action might include targeting schools and placing job fairs in areas of high minority ethnic settlement and including minority ethnic people in sponsorship programmes.

Progress is dependent on learning from initiatives that have been taken to counteract racial discrimination and disadvantage even if their results have been unclear or relatively unsuccessful. However, there is no consistent or comprehensive monitoring of the effects (particularly the benefits) of action taken to address inequalities. Without the systematic collection of information about the participation of minority ethnic people training for or employed in the construction industry, it is difficult to see how the results of efforts made to ensure equality of access to the industry and progress within it can be effectively measured or evaluated. Another important factor is the level of importance placed on the industry's commitment to racial equality by the large construction industry clients, for example the important national retailers, airports and transport authorities, local and national government, and funding organisations, particularly in Europe. Furthermore, the leadership function of industry managers is critical in raising issues at the highest levels and in giving direction and support to the middle managers within the construction companies themselves. When leaders start asking questions, then it is a strong signal to others that policies must translate into results.

Acknowledgments

I am grateful for the assistance of Jamie and June Jackson at CEMS, Royal Holloway University of London, for their help and support under the general direction of Professor K.H. Ansari, OBE, in the preparation of the paper on which this chapter is based.

Notes

1 For example the CITB 'positive images' campaign in 2001 which made greater use of images of young, female and minority ethnic people.
2 The 2001 Census data indicates that BME groups – all 'Asian or Asian British', 'black or black British', 'Chinese', and 'mixed' ethnic groups, excluding 'other ethnic group' – form approximately 2.8 per cent of the total workforce in construction in England and Wales.
3 The CITB website summarises the situation to date: http://www.citb.org.uk/equal_ops/ and contains a copy of 'The Royal Holloway Report' with statements from the CITB Chairman, CRE, and SOBA http://www.citb.org.uk/equal_ops/holloway_2002/default. htm. The most recent work was conducted for the Commission for Architecture and the Built Environment and was published in July 2005 (available from www.cabe.org.uk/ publications).
4 The 'problems' are not just for the construction industry to address but affect higher and statutory education institutions as well with respect, for example, to the quality and quantity of careers and other vocational information, and to the lines of communication that HEIs establish and maintain with construction firms.
5 A higher number of students wished to continue their studies in Construction and the Built Environment on leaving the course than did on entering it. In this respect, the course was obviously a success.
6 What in sports is called 'stacking' – that is the disproportionate concentration of ethnic minorities in certain positions.
7 Frequently confused both in the construction industry and in the media with positive or 'reverse discrimination'. The 1976 Race Relations Act prohibits positive discrimination at the point of recruitment (i.e. no racial quotas or preference), but allows positive action by employers to provide training for employees who are members of particular racial groups and who have been under-represented in particular work.

References

Ansari, K.H. and Jackson, J. (1995), *Managing Cultural Diversity at Work* (London: Kogan Page).

Barker, M. (1981), *The New Racism: Conservatives and the Ideology of the Tribe* (London: Junction Books).

CEMS – Equal Opportunities Consultancy Group (1999), *The Under-representation of Black and Asian People in Construction* (London: CITB).

CEMS (2000a), *The Impact of Race and Gender on Progression from GNVQ Construction into Industry* (London: CITB/FEDA).

CEMS (2000b), *Retention and Career Progression of Black and Asian People in the Construction Industry* (London: CITB).

CEMS (2003), *CITB Case Studies – 'People Profiles'* (London: CITB).

CEMS (2005), *Minority Ethnic Representation in the Built Environment Professions* (London: CABE).

Commission for Racial Equality (1995), *Racial Equality Means Business* (London: CRE). *http://www.constructingexcellence.org.uk/resourcecentre/peoplezone/details/ toolkit.jsp?toolkitID=99* [accessed 22 February 2005].

Construction Industry Board (1996), *Tomorrow's Team: Women and Men in Construction: A Report by Working Group 8 of the Construction Industry Board* (London: Thomas Telford).

Construction Industry Training Board (2002), http://www.citb.org.uk/equal_ops/. http://www.citb.org.uk/equal_ops/holloway_2002/default.htm. http://www.citb.co.uk/research/pdf/ca_survey_2000.pdf [accessed 07 January 2004].

Committee on Women in Science, Engineering and Technology (1995), *The Rising Tide: A Report on Women in Science, Engineering and Technology*. (London: HMSO).

Egan, Sir John (1998), *Rethinking Construction*. London: Department of the Environment, Transport and the Regions.

Grant, B. *et al.* (1996), *Building E = quality: Minority Ethnic Construction Professionals and Urban Regeneration* (London: House of Commons).

Greed, C. (1999), *The Changing Composition of the Construction Professions* (Bristol: University of the West of England, Occasional Paper 5).

Krishnarayan, V. and Thomas, H. (1993), *Ethnic Minorities and the Planning System* (London: Royal Town Planning Institute).

Latham, Sir Michael (1994) *Constructing the Team, Final Report of the Government/Industry Review of Procurement and Contractual Arrangements in the UK Construction Industry* (London: The Stationery Office).

Macpherson of Cluny, Sir William (1999) *The Stephen Lawrence Inquiry* (London: The Stationary Office).

Movement for Innovation/Rethinking Construction (2000), 'A commitment to people, "Our biggest asset": A report from the Movement for Innovation's working group on Respect for People'. http://www.constructingexcellence.org.uk//resources/az/view.jsp?id=290 [accessed 11 October 2006].

Owen, D. *et al.* (2000), 'Patterns of labour market participation in ethnic minority groups', *Labour Market Trends*, Vol. 108, no. 11 (November).

Twomey, B. (2001), 'Labour market participation of ethnic groups', *Labour Market Trends*, Vol. 109, no. 1 (January).

Further reading

Barnes, H., Bonjour, D. and Sahin-Dikmen, M. (2002), *Minority Ethnic Students and Practitioners in Architecture: A Scoping Study for the Commission for Architecture and the Built Environment* (London: PSI).

Barnes, H., Parry, J., Sahin-Dikmen, M. and Bonjour, D. (2004), *Architecture and Race: A Study of Black and Minority Ethnic Students in the Profession* (London: CABE).

CITB-ConstructionSkills (2003), *Construction Skills Foresight Report 2003* (King's Lynn: CITB-ConstructionSkills).

Dainty, A.R.J., Bagilhole, B.M., Ansari, K.H. and Jackson, J. (2002), 'Diversification of the UK construction industry: A framework for change', *Leadership and Management in Engineering*, October, Vol. 2, No. 4, pp. 16–18.

Ellison, L. (2003), *Raising the Ratio: The Surveying Profession as a Career: A Report for the RICS Raising the Ratio Committee* (London: RICS).

de Graft-Johnson, A., Manley, S. and Greed, C. (2003), *Why do Women Leave Architecture?* (London/Bristol: RIBA/University of the West of England).

O'Donnell, L. and Golden, S. (2001), *Construction Apprentices Survey 2000: National Report* (London: CITB).

Pathak, S. (2000), *Race Research for the Future: Ethnicity in Education, Training and the Labour Market* (London: Department for Education and Employment, Research Topic Paper 01).

RIBA (2003), 'Why do women leave architecture? Report response and RIBA action': http://www.riba.org/fileLibrary/pdf/WWLARIBAResponse.pdf [accessed 7 January 2004].

Shepley, C. (2003), 'Tomorrow's planners: It is possible to make a difference', *Planning*, Issue 1537 (19 September).

Spielhofer, T. and Golden, S. (2001), *Construction Apprentices Interim Survey: Draft Report* (CITB): http://www.citb.co.uk/research/pdf/ca_survey_interim_2001.pdf [accessed: 07 January 2004].

10 The gender gap in architectural practice

Can we afford it?

Ann de Graft-Johnson, Sandra Manley and Clara Greed

Abstract

Only 13 per cent of architects in the UK are women. Statistical evidence indicates that the number of female students studying architecture has increased to 37 per cent, but the percentage of women practising has not increased at the expected level. In other words, women are leaving the profession after qualifying.

Research into why women leave architecture revealed some matters of grave concern, which have implications for all the construction professions. Issues such as low pay, poor conditions, macho office culture, sidelining and direct sexism not only have a direct impact on women, but create a working environment that may fail to attract the necessary level and range of people to cover the current skills shortage. This leads to questions about the degree to which the professions are insular, whether the skills of women and minority groups are being over-looked to the detriment of the professions and whether this affects the quality of outputs.

The arguments that women are not fit to be architects through lack of spatial understanding or other shortcomings are identified, discussed and countered. The reasons why the construction professions have been slow to act to rectify the under-representation of women are explored. Strategic recommendations for change are aimed at the professional bodies as well as practitioners to facilitate ways of making the construction professions more attractive to women and create a built environment that meets the needs of the diverse population.

Keywords: architecture, women, under-representation, inequalities and stereotyping.

10.1   Introduction

I guess my experience with my two and a half year old twin daughters who were not given dolls and who were given trucks, and found themselves saying to each other, look, daddy truck is carrying the baby truck, tells me something. And I think it's just something that you probably have to recognize.

(Summers, 2005)

This quotation is from a controversial speech made by Lawrence Summers, the President of Harvard University. The speech sparked a fierce row that spilled out from the academic campuses of America's Ivy League universities to wider society in both the USA and the UK. It reopened an old debate about the innate differences between men and women and the issues of 'different availability of ability at the high end' (Summers, 2005). The interest in Summers' attempts to hypothesise about the reasons for the under-representation of women in science and maths in the high-flying universities of the US lies not in his surprising reliance in parts of his paper on anecdotal information of the 'baby truck, daddy truck' variety. It lies in the fact that his talk concentrated on why women were unfitted to take senior positions, rather than on considering whether there are any actions that could be taken by the universities to promote an atmosphere in which the talents of women could thrive. Not surprisingly, when the Daily Kos published a website article subsequent to Summers' talk, it provoked over 200 pages of responses (Armando, 2005).

It is not the intention of this chapter to debate the latest of a series of controversial views, behaviour and attitudes expressed by Summers (see, for example, Summers, 1991[1]; The Journal of Blacks in Higher Education,[2] 2005). Neither is it the intention to explore the hypotheses raised by Summers and others who have held that the under-representation of women or other groups in certain professions can be explained away by lack of aptitude. The focus of the chapter is to explore ways in which the architectural profession can be proactive in addressing the problem of the gender gap in architectural practice. More particularly it discusses the actions necessary to redress the failure of the profession to retain qualified women as architects. However, it is notable that the research report on which this chapter is based – 'Why do Women Leave Architecture?' (de Graft-Johnson *et al.*, 2003) – provoked similar arguments to those raised by Summers. The gist of these arguments was that women had different types of brain which were not fit to the practice of architecture, particularly regarding spatial understanding and technical ability and that this explained their low participation rates in the profession. Further, opinions were put forward that women did not have the same level of commitment to their profession as men (Phillips, 2003; Glaser, 2003). Unlike Summers, whose talk failed to emphasise the positive actions that can be undertaken to create an environment in which women can reach their full potential, the intention of this chapter is to identify and address issues within the architectural profession that inhibit full participation by women and suggest ways of changing the profession's culture. Many of those issues have wider implications that not only affect other under-represented groups, but may also have repercussions for the whole construction industry.

10.2　The specific problem

From the 1970s, Equal Opportunities legislation, including the Equal Pay Act 1970 and the Sex Discrimination Act 1975 (Equal Opportunities Commission (EOC), 2004a, 2005a), has aimed to improve women's representation in the professions. Despite the extent of legislation, however, the deeper structures of discrimination in the field of construction have proved resilient to change (Fredman, 2002; De Graft-Johnson *et al.*, 2003). In 2002, surveys of the architectural profession and of the student population revealed some worrying statistics. Between 1990 and 2002 the representation of women commencing architectural courses had increased from 27 per cent to 37 per cent, and male and female student drop-out rates were similar. None the less, this noticeable increase in female participation at the educational level was not reflected in the composition of the architectural profession. The percentage of women in practice following qualification had remained relatively static at 13 per cent and indicated skills drainage (Toy, 2001; Mirza and Nacey, 2001 and 2002; ARB, 2003). Toy summarised the problem by remarking that

> If the number of women architects continues to grow at the present rate, their representation in the profession might just achieve parity by the year 3000.
>
> (Toy, 2001)

The problems encountered by such well-known architects as Zaha Hadid and Denise Scott-Brown indicate a continuing malaise in terms of profile and attitudes to women architects. Hadid's talent was recognised early when she left the Architectural Association (AA) with the Diploma Prize in 1977 (Design Museum, 2005) and won the Hong Kong Peak competition of 1983. However, after winning the Cardiff Opera House project, she was dropped as the architect because of local opposition. Undoubtedly, some of the opposition related to the fact that the decision was taken in London and not in Wales and it remains unclear whether additionally the objection related to her design or even her identity as an Iranian woman (ibid.). The vast majority of her work and the related acclaim, which includes being the first woman to win the prestigious Pritzker Prize in 2004 (ibid.), have emanated from outside the UK. It is only recently that Hadid has received more commissions in Britain. In the Scott-Brown/ Venturi partnership, Denise Scott-Brown has fought to win the same recognition as Robert Venturi and has publicly stated that she has not been treated equitably (Toy, 2001; Scott-Brown, 2005).

While it was clear from the research study that the situation was not UK-specific and that there was a problem internationally in relation to women architects' experiences, it is interesting to compare the UK situation with another country where the gender profile is radically different. In Finland, where women have been practising officially as architects longer than in Britain, the representation is more balanced. Wivi Lönn, who received her diploma in architecture in 1896, was the first Finnish woman to set up her own practice (Nikkanen-Kalt, 1999).

Table 10.1 Gender profile of doctors 2001 (in %)

	Males	Females
General Practitioners	67	33
Doctors training for general practice	40	60
Hospital consultants	65	35
House doctors and junior doctors completing specialist training in next 10 years	50	50
Consultant obstetricians and gynaecologists	75	25
Registrars training to be obstetricians and gynaecologists	49	51
Surgery	85	15
Clinical oncology and paediatrics	41	59

Source: The British Medical Association, 2001.

The latest figures from The Finnish Association of Architects (SAFA) show that 38 per cent of Finnish architects are women, and of their student membership, 43 per cent are female (SAFA, 2005).

The Finnish example overturns any argument about correlation between representation and spatial awareness. Finland has long operated an anonymous competition system where entrants have been able to submit work under pseudonyms. This appears to have resulted in a much higher ratio of competitions being won by women. For instance, the system enabled Wivi Lönn to win first prize in an architectural competition in 1904 for the Tampere School of Economics and to go on to win six first prizes and a second prize in another competition, despite some opposition along the way (Suominem-Kokkonen, 1992). In contrast, the first prize by a woman entrant in the UK was not achieved until 1928 by Elisabeth Scott who won the competition for the Royal Memorial Shakespeare Theatre at the age of 29 through anonymous review. Her work is relatively unsung (Matrix, 1984).

It is instructive to compare architecture and other professions such as law and medicine in this respect. In 2003, for the first time more women were called to the bar than men (The General Council of the Bar, 2004). In medicine, women constitute 39 per cent of all medical practitioners (EOC, 2004b). While it is evident, that there continue to be problems for women in both the legal and the medical professions, particularly in relation to progression to senior posts and within some specialisms, progress has been made in reducing the gender gap (Table 10.1).

While the focus here is on the architectural profession, other studies of the whole construction industry show that lack of diversity needs addressing more generally (EOC, 2004d). Nevertheless, it is worth speculating on the reasons why the construction professions have been so slow to react to the need to remove the impediments that limit diversity.

10.3   The arguments and counter arguments

Table 10.2 attempts to summarise why the status quo continues to be accepted in some instances. Each point will then be addressed and challenged in turn. The

Table 10.2 Common arguments for maintaining the status quo

Equal opportunity goals achieved	A belief that the goals have been achieved and equality is no longer an issue
Gendered career choices	The belief that women are naturally disinclined to enter the construction profession and make different career choices to men
Economic disincentive	An assumption that women are not profitable
Commitment/ different priorities	A belief that woman are not committed, have other priorities, and are not willing to work the long hours that are seen to be essential to the practice of architecture
Aptitude	Women do not have the appropriate spatial or technical skills

first justification, which results in little or no intervention to facilitate further change, is a genuine belief by some construction professionals that the question of equal opportunity for women was addressed at the time of the Equal Opportunities Act in 1975 and that, in consequence, an appropriate level of representation has now been reached (Glaser, 2003). In other words, no action is necessary because the problem has been solved. It is worth recognising that some of the people who hold this view are themselves women, including women architects, a point borne out by some of the contributors to the related research. Clearly, the statistics by Mirza and Nacey, highlighting the differentiation of representation between female and male architecture students and qualified architects, indicate that the profession has not achieved its natural balance (2001, 2002). The example of Finland demonstrates possible scope for change.

The second argument is that women do not naturally choose architecture as a career and instead make different career choices from men thus making this difference in gendered choices acceptable and the norm (EOC, 2004b, c, d). This argument may explain the lack of action by the professions to encourage more women to become involved. Alternatively, it may seem an insoluble problem too large for a single industry to address. Proponents of the latter view may recognise that society effectively discourages women from participating because of stereotypical expectations, but see the task as insurmountable. However, it is argued here that the responsibility lies with the construction professions to challenge and change societal attitudes. The example of other disciplines indicates that this is achievable and that attitudes can change. Society has accepted women doctors and lawyers and women have made inroads into other professions. For example, the composition of orchestras has been transformed through wider adoption of 'gender blind' auditions which were first used in the 1970s and 1980s in the USA. By 2003 major UK symphony orchestras were reckoned to be on an average 30–40 per cent female, signifying quite an advance from 1953 when there was just one woman in the orchestra which performed Vaughan William's Sixth Symphony with Sir Adrian Boult (BBC Press Office, 2003).

The next argument is that women are not cost-effective and that career breaks and outside commitments add to employers' costs. The perception is that women (particularly those with children) may be an economic impediment rather than an advantage. Recent research has concluded that 36 per cent of respondents in small- and medium-sized businesses and 22 per cent in larger ones considered that pregnancy places an undue burden on the organisation (Young and Morrell, 2005). However, the majority of organisations were relatively comfortable with managing pregnancy-related leave. What was evident was the lack of knowledge on how to deal with these breaks effectively and economically. The converse of this argument is to ask whether the architectural profession can afford to reduce the pool of qualified experienced practitioners. The answer must surely be no. It is pertinent to consider whether other financial benefits may be gained from making the practice environment more family-friendly. A number of studies have looked at the benefits of flexible working. Case study evidence from a report by the EOC, 'Part Time is no Crime' (EOC, 2005b), sets out the following business benefits that accrue from allowing and supporting flexible working (Table 10.3).

Other economic imperatives can also drive change and it is necessary for architects and other construction professions to pay heed to these and act accordingly. For example, the DIY superstore chain B&Q recognised some time ago that, contrary to stereotypical perception, more and more women were becoming involved in DIY because they tended to have a greater role and interest in the selection and appearance of the home than did men. In order to effect change, women started to learn the skills needed to carry out work themselves. In response, with financial incentive, B&Q started to offer free DIY classes for women in its stores (Woman's Hour, 2002). On the professional side, the number of women recruited to plumbing and other trade skills has increased as these professions have become less attractive to men and demand is not being met by supply (EOC, 2004c, d). Widening the pool of labour is desirable in terms of both financial benefits and quality within the building and design industries. However, the attitude still prevails that women are not cost-effective in the construction sphere, despite the significant percentage of women in the overall UK workforce, particularly those in full-time employment (Table 10.4). The question is, can the industry afford to continue to exclude or effectively expel qualified, skilled people particularly in a time of recognised skills shortages across the entire sphere of the construction professions? An EOC report 'Plugging Britain's Skills Gap' (EOC, 2004d) singled out 27 industries where skills shortages occurred and linked the problem strongly to gender segregation. The construction industry was

Table 10.3 Benefits from flexible working

- Male and female staff are easier to attract and recruitment costs are reduced
- Skilled staff are retained and better returns are gained from training
- Absenteeism and staff turnover decrease
- Staff morale improves.

Source: EOC, 2005b.

Table 10.4 Women in the workforce
Labour market statistics
February 2005
General UK employment statistics during the three months to December 2004

	Men		Women	
	Number in millions	*percentage of total*	*Number in millions*	*percentage of total*
Numbers in employment (full and part time)	15.41	54	13.1	46
Numbers in full-time employment	13.77	65	7.43	35
Numbers in part-time employment	1.64	22	5.67	78
Numbers unemployed	0.83	59	0.581	41

Source: National Statistics February 2005.

cited as having the second highest level of vacancies for skilled workers, while the proportion of women working in construction was only 1 per cent.

> Gender occupational segregation is generally regarded by policy-makers as a 'gender' issue. But they are missing a trick by not recognising that widening the recruitment pool is a major part of the solution to the skills shortage.
>
> (EOC, 2004d)

> The success of our economy depends on our ability to use the talents of all our people.
>
> (Patricia Hewitt in EOC 2004d)

The argument relating to commitment and different priorities is based on a belief that, because women often have other responsibilities, they are not fully committed to the art of architecture. In fact women are deeply interested in the built environment and design partly because often they, and the people they may be responsible for, are so affected by the discriminatory aspects of their surroundings (Weisman, 1992). Television series such as Property Ladder and Grand Designs have profiled women who are extremely involved and interested in building, development and design, and many local campaign groups for change are dominated by women.

The final and perhaps most topical argument in terms of recent commentary is that women are not biologically designed to design. One possible explanation is that the abnegation of responsibility for taking action is justified by the argument put forward by people such as Summers, Glaser and Phillips (Glaser, 2003; Phillips, 2003; Summers, 2005). If it is true that women have neither the desire nor the ability to succeed in the construction industry, then there is no need to proceed and low levels of participation by women can be explained by these factors.

The profession of architecture demands a number of diverse skills. Visual and spatial skills are an obvious requirement. However, technical ability, research and communication skills are also crucial. Without understanding and awareness of people's needs and the way they use buildings and space, an architect is likely to produce poor design not fit for purpose. The RIBA careers' guidance states that

> Architecture reflects the society that builds it, but it also affects the way that society develops. This means we need architects who can respond to the different needs and values of all sections of the community. In the past most architects were drawn from a fairly narrow sector of society but now it is essential we ensure that the profession represents every social and cultural background.

The advice goes on to say:

> The one thing that is constant in architectural work is that it is concerned with people.
>
> (RIBA, 2005)

The assumption that particular traits are exclusive permeates much of the discussion about gendered behaviour and abilities. For instance, it is implicit in many of the arguments that someone cannot be both empathetic and at the same time good at art or possess other visual skills. Academics such as Simon Baron-Cohen and others have argued that the average male and female have different cognitive abilities and that men tend to display more spatial and mathematical skills than women. Conversely, women are considered to be better at languages and communication (Baron-Cohen, 2003a, 2003b). While Baron-Cohen stresses that his findings relate not to a specific female or male but to average trends, others have used his work to support more gender stereotypical assumptions. For example, from the observation that male infants shown a mobile object will focus first on the object whereas female infants will first focus on the person holding it, it is inferred that the female baby cannot be interested in the object. The assumption that these behaviours represent ability is of dubious provenance. It encourages the assumption that women are a homogenous entity; carried through to the wider community, this belief leads to gendered assumptions about ability and fails to recognise diverse skills in the population as a whole. Diversity can enrich society if people's potential is realised and perceived attainment is not based on general normative statistics. In Summers' example of his daughters and the 'daddy' and 'baby' trucks, he draws the conclusion that his twins as females only have the ability to anthropomorphise the trucks within the context of their future role as caring, nurturing women. By implication this appears to exclude any possibility that they might also develop a creative, technical or scientific interest in the world around them. It also implies that a female who is caring and nurturing has only one compartment to her brain and is incapable of other, high-level intellectual activity. Because he was so sure that his individual witness confirmed the stereotype, Summers may well have failed to note his daughters' brains were multi-faceted.

The following questions are pertinent as counters to the argument that no action is necessary and that the current low level of female representation reflects both women's innate abilities and their career choices.

If women do not have the ability or inclination to be designers then why

- Are some women architects high achievers in the profession?
- Are women so much better represented in other countries, such as Finland?
- Are women increasingly taking up self-build activity, development and other construction-related projects?
- Do well-qualified, skilled women leave the architectural profession post-qualification with so much regret?

It is worth exploring the impediments and resistance to change which have maintained the status quo in which women are not widely represented in the profession. Utterback (1974) speculated on the factors that influence the dissemination of an idea into an organisational framework and identified seven key attributes. These are listed in Table 10.5 together with the negative and positive indicators that might be applicable to the architectural profession. In relation to each attribute some suggested actions that may help change the status quo are advanced.

10.4   Why is it important to maintain diversity?

Many of the ills of the contemporary environment can be traced to male assumptions about what constitutes good design.

(Roberts, 1991)

If the majority of professionals come from a relatively narrow band of 'types' of people, and are predominantly white, middle class, middle aged, male, able-bodied, then the construction culture will lack richness and diversity, and there will be a lack of alternative, valid perspectives upon which to draw in the course of professional decision-making, which ultimately will be reflected in the nature of the built environment itself.

(Greed, 1998)

There has been a long-running commentary on the implications of a male-dominated architectural profession (Weisman and Birkby, 1983; Weisman, 1992; Anthony, 2001). The different patterns of people's lives, their gender and other factors impact on the use of the built environment and more specifically buildings. A lack of diversity within the profession may imply a lack of awareness, understanding or dismissal of priorities or issues for groups who are under-represented as architects or are not represented at all. To what extent does the built environment cater for the needs and activities and aspirations of women, disabled people, black and ethnic minority groups, children and young people given the current dominance of males in determining the form and shape of the built environment? At the

Table 10.5 Impediments against and scope for change

	Definition	Negative indicators	Positive indicators	Scope for change
Degrees of compulsion	The nature of the legal framework – the extent to which laws are in place, adhered to and enforced	Employers have poor knowledge of the legal requirements in relation to employment rights of women Some evidence that law has been broken or treated in a cavalier fashion No compulsion to change the curriculum to be more inclusive at either school education or architectural education Onus on aggrieved to take action rather than employers to demonstrate compliance	RIBA have launched a campaign to improve knowledge of workplace law and incorporated equal opportunities into their membership code CABE and other bodies working with schools to promote construction professions	More responsibility should be placed on employer to demonstrate compliance. Need for action to change the schools' national curriculum at all levels to be more inclusive and place greater emphasis on built environment in non-gender segregationist way. Specific requirement for schools of architecture to incorporate inclusive design and curriculum and to embed this approach in project work
Perceived advantages	Enhancement of diversity more likely to proceed if advantages in terms of social and economic factors are clear	Advantages of more gender parity not seen. Women regarded by many employers as high risk in cost terms. Advantages not well championed or disseminated particularly within architectural field	50/50 Campaign by BD Magazine and initiatives by RIBA stress the advantages of greater parity EOC and government actions to promote equality and demonstrate advantages of more diverse skills pool	Need for more research and other actions to focus on the positive contributions women and others can make. Studies needed on added value socially and economically and in quality of output.
Compatibility	The extent to which the idea conforms with current norms, values and structures	Regarded by many as unnecessary, irrelevant and annoying attempt to be Politically Correct Structures not always in place to facilitate full participation by women Male dominated 'gentleman's club' atmosphere has persisted and acted to inhibit change	Some clients require diversity to meet their and/or the user needs and expectations	More diverse client base, more demand for profession which better reflects society at large

Communicability	The extent to which the idea can be explained and identified	Some people think that there is no necessity to change. Resistance to change – it is over 30 years since EO legislation – need can easily be denied	There is still a media profile which advocates the need for continuing action towards equality and stresses value added factors	Equal opportunities training should be reviewed and evaluated to make more relevant to stress business advantages and then more effectively targeted. Outreach strategies to include more celebration of achievement by women and gender mainstreaming of these. Greater profile for female role models in schools and media
Non-pervasiveness	The greater the number of aspects of an organisation or society that are affected by change, the less likely it is that change will take place	Lack of collaboration across the board and across professional divides. Many different groups, from client to contractor to colleagues in other professions, are not convinced. Need for change not recognised in all cases	Positive action by RIBA and AFC may lead to more radical and wide-reaching change	Construction industry to address multiple issues in collaborative way to reinforce the need for change
Reversibility	Idea is more likely to be accepted if it can be experimented with at low cost in terms of time, money and resources and if it can be reversed	Too much effort to change and too costly to dismantle if idea fails		Change is imperative. The cost of no change could be very destructive to the profession in terms of image and future role
Number of gatekeepers	The smaller the number of people involved the more likely the chance of adopting change	Number of potential tiers to gatekeeping from careers advice to architectural education to practice and then career advancement. Glass ceiling in place. 'Queen Bee' syndrome. Successful women may act as impediment to other women's advance in some cases	Actions to identify where gatekeeping may be taking place and to counter this. For instance CABE's 360° schools initiative	More monitoring throughout education and practice to assess performance and identify where and what actions are necessary

macro end of the scale, the design of the urban environment is questioned by many people who feel it does not meet their needs. Many women, for example, would prefer more energy-efficient housing, greater flexibility for working at home, homes that cater for changing family situations, are child friendly and situated closer to amenities to facilitate multi-task journeys (Andrews *et al.*, 2002). There are numerous examples of buildings not meeting the cultural requirements and life-styles of people for whom they are meant to cater. Inclusive design is given a low priority in both practice and education and clients or user needs are often over-looked (Matrix, 1984; Penoyre Prasad *et al.*, 1998; Manley and Claydon, 2000; Preiser and Ostroff, 2000; Imrie and Hall, 2001; Ostroff *et al.*, 2002; Andrews *et al.*, 2002; Women's Design Service, 1998, 2002, 2004). There is a long history of concerns. The Tudor Walters Committee, set up in 1917, put forward proposals for the design of public housing. Raymond Unwin (of Parker and Unwin, the designers of Letchworth Garden City) was a member. A women's sub-group, which was formed in 1918, scrutinised the committee's plans and put forward their findings that women in particular preferred a separate parlour. However, their comments were largely overlooked and, despite Parker and Unwin having already met opposi-tion in Letchworth from residents on this very matter, they nevertheless went on to produce mass housing without parlours in London. The houses provided a tiny scullery in which to prepare food, wash dishes, wash clothes, and bathe. The remainder of family activity took place in a living/dining room and there was no provision for a 'room for best' to cater for special occasions. The Garden City Journal (1906) stated in relation to Letchworth that 'the workmen and their wives...do not take kindly to the innovation; they like the parlour and mean to have it' (Matrix, 1984; Crisp, 1998). Wilson criticises a general attitude that ignores user needs in relation to housing and makes assumptions about roles as demonstrating 'contempt for the residents of the housing estates and their assumptions that a rigid sexual division of labour prevailed' (Wilson, 2001). While Wilson's criticisms in this instance are aimed at planners, the same accusation has been laid at the feet of architects. It is clear that many contemporary architects still continue to ignore the needs and wishes of users and clients to an extent which might be regarded as contempt.

At the micro level, a feature in the BBC's *Woman's Hour* in July 2003 on public toilets is an example of a number of issues which may be encountered in part because of the lack of representation of women as designers of buildings. 'To be able to get into the cubicle and shut the door would be a start', Jenni Murray intoned at the beginning of the piece (*Woman's Hour*, 2003).

The danger in a profession which to some degree could be accused of being self-replicating, is that this self-replication engenders narrow and exclusive thinking and also a lack of ability to develop knowledge and awareness. It is important for clients, users and architects to have an architectural profession which is diverse in its constitution to enable lively and informative exchange and facilitate the design of buildings and spaces which meet user needs and enhance the environment.

The failure to reflect a diverse population has implications for the image of a profession which has already been at the centre of much criticism. The unsatisfactory nature of many post-Second World War developments, particularly housing and shopping centres, whether architect designed or not, has led to much of the blame for this being directed at architects. The portrayal of the architect as arrogant, aloof and dismissive of clients is demonstrated by Ayn Rand's portrayal of Howard Roark, the architect in 'The Fountainhead':

> I don't intend to build in order to serve or help anyone. I don't intend to build in order to have clients. I intend to have clients in order to build.
>
> (Rand, 1943; Penoyre Prasad, 2000)

Prince Charles' 1984 invective against the architectural style of the winning scheme for the National Gallery extension, which he described as a 'monstrous carbuncle', did nothing to improve the reputation of architects, but it was a factor in opening a renewed debate about architecture and a resurgence of interest in urban design (Prince Charles, 1984; Tibbalds, 1988; Department of the Environment, 1994). He has recently launched a new plea for a change in attitudes of Britain's architects and town planners arguing that too many in the professions are still treating cities and towns 'as what Le Corbusier called a "machine for living" – a collection of mechanical parts' and that

> instead of seeing every building as an opportunity to make an ever more imaginative iconic 'statement' – and to indulge our egotistical ambition – I believe we must see each piece of the built environment as part of a living language, connected to a living tradition.
>
> (Prince Charles, 2005; BBC News UK, 2005)

It is clear that Prince Charles endorses opinion by other commentators, such as Jane Jacobs, that cities and the built environment require a much more sensitive response which facilitates the creation of more cohesive communities (Jacobs, 1961).

The Egan Report (1998), Rethinking Construction, which focused on the construction industry in relation to housing, cited customer dissatisfaction as an issue stating that

> More than a third of major clients are dissatisfied with consultants' performance in co-ordinating teams, in design and innovation, in providing a speedy and reliable service and in providing value for money.

In the light of criticism both in terms of the product but also in terms of the public perception of the typical architect and accusations that the profession may be becoming a dinosaur, the challenge is for architects to make themselves indispensable rather than becoming further alienated from the public.

Apart from the issues from a client perspective, there is also the issue of the relationship between architects and other members of the construction professions. The Latham and Egan reports (Latham, 1994; Egan, 1998), which highlighted lack of cohesion and poor team working within the construction industry, have contributed to significant changes in procurement methods which may have resulted in a more cohesive industry. Whether or not there is greater cohesion, the question of whether better value for money and improved design has been achieved is causing great debate, with many arguing that in fact design and quality have been compromised by the new procurement routes (CABE, 2000; Glancey, 2005). There is, however, general agreement that the knock-on effect of the changes in commissioning construction has been that the role and remit of the architect has been diminished to a considerable degree.

> Significant effects of this phase of change became evident in the late eighties, with the spread of methods of procurement other than the 'traditional route': the building fully designed, the contract administered by an architect. Less than 40 per cent of construction (by value) was procured 'traditionally' in 1998, as compared to over 70 per cent in the mid-eighties.
>
> (Penoyre Prasad, 2000; RIBA, 2000)

The RIBA's *Architects and the Changing Construction Industry* (RIBA, 2000) and Penoyre Prasad's *Constructive Change* (Penoyre Prasad, 2000) addressed the Egan report and highlighted the need to respond to change, including countering the erosion of the status of architects.

It is probable that the initiative taken by the RIBA in commissioning and taking action over the retention of women, as well as looking at other aspects of diversity, stems at least in part from concern about the future for architects. The fear is that the loss of qualified women might contribute further to a decrease in the role of the architect making the profession more vulnerable.

10.5   The cost of losing qualified practitioners

It is evident that there are severe financial implications for the taxpayer, the individual and architectural practices associated with the loss of qualified people.

Full cost recovery for education and training has not been reached and students are still being subsidised by the state. Even so, the contributions most students or their parents are expected to make have risen significantly in recent years. For the individual student, the particularly heavy costs associated with training as an architect are at least in part lost if they then embark on another career. For example, information provided to potential students by the University of the West of England undertaking a standard three-year degree course starting in 2005 is that they should allow for an expenditure of £23,000 (University of the West of England, 2005). Architecture students undertake a lengthier, three-part course of training: in general 5 years of academic work with 2 years working in practice, prior to qualifying. Students obtaining Part 2 in 2005 will have spent a minimum of £33,500 including fees and living costs if they have kept within the budgets advised by universities. The RIBA estimates that with the introduction of top-up fees from 2005, students are 'likely to have debts of up to £57k' (Ellis, 2005). If they then subsequently drop out after qualifying as architects, it could be argued that, in addition to the debt, at least two years of potential for earning an income has been lost as well as two years' worth of fees and expenditure on cost of living. These calculations of financial loss do not, of course, take into account the personal cost associated with failing as an architect. The concern is that architecture will become an even more exclusive profession given the combination of high cost of training, the low expected projected income and the fact that fear of debt is particularly strong among people from lower socio-economic backgrounds. These factors, coupled with low salary expectations (the average salary for qualified architects is £36,000) (Ellis, 2005), are likely to act as deterrents to entering architecture as a career.

Aside from issues that affect the individual, due to low salaries and the high cost of training, there is a serious cost implication for practices. The recruitment and replacement of qualified, experienced people when staff leave can be very expensive. A lesser consideration is the extent to which professional institutes lose fee income.

10.6 The research into why women are leaving architecture

The impetus for commissioning research into the reasons for the departure of females from the architectural profession came from concerns triggered by statistical research by Mirza and Nacey. Scrutiny of statistical reports on Architects and architectural education (Mirza and Nacey, 2001, 2002) indicated too small an increase in numbers of women architects post qualification and a failure to reflect the increase in the representation of female students in architectural schools in the architectural profession as a whole. Prompted by the lobbying of Architects for Change (AFC), a pressure group representing men and women from the Women in Architecture group and the Society of Black Architects, the RIBA commissioned the University of West of England to undertake specific research into the reasons for women leaving architecture.

The research undertaken focused on the central question 'Why are women leaving architecture?' The study was not an investigation into the health of the profession as a whole in relation to diversity, but concentrated on women who had already left or were considering doing so. The methodology, illustrated in Figure 10.1, involved a literature review supplemented by the contributions of an expert group drawn both from within the architectural and construction professions and from other professions where more equality inroads had been made. A web-based questionnaire directed at women who had qualified or were studying formed a central core of the research. Women were reached partly through word of mouth and partly through a media campaign that drew attention to the research, as there was no official record that could be consulted of women who had left. A surprising amount of interest was generated and the original target sample of 100 respondents was almost doubled. A series of more in-depth interviews (14 in total, of which 3 were pilots) was undertaken predominantly on

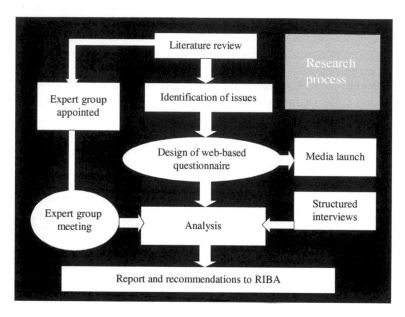

Figure 10.1 Methodology diagram.

Table 10.6 Areas addressed in the online questionnaire and interviews

- Employment including status, advancement, salary and career
- Office culture, ethos and practice
- Support and training
- Professional bodies and codes
- Educational profile and culture
- Attitudes inside and outside the profession
- Media representation.

a face-to-face basis as a follow up to the questionnaire. All but two of the interviews were with women who had left the profession.

Areas addressed in the questionnaire and investigated in more detail are listed in Table 10.6.

10.7   Research findings

Evidence from the research, which established that women were reluctantly leaving architecture after meeting all relevant criteria, qualifying and practising for some time, demonstrates that there is a considerable way to go to achieve parity between males and females. The research revealed issues about the culture and profile of the profession and demonstrated that far from being solved the problem of the lack of equal opportunities is a live issue of serious concern.

Key findings are set out in Table 10.7. What was clear from the findings was that there was no one single overriding factor which contributed to women leaving the profession. For many, an accumulation of factors led to the decision to leave. Consideration of whether the architectural profession might be described as 'macho' provoked a number of responses. Some respondents believed this to be the case and described examples of sexist behaviour that they had experienced. Others denied the existence of sexism but then went on to describe instances of behaviour which would have been unacceptable in other professional areas. Perhaps one of the major concerns was poor employment practice, including inequitable salaries and career advancement, long working hours and lack of opportunities for flexible working. In addition, attitudinal factors played a significant part in women's decisions to leave. For example, the assumption that women, in particular those with family responsibilities, were not sufficiently committed to the practice of architecture, in some cases, led to women being given less responsible tasks and being sidelined. There was also a failure to acknowledge other ranges of skills and understanding that women might develop through their different experience and activities. For example, the personal development of women on a career break or working part-time was assumed to have come to a complete halt. Perhaps one of the more benign but nevertheless obstructive practices was a tendency to 'protect' women from some of what may be perceived to be the 'harder' areas of an architect's work, such as dealing with a contractor or an awkward client or solving construction problems.

While some instances were indicative of a culture that militated against women, perhaps most disturbing was the research finding which revealed examples of illegal practice. This included unfair or constructive dismissal, in particular relating to returners from maternity leave, discrimination in terms of salary and career advancement, harassment, sexist behaviour and treatment, bullying and excessive working hours. Not all examples of unacceptable practice were gender

Table 10.7 Key factors that influenced women's departure from architecture

- Low pay
- Unequal pay
- Long or excessive working hours
- Inflexible/unfamily-friendly working hours
- Sidelining and appropriation of ideas or contribution and lack of attribution
- Restricted scope of work through gender-based allocation
- Glass ceiling
- Protective paternalism preventing development of experience in required areas
- Stressful working conditions
- Macho-culture
- Sexism
- Bullying
- Redundancy or dismissal (including threat of)
- High professional litigation risk and high insurance costs
- Lack of returner training provision
- More job satisfaction and better conditions elsewhere.

specific. For instance, in relation to bullying, instances were cited where males were being bullied or where women were the bullies. Long and excessive working hours were part of the expected conditions for both males and females in many offices, even though evidence suggests that this can be destructive and runs counter to the growing concerns about work-life balance (Bunting, 2004). It is possible that women were more adversely affected than men, for instance through specific personal commitments, particularly through childcare, which conflicted with expectations to work long hours. However, there are signs that gendered commitments are being eroded and that both men and women are increasingly unwilling to comply with the 'excessive hours' culture (Bunting, 2004).

While there was no intention in the research to look further than the UK or to provoke an international debate, responses were received from women in the USA, Hong Kong, New Zealand, Australia, Canada, Germany and Singapore. These responses raised the question of whether or not the problem was solely a British phenomenon.

10.8   Recommendations of the report

Prior to formally submitting the findings, key issues were discussed with the expert panel as part of the framework for establishing recommendations. Over 100 recommendations were made, aimed at the professional bodies, architectural practice, architectural schools and individual practitioners. The intent was to draw up achievable, practical ways in which actions could be taken to address the problems and concerns identified. A summary of the key areas covered by recommendations is outlined in Table 10.8. One important recommendation was that there should be further profession-wide research, including parallel consideration of the experience of males. This was in recognition of the fact that not only were the research objectives premised on one specific question, but also that findings indicated that some issues identified were not gender specific and that men as well as women were encountering problems.

Actions towards diversity since the report

Apart from a small number of negative responses (Glaser, 2003; Phillips, 2003), the general response supported the research findings and advocated the need for change. Even prior to formal publication of the report it was clear that there were parties within the RIBA, such as AFC, with a strong commitment to promoting change (RIBA, 2003, 2004). Since publication there have been a number of initiatives and drives to encourage a more diverse profile for the profession. The 'Diversecity' exhibition, launched in September 2003, coincided with the public launch of the report. The exhibition showcased architectural work both to demonstrate the contribution of practitioners from diverse backgrounds and initiatives and to promote equality in the construction professions. It has now become a travelling exhibition, visiting such cities as Boston, Beijing, Athens, Dublin, Sidney and Auckland, and is receiving widespread international interest and gathering further examples as it proceeds.

Table 10.8 Key areas of recommendations covered to be adopted by architects and architectural practices

---

- Low pay and unequal pay to be countered by better dissemination of employment legislation, e.g. the Equal Pay Act, minimum wage legislation and good practice by RIBA
- Practices to be given guidance by RIBA on Working Time Directive and other models to prevent poor employment practices such as imposition of long working hours
- Information and dissemination by RIBA on flexible working and family friendly employment practice
- More diverse representation of the profession to the public and acknowledgement of the work of women architects and architects from non-traditional backgrounds in the media
- More anonymous competitions
- Practices to recognise the need to acknowledge contribution of all staff
- Practices to monitor work allocation to avoid gender bias including paternalistic attitudes that may hinder personal professional development
- Practices to ensure that equal opportunities policy is in place and implemented to ensure that all staff may reach their career potential irrespective of gender etc. Policies should include terms and conditions of employment and complaints procedures etc. as required by law. Performance should be monitored
- Equal opportunities policy and implementation should be included in the professional bodies' codes of practice
- Information and dissemination by RIBA on good practice and latest legislation on stress at work, harassment etc. Practices to ensure they adopt appropriate procedures and monitor
- Practices to review office ethos and ensure that the working environment is conducive to both males and females to eradicate macho culture, sexism and bullying
- Practices to ensure they comply with legislation to ensure that any termination of employment through redundancy or dismissal is fully justified and legal
- ARB to review minimum requirements relating to Professional Indemnity cover
- Returner retraining to be provided at low cost and regionally based
- More affordable and flexible Continuing Professional Development (CPD)
- Mentoring and advisory/helpline support
- Embedding of equality including gender in both the curriculum and practices of architecture schools and recognition of weight that should be attached to people-centred design
- More diverse staff profile in schools of architecture
- Monitoring of the performance of architectural schools in improving diversity targets and equal opportunities practice
- Advisory practice notes for both architectural practices and schools of architecture by the RIBA
- More careers information in primary and secondary education and again more diverse representation in promoting architecture as a career
- Further profession-wide research on males and females
- Identification of exemplary practices through further research to establish the benefits and provide model guidance on employment and educational aspects.

The RIBA is augmenting the web-based support, advice and professional development that it already offers and is providing more legal advice for practices including workplace law. It is reinstating training for returners and extending the mentoring programme for students. To encourage greater diversity at the point of entry to courses, the Institute has been running projects for school pupils and making arrangements for architects to talk to school pupils about opportunities within the profession (ibid.). Further work on promoting diversity in architectural schools is being undertaken by a special interest group with support from the Centre for Education and Built Environment (CEBE). This has included, in conjunction with ARCHAOS (the association of architecture students in the UK), an investigation of the way in which studio culture can be made more inclusive. All of these actions are in line with the research recommendations.

The 'Why do Women Leave Architecture' report has attracted international attention. In 2003 the research was presented at the 20/20 Vision conference in Boston, two groups of Japanese academics met the research team in the UK to discuss the report in 2004. A presentation has been made in Turkey and future events are planned for Israel in 2005.

BD magazine, after a rather dismissive original response to the research, has launched the 50/50 Campaign to promote equal representation of males and females to radically alter the composition of the profession. Practices have been invited to sign up to this commitment (BD, 2005a; Blackler and Levitt, 2005). A summary of the charter objectives, which is provided in Table 10.9, includes some of the key areas identified by the research and sets out a useful if brief framework for action. The objectives set out highlight some of the key areas of concern and provide a useful if brief framework for action that practices with a commitment to equal opportunities should embrace. At least 240 practices have signed up although others have so far declined to do so (BD, 2005b, 2005c).

The future

The 'Why do Women Leave Architecture' report contains recommendations for practical, achievable actions. The recommendations adopt a 'cradle to grave' approach by taking account of current and future generations of architects. They start with nursery, primary and secondary school education in order to challenge the gender stereotypes and gendered assumptions which are embedded so early. Table 10.10 summarises some of the key remaining questions in relation to gender equality and diversity.

Table 10.9 Building Design's 50/50 charter

- To recruit, promote, pay and allocate work according to experience and ability alone
- To set out maternity and paternity rights in a written contract for every worker and strive to go beyond the statutory minimum
- To offer flexible working to all employees and retraining for returning parents
- To challenge the long hours culture and monitor working time
- To appoint a practice champion to promote and monitor the charter.

Source: BD 2005b.

Table 10.10 Key questions relating to the architectural profession and education

Question	Pre-Higher education (Schools, nurseries)	Higher education: Schools of architecture	Architectural practice	Architectural institutes and bodies
To what extent is the profession abiding by existing law		No concrete evidence concerns raised by anecdotal information	Evidence indicates non-compliance in certain instances	Positive moves by RIBA to encourage compliance and information dissemination
Does the culture of the profession remain in a paradigm which sees women as inherently inferior as architects	School system may still reinforce stereotypical views and discriminatory steering on career directions	Some evidence on gendered treatment. Little evidence of action to address inequalities in attitudes	Respondents raised concern and debate in the media expose opinions which imply lack of ability	RIBA has formally recognised the issue and is taking steps to change the paradigm. Some individual RIBA council members may not be on board
Is the culture of the profession moving towards an acceptance of a positive duty to promote equality	National schools curriculum currently does not include design of the built environment. Scope to improve the curriculum and careers advice on equality matters generally	Some commitment but not universal. Divergence between rhetoric and action in some cases. Little consideration of commitment to inclusive curriculum	50/50 campaign has been endorsed by some practitioners and rejected by others	Formal recognition by the RIBA

10.9   Conclusions

It is clear that despite some initiatives for change in the architectural profession, the climate for practising architecture is not providing adequate scope for women to thrive. Women appear to be voting with their feet and leaving the profession. This unhealthy situation does not bode well for attempts to diversify the profession to make it more reflective of the community it serves. If the working environment is not right for women, it is likely that others, such as disabled people and members of black and minority ethnic groups, will also fail to thrive, even if initiatives to attract a more representative group of people into the profession are successful. There is also evidence that men are being adversely affected by at least some of the problems that women identified as contributors to their decision to leave the profession. Bullying, the macho culture and long hours may prove to be unacceptable to men as well as women as more people seek a healthier and more comfortable work-life balance (Bunting, 2004).

The consequences of the female brain drain are financially damaging for both practices and the individual, but they may also have consequences for the public image of the profession, which is already seen by many people to be out of touch with user needs and predominantly self-interested. The architectural profession needs to respond positively to prevent further problems in the future by adopting the recommendations of both the 'Why do Women Leave Architecture' report and the BD 50/50 campaign and by taking up the opportunities for change. It is clear from Finland's example that it is possible to achieve a more appropriate balance between males and females in architecture, and the progress made in other professions in attaining greater parity reinforces this view. It is likely that a shift in emphasis to build a more inclusive and representative profession will help to re-establish the important role of architects in achieving built environments that can both respond to user needs and also raise public expectations for higher quality design.

Notes

1 An internal Memo World Bank written by Lawrence Summers dated 12 December 1991 put forward economic arguments to support the disposal of toxic waste in developing countries. Greenpeace obtained a leaked copy of the memo and disseminated its contents. The then Secretary of the Environment for Brazil Jose Lutzenburger wrote a highly condemnatory response to Summers after the memo became public in February 1992. He wrote, 'Your reasoning is perfectly logical but totally insane . . . Your thoughts [provide] a concrete example of the unbelievable alienation, reductionist thinking, social ruthlessness and the arrogant ignorance of many conventional "economists" concerning the nature of the world we live in . . . If the World Bank keeps you as vice president it will lose all credibility. To me it would confirm what I often said . . . the best thing that could happen would be for the Bank to disappear' (GS Report, 1999).
2 Following the appointment of Lawrence Summers as President of Harvard in 2001, a controversy arose as a result of series of events relating to Summers' criticisms of the African American Studies programme. Criticisms related to programme content, assessment and the academic credibility of some African-American staff. *New York Times* reporter Jacques Steinberg wrote, 'The dispute cast a cloud over the Summers' administration.'

References

Andrews, L., Reardon-Smith, W. and Townsend, M. (2002) *But will we want to live there? Planning for People and Neighbourhoods in 2020, Housing Corporation.* Available online at www.women2020.com.

Anthony, K. (2001) *Designing for Diversity: Gender, Race and Ethnicity in the Architectural Profession*, University of Illinois Press, Chicago, Illinois.

ARB (Architects Registration Board) (2003) *Registration Statistics: Breakdown of Registrants 2002 Statistics.* Available online at www.ARB.org.uk [accessed October 10th 2004].

Armando (2005) Lawrence Summers: *Foot in Mouth? Or Worse?* Daily Kos, 18 February 2005. Available online at www.dailykos.com/story/2005/2/28/94524/5350.

The General Council of the Bar (2004) *Bar Statistics-December 2003*. LexisNexis available online at www.barcouncil.org.uk/documents/bar_statistics03.doc.

Baron-Cohen, S. (2003a). *The Essential Difference: Men, Women and the Extreme Male Brain*, Penguin UK/Perseus.

Baron-Cohen, S. (2003b) *They just can't help it*, The Guardian 17 April 2003. Available online on www.guardian.co.uk/life/feature/story/0,13026,937913,00.html.

BBC Press Office (2003) *Is the future female for tomorrow's Orchestras?* BBC 11 February 2003. Available online at www.bbc.co.uk/pressoffice/commercial/worldwidestories/pressreleases/2003/02_february/t.

BBC News UK (2005) *Prince's New Architecture Blast*. BBC News UK Edition 21 February 2005. Available online at www.news.bbc.co.uk.

Blackler, Z. and Levitt, T. (2005) *Sexism the real life story. Building Design Issue*, 1665, 7 January 2005.

The British Medical Association (2001) *The changing face of medicine: Today's doctors.* BMA available online at www.bma.org.uk/ap.nsf/Content/Changing+face+of+medicine+-+Today's+doctors.

Building Design (2005a) *Why this change matters, Building Design Issue*, 1665, 7 January 2005.

Building Design (2005b) *BD's 50/50 Charter Explained, Building Design*, 22 February 2005 Available online at www.bdonline.co.uk.

Building Design (2005c) *50/50 Victory in sight*, BD, 4 March 2005, *Issue* 1663.

Bunting, M. (2004) Willing Slaves: How the Overwork Culture Is Ruling Our Lives, Harper Collins, London.

CABE (2000) *Better Public Buildings*, CABE.

Crisp, A. (1998) *The Working-Class Owner-Occupied House of the 1930s*, Oxford Thesis. Available online at www.pre-war-housing.org.uk/internal-planning-services-and-fittings.htm.

De Graft-Johnson, A. Manley, S. and Greed, C. (2003) *Why do Women Leave Architecture?* RIBA. Available online at 'http://www.riba.org/fileLibrary/pdf/WWLAFinalreportJune03.pdf', http://www.riba.org/fileLibrary/pdf/WWLAFinalreport June03.pdf.

Department of the Environment (1994) *Quality in Town and Country – A discussion document*, MSO.

Design Museum (2005) *Zaha Hadid*, Design Museum and British Council. Available online at http://www.designmuseum.org.

Egan, Sir J. (1998) *Rethinking Construction*, London: HMSO.

Ellis, C. (2005) *Gateways to the Professions; The Architectural Profession*, RIBA available online at www.dfes.gov.uk/hegateway/uploads/Chris % 20Ellis' % 20PP % 20 Presentation.ppt [accessed 6 March 2005].

Equal Opportunities Commission (2004a) *Sex Discrimination Act 1975 (As amended)*. EOC available online at www.eoc-law.org.

Equal Opportunities Commission (2004b) *Facts About Women & Man in Great Britain.* EOC available online at www.eoc.orguk/ceng/research/facts_about_2004gb.

Equal Opportunities Commission (2004c) *Britain's Competitive Edge: Women Unlocking the Potential*, EOC, London.

Equal Opportunities Commission (2004d) *Plugging Britain's Skills Gap: Challenging Gender Segregation in Training and in Work*, EOC May 2004. Available online at www.eoc.org.UK/PDF/phase one.pdf.

Equal Opportunities Commission (2005a) *Equal Pay Act 1970* (As amended). EOC available online at www.eoc-law.org.uk/cseng/legislation/epa.pdf.

Equal Opportunities Commission (2005b) *Part Time is no Crime*. EOC available online at http://www.eoc.org.uk/cseng/policyandcampaigns/flexible_working_interim_report.pdf.

Equal Pay Act 1970.

Fredman, S. (2002) *Discrimination Law, Clarendon Law Series*, Oxford University Press: Oxford.

The Garden City Journal (1906) Vol. no 1, p.187, October 1906, London.

General Council of the Bar (2003) *Annual Report 2003*, LexisNexis UK: London.

Glancey, J. (2005) Sweet and Low Down, *The Guardian*, 28 February 2005.

Glaser, K. (2003) *Written in The Womb: Is There a Biological Explanation for Why More Men than Women Succeed in Architecture*, Building Design, 23 May 2003.

Greed, C. (1998) *The Changing Composition of the Construction Professions*: Faculty of the Built Environment, University of the West of England.

Imrie, R. and Hall, P. (2001) *Inclusive Design: Designing and Developing Accessible Environments*, Spon Press/Taylor & Francis, London.

Jacobs, J. (1961) *The Death and Life of Great American Cities*, Random House Inc.: New York.

Journal of Blacks in Higher Education (2005) *Is the Magic Gone From Black Studies at Harvard*? Journal of Blacks in Higher Education. Available online at http://www.jbhe.com/features/45_blackstudies_harvard.html.

Latham, M. (1994) *Constructing the Team-Final Report of the Government/Industry Joint Review of Procurement and Contractual Arrangements in the UK Construction Industry*, HMSO: London.

Manley, S. and Claydon, J. (2000) *Achieving Richness and Diversity: Combining Architecture and Planning at UWE*, in D. Nicol, and S. Piling. Changing Architectural Education, E&FN Spon: London.

Matrix (1984) *Making Space: Women and the Man Made Environment*, London: Pluto.

Mirza, A. and Nacey, V. (2001) *RIBA Education Statistics 2000/01*, RIBA Centre for Architectural Education, London.

Mirza, A. and Nacey, V. (2002) *Architects' Employment and Earnings 2002*, RIBA Journal, July 2002.

National Statistics (2005) *Labour market statistics February 2005*, National Statistics. Available online at www.statistics.gov.uk.

Nikkanen-Kalt, P. (1999) *Female architects in Finland*, ARVHA *October 1999*. Available online at http://www.arvha.asso.fr/archi_fem/arvha_french/info_arvha/document_info/paivi_nikkanen_uk.

Ostroff, E., Limont, M. and Hunter, D. (2002) *Building a World Fit for People: Designers with Disabilities at Work*, Adaptive Environments Center, Boston, MA.

Penoyre Prasad, S. (2000) *Constructive Change*, Penoyre & Prasad Publications. Available online on penoyre-prasad.net/publications/contrsuctive/introduction.html.

Penoyre Prasad with Matrix, Elsie Owusu Architects, Audley English Associates (1998) *Accommodating Diversity: Housing Design in a Multi-Cultural Society*, National Housing Federation: London.

Phillips, P. (2003) *The Gender Gap*, Building Design Magazine, 24 October 2003.

Preiser, W. and Ostroff, E. (eds) (2000) *Universal Design Handbook*, McGraw Hill, University of Cincinnati: Ohio.

Prince Charles (1984) *Speeches and Articles: Architecture*. A speech by HRH The Prince of Wales at the 150th anniversary of the Royal Institute of British Architects (RIBA), Royal Gala Evening at Hampton Court Palace, 30 May, 1984, The Prince of Wales. Available online at www.princeofwales.gov.uk/speeches/architecture.

Prince Charles (2005) *A speech made by HRH The Prince of Wales to the Royal College of Physicians Conference, 21st February 2005*, Clarence House and the Press Association Ltd 2005. Available online at www.princeofwales.gov.uk/speeches/health.

Rand, A. (1943) *The Fountainhead*, Penguin: Putnam.

Ravetz, A. (1974) *Model Estate*, London: Croom helm.

RIBA (2000) (Royal Institute of British Architects) *Architects and the Changing Construction Industry*, RIBA. Available online at www.architecture.com.

RIBA (2003) *Why do women leave architecture? Report response & RIBA action*, July 2003. RIBA. Available online at www.riba.org.

RIBA (2004) *What the RIBA and AFC are doing for equal opportunities*, Available online at http://www.riba.org.

RIBA (2005) *Shaping the future: Careers in architecture*, RIBA available online at www.riba.org [accessed 28 February 2005].

Roberts, M. (1991) *Living in a Man-Made World*. Routledge, London and New York.

SAFA e-mail, 24 February 2005.

Scott-Brown, D. (2005) *Out of Africa Architecture and Academe: Into an ambivalent addiction to the practice of architecture in XXII World Congress of Architecture Istanbul 3–7 July 2005 Abstracts*, International Union of Architects UIA.

Summers, L. (1991) *Internal Memo World Bank December 1991* (report of leaked memo on Toxic Waste) in article '*Treasury Secretary-to-Be Dumps in World* GS Report'. GS Report Available online at www.gsreport.com/articles/art000171.html.

Summers, L. (2005) *Remarks at NBER Conference on Diversifying the Science & Engineering Workforce*, The Office of the President Harvard January 14 2005. Available online at 'http://www.president.harvard.edu/speeches/2005/nber.html.

Suominen-Kokkonen, R. (1992) Wivi Lonn, The Architect, in *The Fringe of a Profession: Women as Architects in Finland from the 1890's to the 1950's*, William Stout Architectural Books, San Francisco. Paper available online at www.kirjasto.oulu.fi/english/wivilonn/literature/wivilonnessay.

Tibbalds, F. (1988) ' *"Mind the Gap": A Personal View of the Value of Urban Design in the Late Twentieth Century'*, The Planner 74(3), pp. 11–15.

Toy, M. (2001) *The Architect: Women in Contemporary Architecture*, Chichester, The Images Publishing Group Pty Ltd.

University of the West of England (2005) *Financial overview for LEA funded students entering higher education in 2005/06*. January 2005 Bristol University of the West of England. Available online at http://www.uwe.ac.uk/csa/saws/prosgeneral.shtml.

Utterback, J. (1974) *Diffusion of Innovations*, Free Press, New York.

Weisman, L. (1992) *Discrimination By Design A Feminist Critique of The Man-Made Environment*, University of Illinois Press: Urbana.

Weisman, L. and Birkby, N. (1983) *The Women's School of Planning and Architecture* in C. Bunch, and S. Pollock (1983) *Learning Our Way*, New York: Crossing Press.

Wilson, E. (1991) *The Sphinx In The City: Urban Life, The Control of Disorder and Women*, University of California Press: Berkeley, Los Angeles, Oxford.

Women's Design Service (1998) *Making Safer Places*, WDS.

Women's Design Service (2002) *Re-moving the Goalposts*, WDS.

Women's Design Service (WDS) (2004) *Disability and Regeneration*, WDS.

Woman's Hour with Murray, J. Greed, C. and Stone, S. (2003) *The Politics of the Ladies Loo*. Radio 4, 10 July 2003. Available online at www.bbc.co.uk/radio4/womanshour/2003_27_thu_04.shtml.

Woman's Hour (2002) *At B&Q warehouses around the country women are taking up their tools and learning how to fulfil their makeover dreams*, Radio 4, 9 July 2002. Available online at http://www.bbc.co.uk/radio4/womanshour/2002_28_tue_04.shtml.

Young, V. and Morrell, J. (2005) *Pregnant & Productive*, Working Paper Series No. 20. EOC.

11 Managing cultural differences in the global construction industry

German and Austrian engineers working in Australia

Karolina Lorenz and Marton Marosszeky

Abstract

Much has been written about cultural issues in doing business around the globe. Well-documented studies have been designed to brief business managers regarding the cultural and business environmental aspects of the market in which they will operate. However, these studies are rarely industry specific and little has been written that specifically addresses issues in the global construction sector.

This chapter reports recent research that has studied, in some detail, differences in culture between designers and contractors from Germany and Austria on the one hand and from Australia on the other, as well as differences in industry regulations and practice between these countries.

Austrian and German companies are well represented in the Australian construction sector, through both joint ventures with local companies and majority ownership of several major Australian construction companies. In these arrangements, it is common to find senior engineers from the European partners/owners working on major projects in Australia; as a result, they need to bridge the cultural and technical gap between practice at home and abroad.

The study reported here revealed significant differences both in the way the industry is regulated and in the way in which organisations and projects are managed in the two different parts of the world.

These differences were evident in relation to safety management regulations and practices on site, the influence of trade unions, industry vs enterprise bargaining and project level agreements, environmental regulatory and management issues, stability of employment and bureaucracy. They have arisen as a result of a diverse range of cultural, regulatory and historical factors. Furthermore, the study revealed differences between the organisational cultures in design and construction firms in Europe and Australia.

The chapter concludes that significant differences exist in both technical and cultural practices between markets, and detailed knowledge of local regulations and practices is therefore important for the successful operation of global construction businesses.

Keywords: intercultural management, organisational culture, and construction management.

11.1   Introduction

Most large construction projects over the world have complex multicultural workforces: workers often come from nearby poor countries and recent émigré populations are usually well represented. Furthermore, with the globalisation of the construction sector, large international design and construction organisations undertake projects in foreign countries either in their own right or through joint ventures with local companies. Hence, the multicultural workforce extends from unskilled labour, through the trades to the professions on such projects, creating a challenge for management seeking to create productive project cultures.

This chapter examines this challenge from several perspectives, including the following:

* the implications of generic national cultural differences for collaboration
* the different organisational cultural traditions that parts of a team bring to the project
* the significance of historical differences in industry structure and regulatory environments for industry practice and organisation of construction activities.

The study approached these broad questions in the context of German and Austrian participation in major engineering and construction projects in Australia. Many large Austrian and German design, specialist manufacturing and construction companies are involved in major construction projects in Australia, often through local joint ventures where they provide specific know-how. Engineers are often sent to fill key management and technical roles within a project management team or to take a senior management position in a subsidiary company. Besides the general cultural differences in day-to-day life (which are not considered here), such transferred staff face numerous differences in relation to the organisation in which they are embedded or with which they are working in close collaboration.

These circumstances created the opportunity for this study and through the examination of factors that affect collaboration in these settings, more general conclusions are drawn regarding the challenges facing large multinational teams on major projects. While it is recognised that this is a limited study of international collaboration, it is argued that at a general level, the observations and conclusions drawn from this analysis are valid for all major projects led by an international management team.

Geert Hofstede's (2001) work was used as the initial basis for the examination of national culture. Then organisational culture was compared between a group of Austrian and Australian contractors and design firms, based on the Competing Values Framework (Cameron and Quinn, 2005). Finally, against this background, seventeen semi-structured interviews were conducted with Austrian and German engineering managers working in Australia to identify in more detail specific differences in the business and regulatory environments in Germany, Austria and Australia.

Definitions of culture abound in the literature and the elements of culture have been widely analysed and discussed; this research takes as its starting point the definitions by Hofstede and Cameron and Quinn. However, there is general agreement that culture is reflected in *the way things are done by a group* and most researchers concur on the six aspects of culture identified by Cameron and Quinn (2005):

1   Cultures are a property of groups of people and not individuals.
2   Cultures engage the emotions as well as the intellect.
3   Cultures are based on shared experiences and thus on the histories of groups of people; to develop a culture takes time.
4   Cultures are infused with symbols and symbolism.
5   Cultures continually change because circumstances force people to change.
6   Cultures are inherently fuzzy in that they incorporate contradictions, paradoxes, ambiguities and confusion.

Hofstede's (2001: 9) generic definition of culture as 'the collective programming of the mind that distinguishes the members of one group or category of people from another' encapsulates all of this.

Most people who have worked internationally have experienced the negative reaction that some behaviour, entirely acceptable or even desirable in one culture, may cause in another. In general, engineers have a reputation for being focused on the technical aspects of their task and they pay less attention to the *softer* or *cultural* aspects of management. This chapter does not set out to investigate whether this stereotype is true or not but rather to identify the cultural differences that managers, especially engineers of Austrian or German origin, should be aware of when working in construction in Australia. It can be generalised from this that most ethnic groups would encounter similar cultural differences to their home practices when working abroad.

Most studies of national culture place Austria, Germany and Australia into the same rough cultural classification of *Western* countries or countries of *European origin* and, therefore, do not distinguish between Europeans and Australians (e.g. Huntington, 1997). However, several more detailed studies (e.g. Hofstede, 2001; Trompenaars and Hampden-Turner, 1997) have identified cultural differences between these countries and argue that they cannot be ignored. Some of these studies indicate that in some *characteristics*, the differences between the two German-speaking neighbours are greater than between Germany and Australia.

Cultural differences have been brought into sharp focus as a result of globalisation; nevertheless, Hofstede's (2001) international study of culture found that although the globe is shrinking as a result of IT, telephony and air travel, the relative differences in national cultures, remain unchanged.

The research conducted in the study presented here consisted of two main parts. The first was based on a survey of organisational culture in a group of Austrian and Australian design and construction companies using Cameron and Quinn's Competing Values Framework (2005). A questionnaire was then developed and piloted, informed by the findings of the initial survey together with the well-known work of Hofstede, Trompenaars/Hampden-Turner and Gesteland who compared, among others, Austrian, German and Australian culture. During 2003, 17 semi-structured interviews were held with managers and engineers of German (11), Austrian (5) or Swiss (1) origin. Fifteen of these interviewees were working on construction projects in Sydney; the remaining two were involved in manufacturing in Sydney.

The inevitable subjectivity of discussions in relation to cultural factors is generally accepted, and readers should therefore recognise that their interpretations will be influenced by their own cultural biases in the same way as the authors bring their particular biases to the writing of this chapter.

11.2 Literature review

Schneider (1989) proposed that an organisation's culture establishes the rules within which people act in addition to the ways and methods in which people communicate. Through an understanding of group culture, employees know exactly what is required of them in any given situation (Deal and Kennedy, 1982) and it also replaces the need to enforce rigid procedures or control mechanisms through rigorously explicit supervision because it 'functions as an informal control mechanism that coordinates employee efforts' (Ross, 2000). As culture is 'the way things are done within the group', it sets priorities and expectations, enabling people to learn and understand what is important, subsequently identifying those actions that lead to punishment and those that lead to reward.

Brown (1995) identified three important benefits derived from a strong culture:

1 A strong organisational culture allows goal alignment as all employees share basic agreement upon organisational goals and also, how best to achieve them. Subsequently, employee initiative, energy and enthusiasm are all channelled in the same direction with obvious benefits in terms of co-ordination, control and internal communication.
2 Strong organisational culture leads to high levels of employee motivation. Operating within a strong and identifiable culture is intrinsically motivating as employees share a common view of how the organisation should work.
3 A strong culture is able to learn from its past as it possesses agreed norms of behaviour, integrative rituals and ceremonies and well-known stories which reinforce consensus on the interpretation of issues and events based on past experience.

While many scholars have commented on the positive effect that a strong organisational culture has on performance, relatively few actually present any strong evidence of a link. However, the empirical work of Kotter and Heskett (1992) surveyed 180 organisations in 19 different market sectors and demonstrated a clear relationship. Economic performance (based on return on invested capital, net income growth and average annual increases on stock price) and cultural strength were evaluated. The strength of organisational culture was found to have positive correlation with long-term economic performance. Cameron and Quinn (2005), from an extensive study of organisational culture and performance, also believe that firms excelling in highly competitive markets owe their success to their strong cultures.

Quinn (1989) argues that the concept of organisational culture is derived from two different disciplines: an anthropological origin, within which the view is held that organisations *are* cultures, and a sociological origin, within which organisations are seen to *have* cultures. These polarised views results in much debate among researchers as to the best way to measure culture (Cameron and Quinn, 2005).

Traditional studies of organisational culture have relied on qualitative methods such as in-depth interviews, open-ended interviewing and ethnographic observations; more recent quantitative approaches are often criticised for their inability to assess the cultural richness of an organisation in depth. However, the quantitative approaches do offer benefits in terms of generalisability and comparability. Chang and Wiebe (1996) state that qualitative methods have limited application in the study of organisational culture, arguing that they are 'too much a product of a social scientists rather than the participant's point of view'. Cameron and Sine (1999) expand on this debate:

> When assessing culture via questionnaires or interviews, is one really measuring superficial characteristics of an organisation – namely organisational climate – rather than in-depth cultural values? Because culture is based on underlying values and assumptions, often unrecognised and unchallenged in organisations, one perspective argues that only by utilizing in-depth qualitative procedures in which artefacts, stories and myths, and interpretation systems are studied over long periods of time in a comprehensive way can cultural attributes be identified. 'One must experience something to understand it' is the philosophical basis of this approach.

> On the other hand, the opposing point of view argues that breadth of comparison is sacrificed by employing a qualitative approach. The investigation of multiple organisational cultures becomes impossible when immersion in each one is mandatory. To conduct comparisons among multiple cultures, quantitative approaches must be used.

Robert Quinn (1981) and his colleagues proposed a framework of four types of culture that has been widely accepted within the organisational research community (Harrison and Shirom, 1999). Described as one of the best models available to help organisations assess culture and implement change (Hooijberg and Petrock, 1993), it has been extensively used in business and enjoyed wide support among researchers.

The framework has been built on empirical studies and is argued to have a high degree of congruence with well-known and well-accepted theories on the way people think, the way they derive their values and assumptions and the way they process information in organisations (Cameron and Quinn, 2005). Psychologists such as Jung (1923), Myers and Briggs (1962), Mason and Mitroff (1973), McKenney and Keen (1974) and Mitroff and Kilmann (1978a,b) are cited among such theorists with whom the framework concurs. Quinn (1988; Figure 11.1) argues that the congruence of the framework exists 'because of an underlying similarity in people at the deep psychological level of their cognitive processes', citing Mitroff (1983) as an example:

The more that one examines the great diversity of the world cultures, the more one finds that at the symbolic level there is an astounding amount of

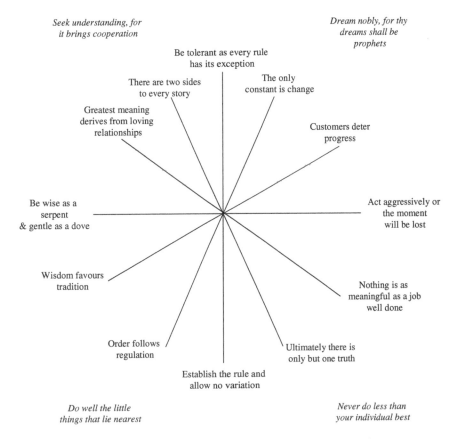

Figure 11.1 Juxtaposed proverbs representing values, beliefs and strategies.
Source: Quinn, 1988.

agreement between various archetypal images. People may disagree and fight one another by day but at night they show profound similarity in their dreams and myths. The agreement is too great to be the result of coincidence alone. It is therefore attributed to the similarity of the psyche at the deepest layers of the unconscious. These similar-appearing symbolic images are termed archetypes.

The framework known as the Competing Values Framework acknowledges the competing, conflicting and paradoxial nature of the modern business environment. Peters and Watermen (1982) suggest that Quinn and Rohrbaugh (1981) recognised that managers in excellent companies had an 'ability to resolve paradox, translate conflicts and tensions into excitement, high commitment and superior performance'. This seminal work conflicted with previous research that neglected to consider such concepts, Quinn proposing that theorists had done so to eliminate contradiction. 'The empirical foundations of traditional social science stands in the way of attempts to build such a theory. Empiricism is primarily a rational-deductive perspective design to answer the question, What is?' (Peters and Watermen, 1982).

The framework was initially developed from studies of the major indicators of organisational effectiveness. John Campbell *et al.* (1974) devised a list of 39 indicators of effectiveness, and Quinn later submitted these to a statistical analysis to establish if patterns or clusters could be identified. Two major dimensions emerged that could split indicators into four clusters. The first dimension distinguishes effectiveness criteria that emphasise flexibility, discretion and dynamism from criteria that emphasise stabilty, order and control. The other dimension differentiates effectiveness criteria that emphasise an internal orientation, integration and unity from criteria that emphasise an external orientation, differentiation and rivalry. The subsequent four clusters of effectiveness represent what people value about an organisation's performance, defining what is seen as good, right and appropriate. 'The four clusters of criteria, in other words, define the core values on which judgements about organisations are made' (Cameron and Quinn, 2005). Each continuum embodies conflicting or competing assumptions, that is flexibilty vs stability and internal vs external. The individual quadrants are conflicting on the diagonal axis, and they are referred to as the Clan, Adhocracy, Hierarchy and Market (Table 11.1). Each name is derived in collaboration with organisational science research that identified precisely matching organisational forms (Quinn, 1988).

In relation to national cultural differences, most of the work in this field has been done by Dutch researchers, possibly because of the Netherlands' multicultural history. Hofstede (2001), Gesteland (1999), and Trompenaars and Hampden-Turner (1997) have each developed a different set of dimensions or indices to characterise and interpret cultural differences and their effects on international business. Of these, Hofstede's is the earliest and best known with initial publication of his research in 1980. Each used a different approach and each dimension represents a specific set of differences in national and business culture. Because of differences in conceptual structure, in order to make a proper comparison between two or more of these frameworks, a single parameter used to describe an area in one of them may need to be compared with two or more parameters in another.

Table 11.1 The four types of organisational culture

The clan culture	The adhocracy culture
A very friendly place to work where people share a lot of themselves. It is like an extended family. The leaders, or the heads of the organisation, are considered to be mentors and perhaps even parent figures. The organisation is held together by loyalty or tradition. Commitment is high. The organisation emphasises the long-term benefit of human resources development and attaches great importance to cohesion and morale. Success is defined in terms of sensitivity to customers and concern for people. The organisation places a premium on teamwork, participation and consensus.	A dynamic, entrepreneurial and creative place to work. People stick their necks out and take risks. The leaders are considered innovators and risk takers. The glue that holds the organisation together is commitment to experimentation and innovation. The emphasis is on being at the leading edge. The organisation's long-term emphasis is on growth and acquiring new resources. Success means gaining unique and new products or services. Being a product or service leader is important. The organisation encourages individual initiative and freedom.
The hierarchy culture	**The market culture**
A very formalised and structured place to work. Procedures govern what people do. The leaders pride themselves on being good coordinators and organisers who are efficiency minded. Maintaining a smooth-running organisation is most critical. Formal rules and policies hold the organisation together. The long-term concern is on stability and performance with efficient, smooth operations. Success is defined in terms of dependable delivery, smooth scheduling and low cost. The management of employees is concerned with secure employment and predictability.	A results-oriented organisation whose major concern is with getting the job done. People are competitive and goal oriented. The leaders are hard drivers, producers and competitors. They are tough and demanding. The glue that holds the organisation together is an emphasis on winning. Reputation and success are common concerns. The long-term focus is on competitive actions and achievement of measurable goals and targets. Success is defined in terms of market share and penetration. Competitive pricing and market leadership are important. The organisational style is hard-driving competitiveness.

Source: Cameron and Quinn, 2005: 58.

Of the three approaches, Hofstede's work is most theoretical. The other two authors place more importance on the practical, more visible outcomes of cultural differences on general business behaviour. Apart from these studies, a great deal has been written in the form of practical guides on how to behave in different business cultures. However, only a few studies are based on detailed research and even fewer compare different countries in relation to each other based on defined dimensions or indices.

Of all the work in the area, the IBM survey conducted by Hofstede in the 1960s and 1970s is by far the broadest with regard to geographic spread and at the same time the most detailed. It looks into many different aspects of culture and not only at those aspects that influence business.

The abbreviations in Figure 11.2 refer to the five dimensions used by Hofstede (2005).

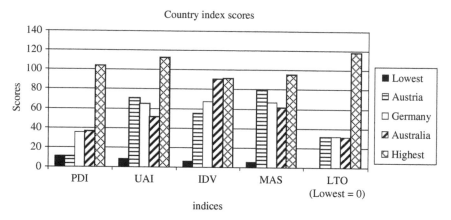

Figure 11.2 Hofstede's indices for Austria, Germany and Australia.
Source: Hofstede, 2001.

1   Power Distance index (PDI): A measure for the degree of power distance in a country's culture, based on the IBM research project.
2   Uncertainty avoidance index (UAI): A measure for the degree of Uncertainty Avoidance in a country's culture, based on the IBM research project.
3   Individualism (IDV): The opposite of Collectivism; together, they form one of the dimensions of national cultures. Individualism stands for a society in which the ties between individuals are loose: everyone is expected to look after himself or herself and his or her immediate family only.
4   Masculinity (MAS): The opposite of femininity; together, they form one of the dimensions of national cultures. Masculinity stands for a society in which emotional gender roles are clearly distinct: men are supposed to be assertive, tough and focused on material success; women are supposed to be more modest, tender and concerned with the quality of life.
5   Long-term orientation (LTO): The opposite of short-term orientation; together, they form a dimension of national cultures originally labelled 'Confucian Work Dynamism'. Long-term orientation stands for the fostering of virtues oriented towards future rewards, in particular perseverance, thrift and adapting to changing circumstances.

In relation to Hofstede's five dimensions, the largest difference among the three countries studied here is found in the Individualism vs Collectivism Index. This index indicates that Australians are more individually focused than group based relative to the other two countries, which suggests that they are more likely to work independently than personnel from either of the German-speaking European countries.

For the PDI, Hofstede found Austrians to be the least hierarchical of all 53 countries included in research. Germany and Australia are both very similar on this measure.

Differences also exist for the Uncertainty Avoidance Index suggesting that Australians are likely to be more tolerant of uncertainty, more adaptive to change and more flexible than people from the two Germanic countries. The Masculinity vs Femininity and Long vs Short-Term Orientation Indices were not considered in the study.

With the exception of Hofstede's PDI, where the two German-speaking countries differ significantly, Germany proves to be relatively close to Austria making it valid to group them and to compare them against Australia in this study.

11.3   Differences in organisational culture

Research method

This first part of the research was designed to identify differences in organisational culture between Austrian and Australian designers, head and subcontractors in the construction sector. Furthermore, the impacts of those differences on construction project management were evaluated.

For this survey, 40 Austrian and 25 Australian companies (designers, head and subcontractors) were contacted, valid responses were received from 13 (33 per cent) and 10 (40 per cent) respectively. The survey instrument was the standard Competing Values Framework Questionnaire (Cameron and Quinn, 2005). The aim was to have independent responses to the questionnaire from a minimum of six employees from different areas of each company. This number was considered to be sufficient to allow for average aggregated and disaggregated results to be compared.

Valid responses were received from 60 Austrian and 38 Australian employees; in each case the figure represented 25 per cent of the people who were contacted. The overall response rates are typical for mailed surveys of this kind. Of the Austrian responses, 24 were from 5 contractors, 12 from 4 subcontractors and 18 were from 4 design firms. From the Australian companies, 16 were from 4 contractors, 7 from 2 subcontractors and 15 were from 4 design firms. For a response to be considered valid, at least 5 of the 6 questions had to be correctly answered; as a result, four responses were invalidated.

The overall numbers in this study are relatively small, especially once the results are disaggregated. The authors are therefore cautious about drawing firm conclusions in the results and discussion sections of this chapter. They argue, however, that the techniques used have great potential and that the general conclusions can be sustained.

Results

The aggregated results for Austrian and Australian companies strongly resemble those found by Cameron and Quinn for the construction industry worldwide (Cameron and Quinn, 2005). This resemblance is shown in Figure 11.3. There is

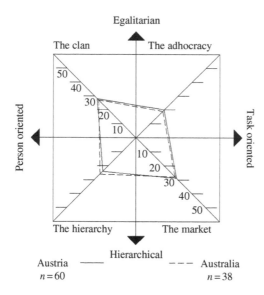

Figure 11.3 Overall results.

no significant difference between the two countries, with both scoring slightly higher on the Clan and the Market Cultures than on the other two axes.

While commonly held preconceptions would lead most people to expect the Austrian firms to be the more hierarchical, this is not the case. However, the result correlates with Hofstede's PDI, which shows that Austria scores lowest of all nations for this index (Figure 11.2).

Although the aggregated results were very similar, once the responses from the subgroups were separated it was found that significant differences exist between Austrian and Australian companies.

The responses from designers suggest that the organisational culture in Australian design firms is relatively more oriented towards Clan and Adhocracy while the culture in Austrian design firms is relatively more oriented towards Market and Hierarchy. Based on one author's experience of using these measures with contracting and design organisations in previous research (Thomas *et al.*, 2002), these differences appear significant (although the relatively small sample size necessitates some caution in making this statement). This finding suggests that Australian design firms are more likely to work in more open, team-based organisations with more independence for individuals and sub-groups within the firm compared to their Austrian counterparts.

The responses from contractors (Figure 11.4) were the reverse to those for designers (Figure 11.5); consequently, the distributions for the aggregated results were similar for the two countries.

The organisational culture in Australian head contracting firms was relatively more oriented towards Hierarchy and Market than in their Austrian

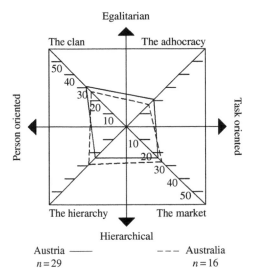

Figure 11.4 Contractors.

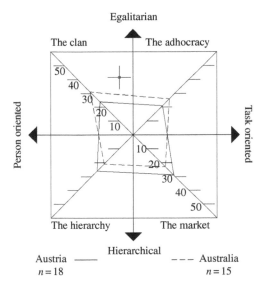

Figure 11.5 Designers.

counterparts; the culture in the Austrian firms was more oriented towards Clan and Adhocracy. This finding suggests that Austrian construction firms are more likely to work in open, team-based organisations with more independence for individuals and sub-groups within the firm compared to their Australian counterparts.

For the subcontractors (Figure 11.6), the sample size is too small to draw any firm conclusions; none the less, the difference on the Hierarchy–Adhocracy diagonal is noteworthy, with the Australian companies significantly biased towards Hierarchy. This bias may well be a reflection of the considerably greater use of formal management instruments in relation to quality and safety mandated by government in Australia.

The results indicate that there can be significant differences between the culture of organisations within similar industry sectors in different countries. In the international construction industry, problems based on differences in organisational culture are most likely to arise when foreign companies expect their local partners to work and behave in a similar way to those in their own country. Hence, the emphasis must not be placed on *how they* (i.e. foreign companies) *compare* with a similar company in the same sector here, but on *what they expect* in a collaborating organisation (i.e. what they are prepared for) and *what they find*.

The lack of significant differences in the aggregated results (Figure 11.2) can be misleading and this example indicates that it is essential to look at the cultural orientation of different players who might work in collaborative relationships. The previous discussion shows that a more detailed analysis that looks at disaggregated results is necessary to obtain a better understanding of the possible impacts of the differences in organisational cultures on contracting relationships.

For managers working on international projects, it is important to know about the organisational cultures of other stakeholders as it influences project management in many ways: how decisions are made, how changes are managed, how goals are set and achieved and the role and style of senior managers.

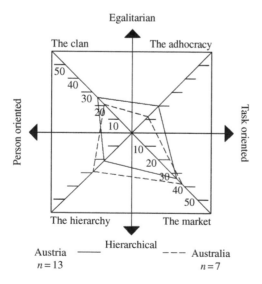

Austria —— *n* = 13

– – – Australia *n* = 7

Figure 11.6 Subcontractors.

11.4   Other differences in the construction industry

Research method

The aim of the second part of the research was to identify specific cultural, organisational and technical differences in project management in the construction sector. This aim was achieved through 17 semi-structured interviews with managers and engineers with German (11), Austrian (5) and Swiss (1) origins or long-term work experiences in those cultures. Fifteen of them were working on construction projects in Sydney; two were involved in the manufacturing industry, working on construction projects in Australia. Because of the relatively small sample size and semi-structured nature of this phase of the research, the focus here was on achieving a rich and detailed insight into differences that might exist between practices in Germany and Austria on the one hand and Australia on the other rather than on the statistical validity of any of the observations.

The questions for this interview were based in some measure on Hofstede's five dimensions, the early findings of the organisation culture surveys and the authors' research experience in similar subject areas. After validation of the survey instrument through three pilot interviews, the questions were refined and the remaining interviews were undertaken. The questions asked referred to culturally and technically based differences in organising, planning, managing, conducting and finishing construction projects.

Although most people do not expect major differences between countries of 'European origin', the results presented in this part of the study indicate that in the construction industry significant differences arise from cultural and historic factors. If a manager understands these differences in advance, they can be relatively simple to deal with and manageable; however, if they are ignored, they may lead to problems arising from project management issues.

While this next part of the chapter provides some guidance for Austrian and German managers working in Australia, the observations can be generalised to indicate the types of problem that any manager faces when working in a foreign country. This section aims to illustrate some of the issues that need to be considered in order to avoid 'negative surprises' when working overseas.

Results

Major differences exist in the influence of unions on construction projects in Germany/Austria and in Australia. Trade unions appear to be significantly more powerful in Australia due to the way that labour-related issues are negotiated at both the enterprise and project level. In Germany 'Tarifverträge' (salary and wage agreements) and in Austria 'Kollektivverträge' are negotiated between employees' and employers' representatives for the whole upcoming year and the whole industry. In Australia in addition to enterprise agreements, through which individual enterprises have to reach an agreement with the unions, site-based project agreements also have to be negotiated for large projects.

Furthermore, there is a significantly greater focus on safety issues on Australian construction sites than in Germany/Austria. The industry safety statistics show that stricter regulations have had a very positive impact on incidence rates. In Australia, everyone who works regularly on a construction site needs to undertake a one-day industry safety induction and must carry an industry Green Card. Furthermore, they have to undertake a two-hour site induction before they can start working on any site. It is the contractor's responsibility to organise and provide these inductions. In Austria and Germany these practices are quite different. In Austria, safety on site is part of the owner's responsibilities. For every site, one person is in charge of first-aid and contractors should only employ workers who have the specific knowledge and skills needed to complete the job for which they are hired.

Although European insurers and governments are increasingly aware of the weakness of safety practices on site (thus increasing the demand for site inspections), basic safety rules – including simply wearing hardhats on site – are still not as widely adhered to in general practice as they are in Australia. This difference is reflected in statistics that show that the incident rate on Australian sites is half of that in the German-speaking European countries.

Another aspect is that Australian companies need to deal with more bureaucracy than do organisations in Austria and Germany: this greater level of bureaucracy is found within the management systems of companies (see below) as well as in reporting to public authorities. In Australia, a greater number of approvals and thus more paperwork are required, which is time-consuming and increases costs.

Furthermore, on an Australian project, management procedures are documented in more detail, especially in relation to the management of safety, quality and environmental issues. It is normal to break down the duties and responsibilities of subcontractors in detail, especially on major projects. In most cases, it is part of the contract that both the head contractor and in turn the subcontractors have to fulfil these procedures and make sure that every employee is familiar with them. This practice is not as common in Austria or Germany.

Finally, there is relatively less stability of employment in the construction sector in Australia, resulting in a lower employee commitment towards the company and *vice versa*. In Australia, personnel tend to be hired for a project and their term of employment is for the project duration. In Austria, employees are expected to be very loyal to their company and employers are expected to ensure that sufficient projects are acquired in order to keep all their employees in work.

11.5   Conclusion

While the research specifically looked at the experiences of a number of German and Austrian engineering managers working in Australia and surveyed contracting and design firms in Austria and Australia, this study has identified a number of significant issues that are not normally discussed in the literature designed for managers taking up international appointments. The issues identified relate to industry-specific cultural differences and differences in regulations and techniques that exist between countries.

The literature review revealed that there are many practical guides for international managers on how to conduct business in an overseas setting. Studies such as those by Hofstede (2001), Gesteland (1999) and Trompenaars and Hampden-Turner (1997) are extremely useful for construction organisations when considering more general management issues in relation to national culture and in areas such as human resource management. However, very little has been published on how cultural and technical differences affect the operating environment of specific industry such as the construction sector.

The most surprising finding of the literature review was that for some national characteristics German culture is more closely related to Australian culture than to its immediate southern neighbour Austria. This finding is contrary to the general perception in Australia that the two German-speaking countries, with very closely related economies, are similar in every way.

In relation to specific construction-related issues, this study found that there are significant differences both in the cultural orientation of sub-sectors in the industry and in relation to regulations, local customs and management practices that have arisen as a result of historical factors in the local industry.

The Cameron and Quinn (2005) Competing Values Questionnaire was sent to Austrian and Australian designers and contractors and, at the aggregated level, negligible difference was found between the construction sector organisations' cultures in the two countries. However, the disaggregated results for designers, subcontractors and contractors showed significant differences between the same types of firms. Significant differences were observed between the organisational culture of Austrian and Australian contractors and similarly between Austrian and Australian design firms. The cultural orientation of Austrian contractors and Australian designers was similar: they were biased towards Clan and Adhocracy cultures relative to their counterparts in the other country, indicating a more open and trusting approach in these parts of the industry.

What needs to be well understood is the difference between what people expect based on their home experience and what they find in the overseas country to which they are posted. Hence the deeper insight into the organisational culture of their collaborators, as advanced in this study, would help a manager to prepare for work in another country. This insight will influence project management practice in many ways: *how* decisions are made, *how* changes are managed, *how* goals are set and achieved and *how* the role of senior managers is defined.

The purpose of the face-to-face interviews was to identify those differences specific to the construction industry. This kind of information can only be identified through interviews with experienced managers already working in the market in question. Significant differences were found in the area of labour-related issues, the extent of union influence on construction projects, environmental, quality and safety management practices on site and also in the detail of government regulations and management practice.

In summary, this research has indicated the existence of differences in relation to culture and practice that go well beyond the existing literature that prepares managers for operations in an overseas market. While this chapter provides some guidance for Austrian and German managers working in Australia, the chapter also indicates some of the issues that companies should examine before moving staff to a new international environment. Once these issues are understood, management can deal with the differences; however, it is important to be aware of these issues as ignoring them may lead to errors in management judgement and the creation of unproductive project cultures.

Acknowledgments

Without the support of the following people and organisations this research would have been much more difficult: Dipl. Ing. Peter Arz (Bilfinger und Berger, Sydney, Australia), Ing. Christian Neumann (Beton und Monierbau, Innsbruck, Austria), the Austrian Foreign Trade Office in Sydney, the German Australian Chamber of Trade in Sydney and the Swiss Australian Chamber of Commerce and Industry in Sydney. Furthermore, the authors wish to thank all the anonymous participants who answered the questionnaires and were willing to be interviewed.

A more detailed description of the results of this study can be found in the diploma thesis 'Intercultural Management for International Construction Projects – A Comparison of Austria and Germany with Australia'.

References

Print documents

Brown, A.E. (1995) *Organisational Culture*. Pitman, London.
Cameron, K. and Quinn, R.E. (2005) *Diagnosing and Changing Organisational Culture Based on the Competing Values Framework*. Jossey-Bass, John Wiley & Sons, Inc.
Cameron, K. and Sine, W. (1999) *A Framework for Organisational Quality Culture*. ASQ, New York.
Campbell, J.P., Brownas, E.A., Peterson, N.G. and Dunnette, M.D. (1974) *The Measurement of Organisational Effectiveness: A Review of the Relevant Research and Opinion*. Minneapolis, Final Report, Navy Personnel Research and Development Center, Personnel Decisions.
Chang, F.S. and Wiebe, H.A. (1996) *The Ideal Culture for Total Quality Management: A Competing Values Perspective. Engineering Management Journal*, 8(2).
Deal, T. and Kennedy, A. (1982) *Corporate Cultures: The Rites and Rituals of Corporate Life*. Addison Wesley, MA.
Gesteland, R.R. (1999) *Cross-Cultural Business Behaviour: Marketing, Negotiating and Managing Across Cultures*. Copenhagen Business School Press, Copenhagen.
Harrison, M.I. and Shirom, A. (1999) *Organisational Diagnosis and Assessment: Bridging theory and practice*. Sage, Thousand Oaks, California.

Hofstede, G. (2001) *Cultures Consequences: Comparing Values, Behaviors, Institutions, and Organisations Across Cultures.* 2nd ed. Sage Publications Inc., USA.

Hofstede, G. and Hofstede, G.J. (2005) Cultures and Organizations: Software of the Mind, 2nd edition. McGraw-Hill, New York.

Hooijberg, R. and Petrock, F. (1993) *On cultural change: Using the Competing Values to Help Leaders Execute a Transformational Strategy.* Human Resource Management, Spring.

Huntington, S.P. (1997) *The Clash of Civilization and the Remaking of World Order.* Touchstone, New York.

Jung, C.G. (1923) *Psychological Types.* Routledge and Kegan Paul, London.

Kotter, J.P. and Heskett, J.L. (1992) *Corporate Culture and Performance.* The Free Press, Macmillan, New York.

Mason, R.O. and Mitroff, I.I. (1973) *A Program of Research in Management. Management Science*, 19.

McKenney, J.L. and Keen, P.G.W. (1974) How managers' minds work. *Harvard Business Review*, 51.

Mitroff, I.I. (1983) *Stakeholders of the Organisational Mind.* Jossey-Bass, San Francisco.

Mitroff, I.I. and Kilmann, R.H. (1978a) *Stories Managers Tell: A New Tool for Organisational Problem Solving. Management Review*, 64.

Mitroff, I.I. and Kilmann, R.H. (1978b) *Methodological Approaches to Social Science.* Jossey-Bass, San Francisco.

Myers, I.B. and Briggs, K.C. (1962) *The Myers-Briggs Type Indicator.* Educational Testing Service, NJ Princeton.

Peters, T.J. and Watermen, R.H. (1982) *In Search of Excellence.* Harper & Row, New York.

Quinn, R.E. (1988) *Beyond Rational Management.* Jossey-Bass, San Francisco.

Quinn, M.P. (1989) Cost of quality and productivity improvement. In: Campanella, J. (ed.) *Quality Costs: Ideas and Applications – A Collection of Papers*, Vol. 2. Quality Press, Milwaukee.

Quinn, R.E. and Rohrbaugh, J. (1981) A Competing Values Framework to Organisational Effectiveness. *Public Productivity Review*, V5, pp. 122–140.

Ross, D.N. (2000) Does Corporate Culture Contribute to Performance? *Journal of Business*, Spring.

Schneider, J.M.A. (1989) *Legacy of quality.* J.M. Schneider Inc.

Thomas, R., Marosszeky, M., Karim, K., Davis, S. and McGeorge, D. (2002) *The Importance of Project Cultures in Achieving Quality Outcomes in Construction.* In: C.T. Formoso and G. Ballard (eds), *Proceedings 10th Conference of the International Group for Lean Construction*, 6–8 August 2002, Federal University of Rio Grande do Sul, Gramado-RS, Brazil.

Trompenaars, F. and Hampden-Turner, C. (1997) *Riding the Waves of Culture: Understanding Cultural Diversity in Business,* 2nd edn, Nicholas Brealey Publishing, London.

Electronic documents

AK Wien. '*Daten und Fakten zum Arbeitnehmerschutz*'. April 2002. <http://www.akwien.or.at/885_8613.htm> (22 May 2003).

Australian Bureau of Statistics, '*Construction trade union membership in the construction industry*'. 11 March 2003. <http://www.abs.gov.au/Ausstats/abs@.nsf/Lookup/A26890 BBDF452285CA256B360003228C> (19 May 2003).

Australian Department of Foreign Affaires and Trade. 12 May 2003. <http://www.dfat.gov.au/index.html> (13 May 2003).

Australian Embassy Berlin. <http://www.australian-embassy.de> (13 May 2003).

Bundesarbeitsgemeinschaft Erste Hilfe. *'Das Unfallgeschehen in Deutschland – Arbeit'*. 2003. <http://www.bageh.org/Tagungen/Forum_2002_Berlin/Forum_2001_ Jena/Fachtagung_Erste _Hilfe_2000/hennef2000-V_03.pdf> (23 May 2003).

Deutscher Gewerkschaftsbund. *'Mitglieder in den DBG Gewerkschaften'*. 31 December 2002. <http://www.dgb.de/dgb/mitgliederzahlen/mitglieder.htm> (19 May 2003).

ExecutivePlanet.com. 2003. <http://www.executiveplanet.com> (13 May 2003).

Hofstede, Geert <http://www.geert-hofstede.com/> (19 September 2005).

Mayer, Beate. *'AUVA – Unfallstatistik 2001 Arbeitsunfälle ohne Wegunfälle, Arbeiter und Angestellte Österreich'*. 23 May 2003. personal email. (26 May 2003).

Österreichischer Gewerkschaftsbund. 2001. <http://www.oegb.at > (19 May 2003).

Statistik Austria. 2000. <http://www.statistik.at > (19 May 2003).

Statistisches Bundesamt Deutschland. *'Erwerbstaetige im Inland nach Wirtschaftsbereichen'*. 25 February 2003. <http://www.destatis.de/indicators/d/vgr010ad.htm> (19 May 2003).

Wirtschaftskammer Österreich. *'Länderprofil Australien'*. November 2002. <http://wko.at/ awo/publikation/laenderprofil/lp_AU.pdf> (28 April 2002).

Part III

Critical perspectives on construction employment and change

12 The impact of eastward enlargement on construction labour markets in European Union member states

David Langford and Andrew Agapiou

Abstract

This chapter concerns the impact of labour migration on the construction industry in the enlarged European Union (EU). It reviews the varying patterns of employment for construction labour in Europe and considers how these different employment conditions influence the labour process. The movement of construction labour within Europe is considered and the chapter argues that diverse economic and geographical conditions will influence the quantity of migrant labour in different European countries. Labour migration was small in the pre-2004 EU and is likely to remain small in terms of the numbers currently involved in the EU construction industries; nevertheless, it is expected to grow in the near future. Finally, the benefits and risks of labour migration are considered from the perspective of the migrant worker and indigenous construction employees in respect of wage levels and productivity.

Keywords: construction industry, EU enlargement and labour migration.

12.1   Introduction

Nearly 10 million people were employed in the construction industry in the EU in 2000 (Lienhardt, 2003). This figure represents around 11 per cent of total employment in Europe – larger than the USA (8 per cent) but smaller than Japan (13 per cent). The added value in the construction industry has been calculated as just over 8 per cent and therefore lower than the share of employment. Clearly, construction labour represents a low-productivity asset. Some comparative data is shown in Figure 12.1.

Another feature of the labour market in the EU is the high proportion of self-employment, although there is great variability in self-employment practices. As Figure 12.2 shows, this proportion is much greater than in manufacturing. This pattern of employment ensures that the EU construction market is dominated by micro firms (41 per cent) and small firms (31 per cent).

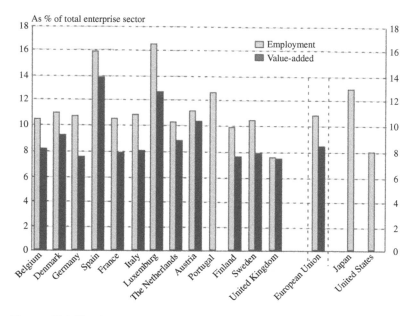

Figure 12.1 Employment and value-added in construction as percentage of total
enterprise sector in some EU countries in 2000.
Source: Eurostat, Japan Statistics Bureau, US Bureau of Labour Statistics

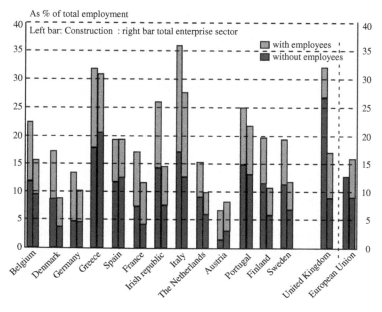

Figure 12.2 Self-employed with and without employees in construction and in the
enterprise sector (2001).
Source: EU Labour Force Survey

In most European countries, the wages of construction workers lag behind those in manufacturing. The evidence on self-employment and wages is indicative of the highly complex structure in the construction industry within Europe. It is now very common (and often encouraged by governments) to make extensive use of subcontractors, although this practice is unpopular with both trade unions and employers. A recent International Labour Office (ILO) report suggests that over-reliance on subcontractors, often employed on harsh contractual conditions, can result in higher final costs for the client. The report also indicated that the abuse of self-employment can lead to safety and health risks and is an impediment to the sustained development of national construction industries, particularly since the workers affected are not often afforded the protection of social security systems, such as sickness and unemployment benefits (ILO, 2001).

The purposes of the chapter are as follows.

- To review the different patterns of employment in the construction industries of the EU.
- To evaluate, in the context of the labour process, the movement of construction labour within Europe and link this to the enlargement of the EU. Regional and sectoral disparities are noted.
- To present a synthesis of the benefits and risks of mobility in terms of the individual migrant worker and the indigenous construction employees.

It is recognised that the data presented in this chapter is merely a snapshot of events at the time of writing (summer 2004). Clearly, it will not be durable data as patterns of migration respond to changing economic and personal circumstances. The intention of this chapter is to present a number of critical issues that will be influential in shaping labour migration within the enlarged EU.

12.2   Regional mobility and labour supply

In any given region, some inward and outward mobility of the skilled workforce can be expected. The net impact for any one region is unlikely to be very large, but for selected occupations such mobility could be an important element in determining labour supply. The UK construction industry already has a long tradition of recruiting workers from Eire.

The recent accession of Central and Eastern European countries to the EU opens the possibility of also obtaining a proportion of Britain's skill requirements from further afield, where pay differentials may make working in the UK an attractive proposition for individuals from the less prosperous regions of Europe. However, much is speculated and little understood about the mechanics and impact of such a move. Problems with persistent skills shortages and future demographic trends suggest that the UK industry may face real difficulties recruiting sufficient numbers of new entrants in a growth economy (Agapiou *et al.*, 1995). Construction companies therefore need to assess whether the employment of individuals from other regions of the UK and Europe is a viable solution to this problem.

The issue of regional mobility of construction workers has been examined by a number of authors and a summary is given in Briscoe (1989). In general, while a few young highly skilled operatives were found to be extremely mobile between regions, most operatives moved around within a given locality and region. In the 1980s, the main movement was from northern regions of Great Britain into London and the south-east, where employment demand was strongest. Large prestige projects, such as the Channel Tunnel, became a magnet for mobile labour because the project's status and the relatively high rates of pay made the inconveniencies worthwhile.

Skilled labour is traditionally more itinerant in some industries than others. Construction is one such industry. In London and the south-east in particular, there has been anecdotal evidence of high levels of workforce mobility, with large numbers attracted from elsewhere in the UK and further afield.

A study of workforce mobility in this region undertaken between May and August 2003 confirmed that London in particular, and the south-east to a lesser extent, draws on labour from a wide geographic region. Of the construction workers interviewed in London, 40 per cent were originally from London and the south-east and 24 per cent were from other parts of England, 17 per cent from other parts of the UK and 19 per cent from outside the UK (IFF Research Ltd, 2003).

12.3   Trends in work organisation and employment practices

It has been shown in the 'Rethinking Construction' report (Egan, 1998) and elsewhere that the procedures and terms under which developers contract with the construction industry have created a sector with employment practices characterised by high-levels of self-employment in place of direct employment (Briscoe *et al.*, 2000; Gann and Senker, 1998; Winch, 1998), a low level of training (CITB, 2002) and a lamentable safety record (National Audit Office, 2001). Traditional competitive tendering combined with the fragmentation of production and the intense use of subcontracting chains also implies that the enforcement of regulations is a key issue for the industry. Health and safety inspections typically result in the temporary closure of up to 50 per cent of sites because of poor and/or dangerous working practices (*Contract Journal*, 2002). There is a tradition of tax avoidance through employing people 'cash in hand' and incorrectly classifying employees as self-employed (Briscoe *et al.*, 2000). These practices are estimated to cost some £2.5 billion each year to the UK government in terms of tax contributions, affecting over 300,000 workers (Harvey, 2001).

Winch (1998) explored the growth of self-employment and the pattern of the drift from direct employment to labour-only subcontracting (LOSC). While geographical shifts are not evident, the pattern of labour migration from one form of employment to another could act as a precursor to migration from one area of

Europe to another. The construction workforce in the UK changed from 10 per cent self-employment in 1961 to 41 per cent in 1997, the largest shift being from 22 per cent to 44 per cent during the Thatcher years of 1979–90. These trends may be explained by labour shortages in an over-heated construction market, the very same driver for the migration of construction labour from east to west. Langford (1975), in a survey of labour and management attitudes to LOSC, found that the practice was commercially advantageous to employers because of the avoidance of government imposts. Examples of these were the Selective Employment Tax of 1966 and training levies, estimated by the Phelps-Brown Report (1968) as comprising a 16 per cent overhead on average wages. Langford found that the use of LOSC was detrimental to the orderly working of the industry as safety standards, quality of training and materials wastage were, in the opinion of the workforce, all adversely affected by the practice and managers believed that they had greater difficulty in exercising managerial control. Similar arguments are now being heard about the use of migrant labour on sites in London but there is no evidential basis for such claims.

Increasingly, the UK government has in mind to regulate the use of LOSC. Although the level of regulation is low and is exercised through the taxation systems, it is still present. There are, however, no barriers to entry to restrict any company or individual from establishing a business or working as a professional in the industry, although this weakness is counterbalanced in UK law by company liability.

The situation contrasts significantly with such countries as Germany and the Netherlands where strict regulations linked to corporatist structures have governed entry to the industry. However, these regulations have been challenged by the provisions of the EU 'Posted Workers' directive, designed to further promote the free movement of labour between member states (Hunger, 2000).

12.4   Employment trends in Europe

Construction is one of the EU's key economic sectors: output in the construction industry accounts for almost one-tenth of its combined GDP. With a combined workforce of 10 million people, the European construction sector is as important as the agricultural sector (ILO, 2001). Small- and medium-sized firms, which number almost a million, predominate in the EU. However, the nature of employment in the European construction industry is unclear as is the extent and variety of employment for foreign construction labour within individual European countries. Current trends in employment practices are discussed below, including an analysis of various forms of employment and the implications of these practices for free movement of construction labour.

Various form of non-permanent employment, in particular temporary work and contract labour (including migrant labour), play a significant and rapidly growing role in the European construction industry (Hellman, 1992). The extent of this phenomenon is not very clear. However, the European Federation of Building and Woodworkers (EFBWW) has identified the main trends in employment within Europe (EFBWW, 1991). The main findings of the EBFWW study were as follows:

- Non-permanent employment in the construction industry was widespread and on the increase.
- Flexible and insecure forms of employment, and contracting work to subcontractors, were also on the increase.
- Sole independent trade subcontractors were to be found in insulation and acoustics; scaffolding; stonemasonry; wood craft; stone-floor, tile and floor-laying work; and a wide range of assembly jobs for so-called independent kitchen furniture fitters.
- Construction activity has historically been based on project-oriented work and changing work-sites. As a result, the construction industry has traditionally employed temporary workers engaged on fixed-term contracts. However, during recent years, this trend has accelerated at an alarming rate, whilst the protection for workers has not kept pace.

There are a number of different categories of permanent and non-permanent employment in the construction industry, and definitions vary between different European countries. Hellman (1992) usefully described the various forms (Table 12.1).

Table 12.1 Categories of employment in the construction industry

Category of employment	*Description*
Normal work in the formal sector	A full-time job paid on a monthly basis with a normal distribution of working hours performed in companies that apply statutory provisions and protection against unfair dismissal. This category of work provides a small degree of job security and financial benefits depending on the worker's ability and/or company seniority and when job contents are determined by collective provisions, particularly collective agreements.
Atypical or precarious work	All forms of employment that differ from the traditional patterns in respect of the number or distribution of working hours, organisation and location of production and the way in which wages and work regulations are determined. Specific examples include part-time workers, the majority of whom are not protected by unemployment, health or pension insurance schemes.
Self-employment	Sole independent tradesmen accountable to themselves, owning their own means of production, and responsible for planning and carrying out work often without supervision. The vast majority of workers in this category are in reality non-permanent employees, working under someone else's direction without their own tools and according to working hours and at a pay rate established by a subcontractor. The extent to which these

workers are artisans working as independent tradesmen or as labour-only subcontractors is difficult to determine. Self-employed workers often have to forego important aspects of protection afforded by labour legislation, social security and collective agreements.

Work in the informal sector | Workers are informal in the sense that for the most part they are unregistered and unrecorded in official statistics. Informal workers are often beyond the reach of social protection, labour legislation and protective measures in the workplace.
The informal sector provides workers with low and irregular incomes and highly unstable employment.

Source: Hellman (1992).

12.5 The construction labour process in Europe

The increase in both temporary or contract labour and self-employed workers has often been associated with the recession that affected construction throughout Western Europe during the late 1970s and early 1980s. However, the increasing use of such flexible and precarious forms of employment is in part due to new personnel policies which construction companies have been implementing since the mid-1970s.

The widespread use of flexible employment practices has been made possible by government policies to deregulate national labour markets and reform social security systems. Attempts to reduce the coverage and value of state welfare benefits have been widespread. Many European governments place higher priority on the control of inflation than on maintaining full employment. There has been a general trend towards extending market pressures, privatisation of nationalised industries and cuts in government subsidies to loss-making firms and industries.

Changes in government social and labour markets policies during the 1980s were accompanied by the adoption of new business practices by the construction industry. These had a profound effect on the structure of employment in the industry throughout Europe. Changing construction methods, on-site working conditions and alternative working arrangements allowed construction firms to react flexibly to the demands of the relevant market. This flexibility was due in the first instance to a well-defined division of labour, embracing such elements as:

- supply and market orientation
- specialisation of the construction sector and trade segregation, coupled with diversification of services offered
- reduction in the levels of employment, accompanied by the increased use of contract labour.

Growth in construction is commonly associated with higher levels of employment in the industry. On the other hand, reductions in employment levels, as seen in German construction during a period of sustained growth in the early 1990s, are considered indicative of a crisis in the industry. Yet they may also, in this instance, indicate the reliance on foreign contract labour employed on German construction sites. Peaks in construction demand in the early 1990s were addressed through the engagement of contract labour from Eastern Europe. The reliance on this source of labour has remained unchanged in the intervening years. In 2002, for example, an estimated 70,000 workers from Poland were employed in the German construction industry (Kus, 2004b). Anecdotal evidence suggests that the number of Polish workers employed illegally in the sector greatly exceeds this figure, but it is difficult to estimate. The available research plays down this evidence. Despite 8 per cent unemployment in construction against an overall employment rate of 20 per cent (Kus, 2004a), the propensity for Polish workers to relocate in developed Western countries is relatively low. Research by the European Foundation for the improvement of Living Conditions (Krieger, 2004) found that only 1 per cent of the population would move into 'old Europe' – the 15 countries in the EU prior to enlargement. The main migrant group would typically comprise young, well-educated workers, mostly graduates. This small proportion with a 'firm intention' of leaving Poland is reinforced by those with a 'basic intention' and a 'general inclination'. Figure 12.3 shows the research referred to earlier.

Now, if the construction workforce in Poland numbered at 360,000 decided to migrate at the rate of all three categories in Figure 12.3, this would add 22,680 construction workers to the EU-15 countries if all applicants could get work permits. This adds 0.002, 268 per cent to the EU construction labour force!

However, at the level of the individual, Poland has the highest migration potential. The income gap, the high unemployment rate, the short distance of Poland from the rest of the EU-15 and the Polish tradition of emigration are key factors in the determination of high migration rates (Wisniewski and Oczki, 2001).

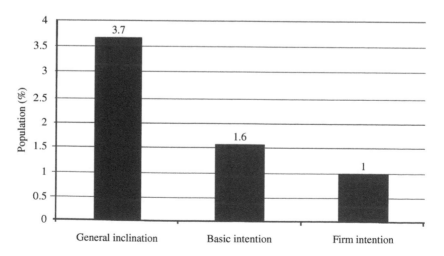

Figure 12.3 Intention of Polish workers to go and work in EU '15'.
Source: Research Report by European Foundation for the Improvement and Living Conditions in 2004

12.6   The movement of construction labour within Europe

One of the potentially dramatic shifts in the labour process of the construction industries of Western Europe is the migration of construction labour from Eastern to Western Europe as part of the enlargement of the EU. On 1 May 2004, ten new countries were incorporated into the EU. The reasons for the enlargement were largely political and cultural. Enlargement returned Europe to its pre-'iron curtain' shape. It also expanded the economic area to around 450 million people, thus representing the largest single market in the world. This is a great asset in a context of economic globalisation (European Policies Research Centre, 2003).

According to Tupy (2003), 'many of the barriers to trade and investment and movement of labour will disappear'. This envisaged movement of labour has some caveats. For example, the European Commission (2004) noted that the requirement for work permits in the pre-2004 states would continue for a transitional period of five years with an option for member states to extend this requirement for a further two years if disruption to their labour market was feared. (Malta and Cyprus were excluded from the prescription.) The freedom for persons to move and settle throughout the EU will have the biggest impact on migratory patterns and the differences in wages will be the main prompt of each movement. In the case of the construction industry, the nomadic traditions and the relatively casualised labour open up the construction labour force to an even greater extent.

Yet the metrics of such mobility are unclear. The poor coverage of available statistical information makes any description of construction labour mobility towards and within Europe problematic. The extent and variety of such movement is not sufficiently recorded by the available statistics of the EU, which have been based on national statistics. Some very useful but limited information is available. These figures are often outdated and only collected on an *ad hoc* basis.

Employment in the European Construction industry is already characterised by a large number of foreign workers, though their numbers differ from country to country. In the UK, foreign craftsmen represent only a small proportion of the construction workforce. During 2000, an estimated 70,000 workers from Eire were employed in the UK construction industry (Dobson *et al.*, 2001): this figure only represents around 5 per cent of the workforce. The situation is different in other European countries. In Germany and Switzerland, significant numbers of foreign workers are employed in construction (Dobson and Salt, 2002). During 2000, for example, 11 per cent of workers employed in the German construction industry were foreign nationals; in western Germany this figure was around 14 per cent, if only those employees subject to social security contributions are taken into account (ibid.). In principle, these latter employees worked under the same conditions as their national colleagues and are in this respect integrated into society.

The 1996 Posted Workers Directive, implemented via national legislation in all EU countries, required that the terms and conditions of employment in the posting country be directly applied to the workplace in the recipient state. Nevertheless, in both the UK and Germany there has been widespread concern that this legislation has led to the erosion of employment conditions of native employees (European Industrial Relations Observatory, 1999; Faist *et al.*, 1999).

The extent of irregular employment of foreign labour in Europe is difficult to estimate. Construction is a labour-intensive industry and, next to the hotel and catering sector, it is the sector where workers foreign to the region and the specific occupation are most readily able to find employment. Despite the removal of many administrative constraints and legal barriers on worker mobility (e.g. the removal of border controls), labour mobility among the old EU countries is insignificant (Langwiesche, 2000a). The vast majority of irregular construction labour originates from countries outside Western Europe.

Whereas the northern European countries, especially Germany, employ labour that originates from Central and Eastern Europe, increasing flows from the North African region can be observed in the southern EU countries. The majority of irregular workers in Portugal are employed in the construction industry. In Italy, and even in Greece, irregular workers are employed as seasonal or occasional workers on construction sites. One reason is that in these countries national labour markets are not sufficiently organised to cover the seasonal shortages of mainly unskilled construction workers. Irregular employment also avoids payment of social security contributions, in addition to considerably lower wages already paid to foreign workers.

12.7   Migration potential and EU enlargement

Both construction companies and trade unions from the old EU states have consistently warned that the number of migrants from the acceding countries could lead to serious distortions of labour markets. In fact, the income differential in the case of Central and Eastern Europe is significantly larger than in previous enlargement rounds (Alvarez-Plata *et al.*, 2003). Over the past five years very many studies have therefore tried to estimate the migration potential for sectors and regions (e.g. Langweische, 2000b). Although all these figures should be treated with a degree of caution, it is safe to say that the predicted migration level was higher in the former studies than in more recent ones.

The European Integration Consortium (2001) forecasts that the number of foreign residents in the EU-15 from the acceding countries (including Romania and Bulgaria) will increase by around 335,000 people per annum immediately after the introduction of freedom of movement. About one-third of them will be employees. These figures sound significant. In practice, however, this number of potential migrants only represents 0.1 per cent of the EU-15 population (Langweische, 2000a). Furthermore, as almost all present EU members will make use of the transitional periods for the free movement of workers, real figures should be somewhat lower in the first years after the enlargement. In order to effect the transitional arrangements the UK government capped the number of work permits issued to Romanian's at 20,000 per year. In other words they are free to come to the UK but could not work here.

The study also expects the migration figure to fall below 150,000 people within a decade. After thirty years the share of the foreign population originating from Central and Eastern Europe in the EU-15 population may reach 1.1 per cent (European Integration Consortium, 2001).

Regional and sectoral disparities

All studies agree that the impact will be felt disproportionably according to the region and the industry. Obviously, EU-15 countries and regions closest to the new member states are more likely to be affected than others. For example, Germany is expected to receive 220,000 people immediately after freedom of movement is introduced, representing two-thirds of total migration (Langweische, 2000b). After thirty years, 3.5 per cent of Germany's population could originate from the acceding countries. Other countries bordering the acceding countries, such as Finland, Austria and Italy, will also see higher migration figures than the rest of the EU-15 (Hunger, 2000). Border regions will also experience considerable short-term (even commuter) migration for work. Construction is regularly mentioned among the industries considered most likely to be affected, along with tourism and agriculture. In this context, it appears logical that the accession treaties should contain safeguard clauses suspending the free movement of labour.

Wage differentials not enough to stimulate migration

Differences in wages and labour costs are usually considered as the major drivers for the migration of workers and self-employed service providers. Indeed, the differences are substantial, even between the most advanced accession countries and the EU-15. For example, in the Czech Republic, the total labour costs in construction (Euro/month/employee) amounted to €710 in 2002, well below the level of the EU neighbours (Geddes and Balch, 2002). On the other hand, the turnover per person employed in construction is also considerably lower (€28,000 per annum in the Czech Republic, 38,000 in Hungary and only 12,000 in Lithuania). Lower labour productivity accounts for a significant part of the low costs of labour, thus reducing the competitive advantage of construction firms from the acceding countries. Neither construction workers nor businesses will move westwards at any price. Many other criteria are involved in the decision process and may, at least to some extent, outweigh differences in labour costs:

- Business opportunities and unemployment in the acceding countries – construction markets will grow rapidly, and unemployment among construction workers is likely to diminish.
- Business opportunities and unemployment in the EU-15 – growth is predicted to be sluggish over the next decade in most EU-15 countries, including the two main 'target countries' Austria and Germany.

- Indirect factors such as distance to families and homes, cultural differences, language and so on.

High specialisation versus low labour costs

When determining the impact of migration on the construction industry, it is not enough to look at the geographical situation. One can also expect that the more complex and specialised the activity, the lower the pressure from foreign job seekers or service providers. This means that architects, engineers and technical building contractors are likely to see fewer new competitors than, for example, painters or masons. Obviously, the provision of services involving a high degree of expertise, quality, specialisation and knowledge of state-of-the-art technologies and standards provides a counterbalance to low labour costs. In any case, it seems pointless to wait until wage differentials with the acceding countries have disappeared. Alvarez-Plata *et al.* (2003) estimate that it will take about thirty years to half the income gap between the Central and Eastern European countries and the EU-15. During this time equilibrium will be found with equalisation of wage levels. The 'old' EU is abundant in capital and the accession countries are abundant in labour. Boeri *et al.* (2002) see the following consequences.

- A fall in the price of a labour-intensive good (construction) relative to the price of the capital-intensive good (say, electricity generation).
- An increase in the production of the capital-intensive good (e.g. electricity) and a decrease in the production of the labour-intensive good (construction).
- A decline in wages relative to the price of capital, which will continue until factor prices have equalised.
- A movement of labour from the labour-intensive sector to the capital-intensive sector, and a rise in labour intensity in both sectors, since the price of labour has fallen.
- A constant total employment of labour, since the effects of a declining production of the labour-intensive good (construction) and the increasing ratio of labour to capital in both sectors cancel one another out completely.

The result is a movement of labour from the labour-abundant country and a movement of money from a capital-abundant country until prices have stabilised. This of course assumes that mobility is perfect across the countries and that skills of either a professional or a trade type are portable with techniques, standards and levels of workmanship and that quality expectations are uniform among the giving and receiving countries in respect of labour. With regard to capital movement the assumption is that it can be harnessed to similar effect in the new European country as in the pre-2004 EU countries.

Construction companies in the present member states, and in particular in the border regions, should therefore carefully analyse developments in customer requirements. Service packages should be developed and market niches secured.

12.8 The benefits and risks of labour mobility

The employment of foreign nationals in individual European countries can be distinguished by different rights of access to national labour markets, by different qualification levels and by varying employment conditions (Agapiou, 1996). Foreign construction workers in regular employment often receive lower wages than similarly qualified nationals, despite prescriptions for regular treatment. However, there tends to be little or no distinction in working conditions on the basis of nationality. The current trend towards forms of dispatched employment (i.e. workers on contract for services) and the irregular employment of foreign labour offers both benefits and risks for the construction industry in individual European countries.

Employers and employees recognise the benefits of labour migration. In the short term, workers moving from countries with low-wage levels and a lack of employment opportunities to countries with higher income levels are able to improve their individual living conditions, even if their income is lower than that of host country employees. This is especially the case if they transfer part of their income to their home country.

The employer from the host country can benefit financially from the use of foreign workers, even those employed on regular contracts, provided that services are costed on the basis of lower foreign wages. This practice avoids employer contributions to social security benefits and premiums provided for by private occupational schemes, such as entitlement to holiday pay or time lost due to unfavourable weather or economic cycles. Employers openly express an interest in employing foreign nationals, and even a preference for irregular employment in order to avoid such high incidental costs (Balch *et al.*, 2003).

The employment of foreign labour can also be used to ease regional skill shortages, particularly in countries where labour-intensive construction work is considered unattractive. However, this can also impede the development of new construction techniques and methods. Construction still remains a labour-intensive industry throughout Europe. Irregular and low-wage employment can slow down the substitution of labour by capital-intensive methods. This in turn can restrict the growth in labour productivity and the development of efficient practices.

12.9 The influx of irregular foreign migrants into the construction industry: A UK perspective

The impact of irregular migrants working in the UK construction industry is predominantly assessed on the basis of anecdotal evidence, but raises important issues regarding labour and skills shortage within the sector. While migrants may bring much-needed skills to the industry, the existence of an irregular workforce can have an impact upon the labour market conditions, public finances and health and safety measures. For instance, the Inland Revenue has no figures on lost duty or the tax and national insurance fraud that accompanies illegal employment, but it is estimated that it contributes a significant proportion of the construction industry's £10 billion informal economy (Fairs *et al.*, 2001). However, health and safety issues have been identified as the most immediate area of concern (*Contract Journal*, 2002). Many migrants do not have the necessary safety qualifications required for UK construction work (i.e. the Construction Skills Certification Scheme) or the language skills to communicate effectively in English (Broughton, 2001). The inability to understand written or spoken instructions and warnings can have dire consequences not only for migrant labour, but also for the indigenous workforce.

Such arguments derived from anecdotal evidence have some credibility, but they further emphasise the lack of knowledge regarding the scale of the problem and its implications. The lack of hard evidence makes it difficult to demonstrate accurately the extent of the problem in terms of numbers, and only encourages wild speculation.

12.10   Conclusion

The construction industry requires a high degree of mobility from its employees. Contracts of employment tend to be short term, people work for many different employers, and there is poor job security with alternating periods of high and low employment. Throughout Europe employers and employees in the construction industry recognise the benefits of labour mobility, at least from an economic perspective. However, there are risks involved in such movement. Construction is a very labour-intensive activity and wage differentials have a significant impact on the price a construction firm can offer. The competitive advantage of firms established in new EU countries is counterbalanced by lower labour productivity, additional costs due to cross-border activities and, in a number of cases, lower skills.

When looking at the effects of EU enlargement, it is evident that some regions and trades will be more affected than others. The regions along the border between the present and the new member states (Finland, Germany, Austria and Italy) are without any doubt more likely to see new competitors than are Portugal or Ireland. Construction firms in the border regions are particularly concerned about commuters, who, if they come as independent service providers, will not be covered by the safeguard clauses regarding the free movement of labour as stipulated in the Accession Treaties. Taking account of the particular exposure to new, low-cost competitors, the Accession Treaties also contain safeguard clauses for Austria and Germany concerning the provision of services. However, the enlarged Union will have to guarantee the free movement of people, goods, services and capital. Safeguard clauses can therefore only have a transitional character.

The challenge of the latest enlargement is to ensure that subsequent labour regulation is equitable to the migrant worker. Currently, legitimate movement of capital is freer than the legitimate movement of labour and, as such, rogue employers may be tempted to use low-cost, flexible labour to undermine working conditions and wages in the construction industries of the pre-2004 EU countries. This is a challenge to ordered industrial relations in the EU construction industry.

From a critical perspective, the free movement of workers has several implications. Like the free movement of ship registrations, employers may be tempted to record workers as being allied to a nation which offers the most free and deregulated labour market and post them to where their skills are required. This has grave implications for the free movement of labour because workers would be shackled if they were to lose local benefits when moving within the community. Students of the construction labour market should be alert to the dumping of labour by posting workers from countries with lower costs to those with higher labour costs. A labour dispute in Sweden in 2004 offers an example of the dangers. Swedish construction workers blockaded a site because 'a Latvian company was employing low-waged Latvian construction workers on the site' (Woolfson and Sommer, 2005). This example illustrates the central challenge to be overcome, which is to preserve a social model with minimum standards for workers' rights within a global climate of economic liberalism and European enlargement.

Perhaps this is a redundant fear. The expectation that wider enlargement to incorporate Spain, Portugal and Greece would lead to mass migration has been confounded. As Cremer (2005) points out, 'it takes a lot to leave your home and your soil' and if the new European states now have a perspective for social and economic development, then the motivation to migrate diminishes.

One frequently heard argument is that if construction, as an industry, is to be attractive to young people, it should develop a self-image which shows that it is intrinsically satisfying as a career – whether in the professions or the trades. It is further said that over-reliance on migrant labour diminishes the image of the industry and so it becomes less attractive to the indigenous population. Such an attitude is xenophobic and potentially racist.

If construction is to be modernised, then its culture of being suspicious of 'outsiders', whether women, ethnic minorities or, as now, workers from the new EU countries, has to be reformed. The slogan 'equal treatment on site' needs to apply to all whether they are new Europeans or old Commonwealth workers.

References

Agapiou, A., Price, A.D.F., and McCaffer, R (1995) Planning future construction skill requirements: Understanding the labour resources issues. *Construction Management & Economics*, 13(2), 149–161, ISSN 0144–6193.

Agapiou, A. (1996) Forecasting the supply of construction labour. Unpublished PhD Thesis, University of Loughborough.

Alvarez-Plata, P., Bricker, H. and Silverstos, B. (2003) *Potential Migration from Central and Eastern Europe into the EU-15 – an update*, DIW, Berlin.

Balch, A. and Geddes, A. (2002) UK migration policy in light of sectoral dynamics: The case of the construction sector. Paper presented to the 3rd meeting of the UACES sponsored study group on evolving EU migration law and policy.

Boeri, T., Bertola, G. and Bruicker, H. (2002) Who is afraid of the big enlargement: Economic and social implications of the EU's perspective eastern expansion. CEPR Policy Paper No. 7 Centre for Economic Policy Research.

Briscoe, G. (1989) *Construction Occupations. Review of the Economy and Employment*: Institute for Employment Research, Coventry. University of Warwick, 212–239.

Briscoe, G., Dainty, A.R.J. and Millett, S.J. (2000) The impact of the tax system on self-employment in the British construction industry. *International Journal of Manpower*, 21(8), 596–613.

Broughton, T. (2001) Wilson in Talks to Legitimize Illegal Immigrants, Building, 02/11/01, p. 11.

Contract Journal (2002) New safety alert over foreign labour influx. 29 November, *Contract Journal.*

CITB (2002) *Skill Foresight Report. CITB Research Department.* Bircham Newton, Norfolk, England.

Cremer, J. (2005) Free Movement Revisited. CLR News, no 2/2005. European Institute for Construction Labour Research, pp. 3–10.

Dobson, J., Koser, K., McLaughlin, G. and Salt, J. (2001) International migration and the United Kingdom: Recent patterns and trends Research, Development and Statistics Directorate (RDS) Occasional Paper No. 75.

Dobson, J. and Salt, J. (2002) Review of migration statistics, the political economy of migration in an integrating Europe working paper 7, Migration Research Unit, Department of Geography, University College London, London.

EFBWW (1991) Conditions of Employment in the European Construction Industry, Brussels, March.

Egan, J. (1998) *Rethinking Construction: The Report of the Construction Task Force.* Construction Industry Council, London.

European Commission (2004) Free Movement for Persons: A Practical Guide for an Enlarged EU, Brussels.

European Industrial Relations Observatory (1999) Analysis of the impact of the Posted Workers Directive on the UK: (http://www.eiro.eurofound.ie/1999/09/word/uk9907122s.doc).

European Integration Consortium (2001) The impact of eastern enlargement on employment and labour markets in the EU Member States, Employment rand Social Affairs Directorate of the European Commission, Berlin.

European Policies Research Centre (2003) *Impact of the Enlargement of the European Union on Small and Medium Sized Enterprises in the European Union*, University of Strathclyde, Glasgow.

Fairs, M., Madine, V. and Black, S. (2001) The invisible men, *Building Magazine*, 26/10/01, pp. 26–29.

Faist, T., Sieveking, K., Reim, U. and Sandbrink, S. (1999) *Ausland im Inland. Die Beschäftigung von Werkvertragsarbeitnehmern in der Bundesrepublik Deutschland*, Baden-Baden, Nomos.

Gann, D. and Senker, P. (1998) Construction skills training for the next millennium. *Construction Management and Economics*, 16(5), 569–580.

Geddes, A. and Balch, A. (2002) The political economy of migration in an integrating Europe: Patterns, trends, lacunae and their implications. The political economy of migration in an integrating Europe working paper 6/2002, University of Liverpool, Liverpool.

Harvey, M. (2002) Undermining Construction: The Corrosive Effects of False Self-employment. London: Institute of Employment Rights.

Hellman, M.F. (1992) Non-permanent employment: The IFBWW view. Proceeding of the Bartlett international Summer School, Brussels, 5–10 September, 110–122.

Hunger, U. (2000) Temporary trans-national labour migration in an integrating Europe and the challenge for the German Welfare state, in M. Bommes and Geddes, A. *Immigration and Welfare: Challenging the Border of the Welfare State*, Abingdon, Oxfordshire, Routledge.

IFF Research Ltd (2003) Workforce mobility and skills in the construction sector in London and the South-east. Research report for the CITB, ECITB and SEEDA.

ILO (2001) *The Construction Industry in the Twenty-first Century: Its Image, Employment Prospects and Skill Requirements.* International Labour Office, Geneva.

Krieger, H. (2004) *Migration Tends in an Enlarged Europe.* European Foundation for Improvement of Living Conditions, Dublin.

Kus, J. (2004a) Polish Ministry of Economy Labour and Social Policy (2003) The list of major foreign direct investors in Poland in 2002, http://baltic.mg.gov.pl/invest/directinv.htm.

Kus, J. (2004b) The Polish construction labour market in 2003 – fourth year of crisis, *Construction Labour Research News*, no. 1, Brussels.

Langford, D. (1975) Labour only subcontracting in the construction industry. Unpublished MSc thesis. University of Aston.

Langweische, R. (2000a) EU enlargement and the free movement of labour, in Gabaglio, E. and Hoffman, R. (eds), *European Trade Union Yearbook 1998*, ETUI, Brussels.

Langeweische, R. (2000b) Mobility, migration and freedom of movement in East-West Integration, *Construction Labour Resource News*, 2–3.

Lienhardt, J. (2003) *The Construction Industry in the EU*, European Business: Facts and Figures. Part One, Energy, Water and Construction, European Commission, Luxembourg.

National Audit Office (2001) Modernising Construction. A report by the Comptroller and Auditor General. HC 87. Session 2000–01. National Audit Office, London.

Taylor, D. (2004) On call for the eastern flood. *Construction News*, 29 April.

Tupy, M.L. (2003) EU Enlargement: Costs, Benefits, and Strategies for Central and Eastern European Countries. Policy Analysis, no. 489, 18 September, Brussels.

Winch, G. (1998) The growth of self-employment in British Construction. *Journal of Construction Management and Economics*, 16(5), 531–541.

Winiewski Z., and Oczki, J. (2001) Migration effects of Poland's EU membership, *Intereconomics*, 36(4), 174–179.

Woolfson, C. and Sommer, J. (2005) European Mobility in Construction: The Swedish Trade Unions and the Latvian Construction Workers' Dispute. CLR News, no. 2/2005. European Institute for Construction Labour Research.

13 Developments in construction supply chain management

The impact on people and cultural change

Simon A. Burtonshaw-Gunn and Bob Ritchie

Abstract

Supply chain management in the UK construction industry is a relatively recent phenomenon, presenting interesting challenges to the embedded cultural norms, inter-organisational relationships and individual behaviour patterns. The growing trend for contractors and clients to embark on strategic relationships aimed at improving efficiency, effectiveness and risk management are examined. Such strategic developments between the primary supply chain members can provide a number of benefits for both the client and the primary contractor organisations alike. They may also pose unique difficulties for the primary members as well as the sub-contracting organisations. A review of the key professional/practitioner and academic literature provides the basis for the development of a conceptual model of the partnering relationship. The central features of this model are examined through a case-based empirical approach involving two major contracting organisations and a blue-chip client. Conclusions are drawn about the implications for supply chain management within the construction industry at the strategic and operational levels. The impact on the cultural norms, behaviour and practices are examined and the actions taken to overcome such potential barriers to successful implementation are evaluated. The consequent strategic development from partnering to the lead supply chain management role associated with 'Prime Contracting' is examined in the light of the evidence, including the impact of this change for both the industry's and the client's personnel.

Keywords: construction, partnering, cultural change, prime contracting and supply chain management.

13.1 Introduction

The construction industry has, like other parts of the UK economy, witnessed major developments aimed at improving profitability and performance. These developments have occurred in particular phases, for example the emphasis on sales and marketing in the 1980s followed by human resource management and organisational design in the early 1990s. This latter phase resulted in downsizing, de-layering and re-engineering with evident consequences for employees within the industry. Since the mid-1990s, mainly due to the publication of the Latham Report (1994), the construction industry has experienced increasing interest in the subject of partnering. The Latham Report acted as a catalyst for greater attention to supply chain management and partnering between the client and the main contractor. Although this bipartisan strategic relationship is the principal focus of this chapter, it is recognised that such relationships will result in greater emphasis on ensuring effectiveness and efficiency across all dimensions of the total supply chain, both horizontally and vertically. Equally, there will be potential consequences for the relationships of other supply chain partners at the organisational and individual levels.

The chapter aims to develop an understanding of the factors influencing the emergence of strategic relationship building between partners, the key elements of such relationships and the implications for people and culture within the construction sector. The key objectives are designed to

- demonstrate the key drivers to strategic relationship development (e.g. partnering) within the construction sector
- identify the key components and associated processes through the development of a conceptual model
- evaluate the practical issues in developing and sustaining this strategic position
- determine the future developments in the construction sector.

From its inception, partnering was primarily designed to reduce or eliminate the established culture of adversarial relationships, replacing these with long-term relationships based on mutual trust and mutual benefit. The premise underlying this aim was that important and complementary opportunities exist between the two companies which could only be realised through developing a robust and permanent relationship. Although powerful barriers may obscure the opportunities and often preclude their realisation, Latham (1994) argued that the development of appropriate structures, processes and training ought to overcome these barriers, enabling the benefits of partnering to be achieved. Research has cited many examples from the USA that have resulted in cost savings from both one-off project partnering and long-term strategic partnering arrangements (e.g. Bennett and Jayes, 1995). In addition to reducing costs and conflict there is evidence that partnering can also 'improve service quality, deliver better designs, make construction safer, meet earlier completion deadlines and provide everyone involved with increased profits' (Construction Industry Board, 1997a: 5).

Previous research (e.g. Bennet and Jayes, 1995; Critchlow, 1998; CBI/MoD 1998; Galliford, 1998; Construction Industry Council (CIC), 2002; Armitage, 2002; Green *et al.*, 2004) suggests that the most important benefits identified by those engaged in partnering are not directly related to the 'hard' project cost factor; instead they cover a number of 'softer' factors including enhanced team-working, identifying mutual objectives, reduced risk and more efficient problem-solving, all of which may indirectly influence project cost. Davidson and Trinick (1997: 6–7) suggest that the benefits of partnering are

> improved productivity, efficiency, quality, safety, research solutions, innovative design work to speed up the design and construction process, reducing overhead and project costs, and improving profitability. Above all, the change to a non-adversarial culture will improve relations and subsequently reduce disputes, claims and litigation.

The literature also demonstrates that establishing a *partnering* relationship is not a simple operation. For example, the UK government–sponsored report (Egan, 1998: p. 24) stated that 'there is already some evidence that [partnering] is more demanding than conventional tendering, requiring recognition of interdependence between clients and constructors, open relationships, effective measurement of performance and an ongoing commitment to improvement'.

This chapter begins with a review of the operation of the construction industry from the perspective of a supply chain, identifying some of the key strategic changes and their implications. A conceptual model of the supply chain and its key components is derived from the literature and the contemporary analysis of the construction industry (Burtonshaw-Gunn, 2001). Key features of the model including human resource issues and their links to other components are highlighted. The particular component – partnering – is examined in some depth identifying the strategic consequences for partner organisations and evaluating the implications for cultural change and human resource management policies and practices. The evidence from an empirical study involving two major organisations is presented. Conclusions are drawn for the industry as a whole in terms of the impact of partnering and the implications for the organisational relationships and human resources involved.

13.2   The UK Construction Industry: Practices and attitudes

Construction involves assembling materials and components designed and produced by a multitude of suppliers working in a diversity of disciplines, technologies and services, in order to create the built environment. Associated tasks include planning, regulation, design, manufacture, construction, maintenance and eventual decommissioning of buildings and other structures. The scale, complexity and intricacy of individual projects vary enormously, ranging from work undertaken by 'jobbing builders', to multi-billion-pound schemes such as power station construction or such exceptional projects as the Channel Tunnel. The UK industry statistics supplied by the Department of Trade and Industry (DTI) indicate employment close to 1.6 million employees with over 122,000 Small to Medium Sized Enterprises. Its turnover in 2003 was *circa* £85 billion with the top

30 companies accounting for 17 per cent of the supplier output. Other differentiating features of construction are that the activity is temporary in nature, mainly carried out at the client's premises, and often requires special arrangements to enable the client organisation to conduct its business as normal. These characteristics generate such problems as low and discontinuous demand, poor productivity and low profitability. Although a large industry in terms of turnover and number of employees, its structure is seen as fragmented.

The construction industry has a long-standing reputation for its adversarial approach (e.g. Simon, 1944; Emmerson, 1962; and Banwell, 1964), and the amount of time and money spent on litigation. Since the early 1990s the industry has reduced and deskilled its workforce, and largely abandoned apprentice and other training schemes, while sustaining its conflict-ridden competitive tendering culture. The consequent adversarial working relationships persist throughout the whole supply chain, impacting on suppliers, contractors and clients, and often manifesting themselves at the level of the individual. The government-initiated industry review conducted by Sir Michael Latham (1994) recommended changes specifically intended to move the culture away from the traditional adversarial client/supplier relationship. Subsequent development of these recommendations by Egan (1998: 10) identified a number of areas that the industry needed to address, concluding that

> too many clients are undiscriminating and still equate price with cost, selecting designers and constructors almost exclusively on the basis of tendered price...[partnering] envisages a very different role for the construction supply chain...the supply chain is critical to driving innovation and to sustaining incremental and sustained improvement in performance.

The last ten years have seen a steady increase in interest and support for partnering and supply chain management within the construction industry (e.g. Latham, 1994; CIB, 1997b; Critchlow, 1998). However, the term 'partnering' can, and often does, mean different things to different people (Cousins, 1995). Davidson and Trinick (1997: 6) believe the term has perhaps been used 'too loosely and as a consequence it is in danger of becoming debased'. The most quoted and probably now universally accepted definition is that from the National Economic Development Council (NEDC, 1991), which states that 'partnering is a long term commitment between two or more organisations for the purpose of achieving specific business objectives by maximising the effectiveness of each participant's resources. The relationship is based upon trust, dedication to common goals and an understanding of each other's individual expectations and values.' Implicit in this recommended strategic change and the adoption of partnering is the requirement to modify the culture and behaviour patterns that have become entrenched in sustaining the adversarial regime in the past.

The principal barriers inhibiting the adoption of partnering fall into three groups: corporate culture, the traditional client/contractor roles and the time required to develop the necessary relationships. Accustomed to more traditional business relations and corporate culture, construction industry managers can find partnering relationships threatening to their company and to their personal *modus operandi*. Such managers may become uncomfortable with the idea of partnering because of their unwillingness to relinquish control and reluctance to share company information which they consider to be of a confidential nature. Secondly, the traditional industry roles and relationships between client and contractor have in the past led to a more adversarial climate generating significant perceptual hurdles to the successful implementation of partnering. The third barrier to partnering concerns the time and effort required to develop a partnering relationship. Finding the right partner(s) and developing an effective partnering agreement and relationship require a significant investment of senior management time and effort, as each organisation must assess how the other functions and the organisation itself will need to change under the partnering agreement. In recent years Green *et al.* (2004: 29) have argued that 'whilst trust is widely held to be central to effective supply chain management, it is rarely considered in any great detail in the literature'. The authors go on to cite Korczynski (2000) who notes that 'trust is a consequence of the basic premise that one party has confidence that another will not exploit its vulnerabilities...such factors include organisations' reputations, existing interpersonal relationships, the extent to which the organisations are perceived to be interdependent and the likely continuity of future workload'.

Both Latham (1994) and Egan (1998) sought to make operational the concept of partnering, with the former identifying three crucial activities:

- agreeing mutual objectives
- creating a process to deal swiftly with problem resolution
- gaining a commitment to continuous improvement.

It is widely accepted that a relationship lacking any one of these activities would not be considered 'partnering' in the spirit of the NEDC definition, though it might still possess some 'friendly' qualities. A looser relationship might improve collaboration through a less adversarial approach, but would lack the supporting operational framework required to set the boundaries of the relationship. A partnering framework provides visibility to the project participants, formalises the agreed expectations surrounding mutual objectives, establishes problem resolution mechanisms and ensures all parties have a commitment to continuous improvement. Indeed in recognising both the lessons from other industries and the unique nature of the construction industry, the Construction Industry Board (1997a: 5) advises that partnering 'empowers people and encourages them to work together but without rigorous management (measurement, benchmarking, goals) this can lead to "cosy" relationships (even fraud), reduced exposure to raw market forces and overlong carrying of non-performers'. The Australian Construction Industry Development Agency (1993) viewed partnering as a management strategy offering a new way of working for owners, consultants, contractors and subcontractors where all agree from the outset to a formal structure, focusing on creative co-operation and avoiding adversarial confrontation. The Agency (1993, p.11) added that partnering establishes a moral charter among the project team members which binds each party to act in the best interests of the project and the project team members. The main aim is to meet the project objectives by working together rather than by confrontation.

13.3 Culture, people and change

These developments in partnering throughout the construction sector internationally imply a radical change in the existing structures, processes, behaviours and relationships. People at every level in the sector will feel the impact of such developments even if those operating in managerial roles will be exposed more widely and more immediately than others. The demand is not simply for changes to the 'way things are done' but also for changes in attitudes, communications and respect between those involved in the client and contractor organisations: in short, a change to the 'culture'. Embedded in this notion of a change in culture is the need for a greater degree of trust.

Seeking a universal definition for the term 'culture' is fraught with difficulties (Baskerville, 2003) since academic disciplines differ in their perspectives. Several authors (e.g. Guest, 1997; Purcell, 1999; and Gerhart *et al.*, 2000) have posited that the development of a positive and supportive culture will yield benefits in terms of interpersonal working relationships and consequently improvements in individual and organisational performance. In addition to the focus on human capital as the basis for defining organisational culture, O'Keeffe (2002) argues the importance of core competences as a major influencer. It would be possible to envisage the matching of core competences between client and contractor as providing the basis for a mutually supportive culture. Hofstede's (1998) seminal work sought to identify and quantify culture and its link with different dimensions of national culture as a means of explaining differences in development and practices in the Accounting discipline. Other writers (e.g. Baskerville, 2003) have suggested that the concept is far more ephemeral, varying between organisations in the sector and within the same nation. Moreover, the 'general lack of confidence in the assumption of stability of cultural differences...and [c]ultural diffusion and the dynamism of both national and ethnic shifts may be problematic' (ibid.: 1).

If the term 'culture' itself is insufficient to explain fully the change in behaviours and attitudes required in partnering, then the term 'trust' may provide a more substantive basis on which to build such changes. Ideally, what is sought within partnering in the construction sector is the creation of trust in terms of developing long-term strategic relationships between suppliers and clients in terms of the supply chain. Sako (1992) provides an interesting categorisation of trust, appropriate to the construction sector:

- Contractual trust – the adherence to terms of the contract or agreement, incorporating both explicit and implicit terms.
- Goodwill trust – the expectation that the other party will perform tasks beyond these necessarily specified in the contract terms.

- Competence trust – the belief in the ability of the partner to be able to deliver what is contracted.

The presence of embedded adversarial relationships in the construction sector, fuelled by low profit margins, with dispute resolution conducted through due legal process, provides little resonance with Sako's three dimensions of trust. The development of the partnering approach has encouraged the adoption of cultural change and ushered in more tangible changes that reflect more these dimensions of trust. For example, the launch of new types of contract (e.g. 'PPC 2000') was specifically aimed at fostering greater trust in partnering contracts between contractors and clients. Critchlow (1998: 53) addresses the issue of *contractual trust* indicating that

> [c]ontracts serve two main purposes: they identify the project in terms of scope, quality, cost and timing: and they apportion risk. Partnering projects on the other hand are largely concerned with the manner in which the projects are executed i.e. improving efficiency by promoting a culture which accentuates the identification of mutual interests and minimises conflict.

Adoption of the practice of agreeing mutual objectives, negotiating a joint problem resolution system at the commencement of the project and the production of a 'Partnering Charter', supports the second dimension of *goodwill trust*. The third dimension, *competence trust*, can be promoted by more regular and open communications between clients, contractors and their supply chain members. These partners will have been selected against various criteria in terms of their competence to perform the necessary tasks and meet the project requirements. The key feature of trust in this respect is allowing each party to execute their part of the project using their skills and experience without the need for continual approval or justification of actions. The results from a recent study of construction partnering (Armitage, 2002: 14), which surveyed 19 construction companies each with an annual turnover of between £100 million and £1 billion, indicated a uniformly expressed view that the 'absolute requirements for success in partnering are trust and commitment, based on relationships rather than a rule book'.

Green *et al.* (2004: 29) undertaking a comparative study of Supply Chain Management in the construction and aerospace sectors, concluded that

> No amount of best practice initiatives advocating the need for greater trust can escape the consequences of industry structure. However, there are emerging niche markets within the construction sector that approximate towards the characteristics of a high-trust economy. Within these contexts supply chains are highly interdependent. Hence collaborative working becomes a commercial necessity.

13.4   Conceptual model development

Two conceptual models were evolved as part of the research process; in essence, however, they are interrelated and simply provide differing perspectives of partnering. The first model represents the portfolio of components necessary to ensure effective adoption of the partnering process within the construction industry. This initial model was derived from the literature and key studies in the industry, for example 'Constructing the Team' (Latham, 1994), and further refined through discussions with academic and industry representatives. The second model was designed to focus on the phase after the initial selection of the partners, addressing the management processes that the contractor and client need to successfully engage in if the partnership approach is to achieve ongoing success.

The literature suggested a number of key components for the first model (Figure 13.1), the structure of which may be viewed as analogous to the construction of a building. Those components represented below the ground level are the foundation components necessary to support the partnering relationship for the duration of the project and potentially beyond, especially if considering a more strategic partnering arrangement. It is suggested that the absence of sufficiently

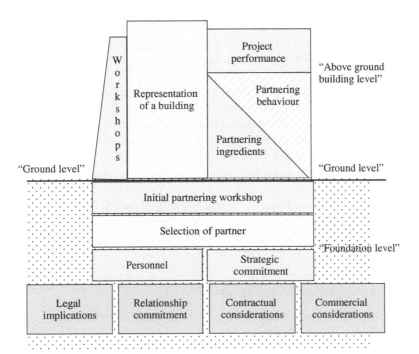

Figure 13.1   Model of construction partnering components.

robust foundations would constrain the success of the partnering relationship and the achievement of the potential benefits from the outset.

The foundation components possess a common theme: the role of the industry personnel, from both the customer and the construction contractor organisations. The main groups of components include the following:

- Legal implications – for example, changes to the contractual relationship consequent on partnering.
- Relationship commitment – to achieve full potential (Bennett and Jayes, 1995) is more than simply making a longer-term commitment to work together. Realising the full potential in terms of risk reduction and joint profit margins requires a more immediate and also a lasting commitment by both organisations to developing a robust relationship.
- Contractual – recognition and resolution of disputes by a mutually agreed process, which only as a last resort refers to the contract, is viewed as one of the cornerstones of the partnering philosophy (Latham, 1994; Bennett and Jayes, 1995).
- Commercial – performance incentives incorporating rewards and penalties (e.g. maximum value for money, improved quality and safety, timely completion, reduction in fixed overheads and shortened learning curves) are 'an essential aspect of partnering...the opportunity for participants to share in the rewards of improved performance' (Egan, 1998: 45).
- Personnel – one of the most significant consequences of partnering is the impact on people and the change in behaviour that such a closer relationship necessitates. The personnel must possess the appropriate technical skills, experience, attitudes and a genuine commitment towards the partnering approach and philosophy.
- Strategic commitment – to develop a strategic relationship, fully endorsed by the senior management of both organisations.
- Selection of partner – most partnerships originate through the competitive tendering process; however, Cousins (1992) emphasised that the partnership philosophy should not look merely at price and delivery, but concentrate on other attributes such as culture, level of technology, production flexibility, commercial awareness, innovation and ease of communication.
- Initial partnering workshop – provides the opportunity to understand and agree how the relationship will work, how the parties will interact with one another and what each expects of the other, at the strategic and operational levels.

13.5   'Above ground' components

The 'above ground' components (Figure 13.1) comprise the following:

- Partnering ingredients – incorporating three crucial ingredients: establishment of mutual objectives; development of a process for problem resolution and an all-party commitment to continuous improvement (Latham, 1994; Bennett and Jayes, 1995; Construction Industry Board, 1997a).

- Partnering behaviour – gaining agreement on the behavioural attributes (e.g. commitment, trust and mutual advantage) to support and engender success between the organisations and the individual team members. Critchlow (1998) emphasised the need for a commitment to a long-term relationship through the use of these attributes to achieve an amicable 'partnering relationship'.
- Project performance – reflects the composite improvements in performance from all of the components (e.g. lower costs, improved quality, fewer overruns, fewer disputes and improved working experience for employees).

Prior to the commencement of the project, the client organisation is primarily in control of the foundation components; but the relationship should evolve to become bilateral, or multilateral, such that client and contractors have more equal influence on the overall project development and performance.

This initial representation of partnering (Figure 13.1) was designed to highlight the key components necessary to establish a successful partnering relationship. The focus of the second iteration of the model (Figure 13.2) relates to the phase following the award of a contract where the partnering philosophy is being utilised – 'partnering in action' – providing a more interactive representation of

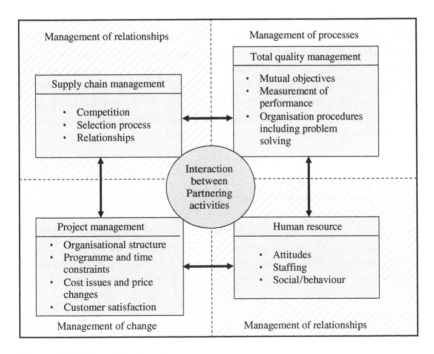

Figure 13.2 Partnering in action.

the issues likely to be encountered in the implementation and continuing management of the process. Many of the key contributors in this field (e.g. Construction Industry Development Agency (CIDA), 1993; Bennett and Jayes, 1995, 1998; CIB, 1997a) concentrate on issues surrounding the establishment of the partnership and fail to fully address the issues involved in implementing and sustaining the partnering relationship.

Each of the four primary components – Project Management, Total Quality Management, Supply Chain Management and Human Resource Issues – is represented by a quadrant together with the associated main sub-components. The discussion of the sub-components is focused towards the implications for closer collaborative working practices and relationships.

Organisational structures and processes are important issues to be settled at the outset to clearly define people's roles and responsibilities and to establish agreed communication channels. The project manager, as part of the contractor's team, remains primarily responsible for delivering the project within the requirements of specification, time and budget, often regarded as the cornerstones of project management, together with achievement of client satisfaction. Establishing effective structures, processes, communication channels and relationships should discourage the adversarial approach, providing benefits in terms of efficiency (e.g. costs, work flow and flexibility) and effectiveness (e.g. meeting the client's needs and funding additional project requirements).

Human resource management

The emphasis on relationships within the partnership suggests the need for effective Human Resource strategies and policies, designed to inculcate the appropriate attitudes and behavioural traits among those employees directly involved. Selection and training of the key senior staff concerned may be critical to the partnering relationship's eventual success. These requirements apply to both contractor and client organisations and there may be merit in hosting joint staff development programmes (e.g. team building).

Total quality management

Partnering and Total Quality Management may be viewed as highly complementary philosophies. Bennett and Jayes (1995: 14) suggest that 'partnering is now being increasingly seen as essential to the implementation of Total Quality Management (TQM)', including the establishment of mutual objectives, the shared commitment towards continuous improvement and the need for mutually agreed measures of performance.

Supply chain management

The characteristics of the evolution towards strategic partnering relationships in the construction industry, was noted previously in Section 13.2 (e.g. the move away from selection on lowest tender price to consideration of life cycle costs and value for money). Important features of Supply Chain Management include the initial selection of partners in the chain, building and developing effective relationships, and agreeing arrangements for the nature of competition both within the partnership and with other suppliers/customers (Ritchie and Brindley, 2004). In the case of the construction industry, it is important that relationships are forged not only between the customer and the first-tier main contractor but also between the second-tier subcontractors and the prime contractor. Evidence from the manufacturing sector suggests that maximum benefit is gained by early and full involvement of all the key partners in the supply chain. Experience to date suggests that this has yet to occur in the construction industry, although there is some evidence that large first-tier prime contractors have already instituted limited framework and supplier management processes.

The other important, if implicit, feature of the Partnership Model is the dynamic interaction between the sub-components within each of these four main groups and between the sub-components in the other groups. For example, staff selection and inculcating appropriate attitudes (i.e. Human Resource Issue sub-components) will influence simultaneously relationship development (i.e. a Supply Chain Management sub-component) and the effectiveness of continuous improvement (i.e. a Total Quality Management sub-component). However, tensions between the various sub-components are likely to arise at different times during the partnership as they are not always mutually compatible. For example, project management imperatives, involving the expedition of particular tasks, may conflict with Total Quality Management objectives, practices and procedures.

The final significant feature of this conceptual partnering model, in this context, is the integration and aggregation of all of the sub-components to achieve effectiveness in relation to the three strategic themes: Change Management, Process Management and Relationship Management. The proposition from the case study research conducted is that Change Management and Process Management are largely, though not exclusively, internally oriented and managed by each organisation in the partnering arrangement. Engaging these two themes with other organisations necessitates effective Relationship Management, both internally and externally. This means, internally, undertaking relationship management between employees in all functional areas and, externally, between these employees and counterparts in other contractors or the client organisation. Effective management across these three themes should lead to an effective, robust and enduring partnership. The contingency dimensions of this conceptual model need to be appreciated and understood, especially those relating to the risks and the impact of relationship development in ameliorating such risks (Ritchie and Brindley, 2004). In essence, the composition and importance of the components and sub-components will vary between different partnering situations, depending, for example, on the length of prior engagement between the partners, the number of partners involved, the nature and scale of the project, the experience of the staff in the organisations and so on.

13.6   Research evidence

The purpose of the research study was twofold: to evaluate the dichotomy, evident in the literature, between the reported benefits and the challenges of partnering, and to evaluate the appropriateness of the conceptual models presented earlier.

A preliminary stage of the empirical research involved an industry-wide survey combining secondary research sources and primary research. The secondary research chiefly involved accessing and reviewing the professional literature, publications and presentations at conferences and seminars. The second stage of the research study involved direct engagement with key representatives in professional agencies, contracting organisations and clients which provided the basis for the empirical work and the development of a case study. The study focused on the non-residential and infrastructure building sector and particularly on projects where the main contractor engaged more directly with the client. It was considered that the scope of this type of project in terms of complexity, physical size, financial scale and construction timescales was most appropriate to the application of the partnering philosophy articulated in the literature. Apart from providing the opportunity to assess most of the possible dimensions of the relationship between the client and the end user, this sector case provided greater potential for the development of a strategic relationship between the two groups than did construction sectors such as housing, where there is much less of an industry/purchaser interface. In many respects the case study (Burtonshaw-Gunn, 2001) itself was exploratory, seeking not only to understand how partnering was operating in this case but equally why the partners had chosen to evolve the partnering relationship in a particular manner, that is, reflecting the contingency nature of the conceptual models developed.

Initial data collection in the form of two questionnaires was undertaken alongside the selection of a main contractor for a major infrastructure project. This approach provided the opportunity to use the tender selection process to draw out knowledge from the many potential major contractors, through a shortlisting interview process and by collaboration with the selected contractor and client organisations. As a result of this process, another contractor, although unsuccessful in being awarded the immediate infrastructure project contract, expressed their desire to be involved in the study, considering it to be of mutual benefit.

Two in-depth case studies were selected for investigation of the models, focusing primarily on the contractor and client perspectives. The two major contractor organisations in the studies differed in their culture, level of partnering experience and overall work philosophy. One of these was and remains a family-run business in the UK and the other has an international involvement in addition to its UK construction operations. The organisations involved all had prior experience in partnering and the evidence drawn related to the prior and current experiences of the managers interviewed. Different collaborative relationships were identified and discussions with senior managers in the organisations revealed a number of factors seen as important to the success of their collaborative relationships.

13.7   Implications for culture and people

The findings from the preliminary industry-wide research indicated a general awareness of partnering and supply chain management concepts within the construction industry as a whole, though evidence of full-scale adoption was limited. Although there were many examples of contractors and clients embarking on the two types of partnering relationship – project and strategic partnering – their commitment to the three key themes of mutual objectives, problem resolution and continuous improvement (Figure 13.1 – Partnering Ingredients, and Figure 13.2 – The TQM quadrant) was mainly limited to establishing the first of these: the project objectives.

The empirical research evidence based on the case studies may be summarised as follows:

- The individual components in the model (Figure 13.1) and their interrelationship have an important contribution to make to the successful outcome of a partnering relationship and project delivery. The foundation components are primarily items for initial consideration by the client organisation, though they may consider modifying these for the benefit of the project. These foundation levels have a particular impact in terms of team selection, attitudes, behaviour and strategic level support.
- Selection of the partnering organisation is an important task which should be undertaken in the light of the partnering philosophy. Interestingly, the case study organisations expressed some caution concerning the amount of partnering work in total that contractors will consider undertaking in terms of their total portfolio at any one time. This is not to suggest a rejection of this approach but to recognise that effective partnering demands a significant and intensive amount of senior management time; this potentially scarce resource might not be available for every contract. The availability of senior management time and support may be a factor for contractors and client organisations equally. In addition to the demands on management time, partner selection may also encounter constraints in terms of suitable candidates to engage in the process. On the one hand, the client may wish to select a contractor with prior partnering experience, quality standards, a good safety record, a track record of project performance, good quality resources and a sound financial standing. On the other hand, contractors with experience of partnering will themselves be looking to enter a relationship with an 'educated' client who will have some experience of the partnering philosophy and expected actions. The 'learning curve' costs associated with one or both novice partners being new to partnering may prove a significant deterrent. Clearly then, as clients try to select the 'best in class' contractors with which to partner, contractors too will look at their experience of undertaking projects on a partnering basis. However, the literature (e.g. NEDC, 1991; Latham, 1994 and CIB, 1997a) advises that some caution is required since there may be a difficulty in determining the genuine level of commitment

to partnering. Contractors may promote their experience of the partnering concept purely as a 'marketing tool' to gain new work.

- Professional facilitation of workshops, ideally at project commencement, is a view fully endorsed by both the contractors and the client organisations, seeking to develop mutual understanding of each member's roles and responsibilities. In addition, an ongoing programme of workshops can play an important part in keeping all parties committed to the partnering relationship through active communication, resolution of problems and introducing any new staff (client and contractor alike) to the agreed project objectives.

- All interviewees recognised that the management style displayed by both parties must be non-adversarial in nature and create an open forum for communication. This degree of change to the previous management style may require changes in behaviour among both senior and middle management, an issue the case study companies were actively addressing through appropriate training and development.

- The industry now sees the Supply Chain Management component as offering a positive contribution to the construction process. As discussed earlier, the construction industry has been relatively slow to engage with the developments in Supply Chain Management and the associated changes in client/contractor relationships. Arguably, partnering is a key component of developing the supply chain and the need to engage more participants in collaborative, or at least less adversarial, ways of working must be recognised.

- The Project Management component generated the most agreement among those interviewed. Representatives of both contractor and client organisations viewed the role and individual attitudes of the Project Manager as the key factor in achieving partnering success and as responsible for the effective management of the relationship between the contractor and the client organisations.

- While both clients and contractors recognise the need to change the nature of their relationships, the research undertaken suggests that there is less clarity on how best to achieve this.

13.8   Limitations of the study

Although a considerable period of time has passed since the publication of the Latham Report (1994) and the subsequent recommendations by Egan (1998), there is still a paucity of organisations in the construction industry actively engaging in the full partnering process. The preliminary survey, although not systematic or representative, indicated a high degree of awareness and general acceptance of the partnering model within the management of major contractors and clients. The survey did not address directly the reasons for the non-adoption of partnering, though some of these may be inferred by the responses given. The in-depth investigation by way of the two case studies provided invaluable insights into the process, issues and tensions within a given context. In both cases the partners were highly committed to the achievement of a successful outcome.

It would be unwise to generalise too much from the experiences in these two cases as the nature of the partnering relationship may vary based on the size of the organisations involved, their level in the contracting hierarchy on a particular project, the extent to which this collaboration was a contractual requirement, together with other contingency variables.

13.9 Conclusions

Evidence from the literature and this study suggests that many clients and contractors believe that there are benefits which partnering can offer. Primarily, the more closely engaged and synchronised the approach to supply chain operations facilitated by partnering, the greater the potential benefits offered in terms of cost savings, reduced timescales, improved productivity and better working relationships. These improvements are achieved through the elimination of the previous adversarial behaviour typical of the construction industry.

The research confirmed that clients and contractors recognise the need to change the nature of their relationships. However, there is much less clarity on how best to achieve this, suggesting a need for more detailed and practical guidance following the award of the contract. This guidance could provide practical assistance to those organisations engaging in a partnering relationship for the first time. An example would be the recent acceptance of some newer standard forms of contract specifically tailored for the partnering approach. Equally, partners with prior experience may benefit from a different level of support and guidance as they will often have new staff engaging in the process for the first time. This partnership support reinforces the contingency approach articulated in the conceptual model.

Continuity following the initial phase of contract start-up is critical. Emphasis should be placed on the importance of continuing partnering workshops, as the evidence suggests that ongoing interchange is necessary to sustain and develop the relationship into a strategic partnership stretching beyond an initial immediate project.

Regarding future trends in construction supply chain management, the evidence suggests that the evolution from partnering to Prime Contracting may be a natural development for those organisations with the necessary understanding and partnering experience. Other organisations might find such a move fraught with problems and risks, presenting them with a notable degree of challenge. There is support for this view of future trends, as evidenced by the commitment of the UK government's major capital spending departments (e.g., the Ministry of Defence, Defence Estates Organization and the Department of Transport), which have expressed a preference for infrastructure projects to be undertaken on a Prime Contractor approach and with a contractual undertaking typically of five years' duration. The recommendation by the UK Treasury and the National Audit Office for the procurement of central government construction projects through the Private Finance Initiative (PFI) and the Public Private Partnership (PPP) promotes the development of a Prime Contractor approach. None the less, it has been suggested (Burtonshaw-Gunn, 2004) that both of these procurement methods introduce new considerations of project risk especially when applied to international projects.

The evidence available from the policy imperatives directing the construction sector together with the empirical evidence provided through the two case studies demonstrates that the sector is undergoing fundamental changes. While such changes may be portrayed as structural changes and developments to contractual relationships it is evident that the required changes must run much deeper. The term 'cultural change' is insufficient to capture the nature of the changes needed. People engaged in the industry and especially the management of both clients and contractors, including other subcontractors in the supply chain, need to adopt new patterns of behaviour, engage in more responsive modes of interaction and recognise the full implications of the partnering process. Underlying these requirements is the importance of developing trust: a major challenge for a sector that has long evinced mistrust in all its activities and relationships. The development of trust is a function of commitment, long-term engagement, willingness to share information, and recognition that partners will work towards their mutual benefit.

References

Armitage, A., 2002, *Construction Partnering in Practice: A Survey of Attitudes and Experience Among Senior Executives in Construction Contractors.* Construction Forecasting and Research Limited, London.

Australian Construction Industry Development Agency, 1993, *Partnering – A Strategy for Excellence A Guide for the Building and Construction Industry*, Commonwealth of Australia and Master Builders Inc., Australia.

Banwell, H.,1964, *Placing and Management of Building and Civil Engineering Work*, HMSO, London.

Baskerville, R.F., 2003, Hofstede never studied culture, *Accounting, Organisation and Society*, Vol. 28, No. 1, pp. 1–14.

Bennett, J. and Jayes, S., 1995, *Trusting the Team – The Best Practice Guide to Partnering in Construction*, Thomas Telford Publishing, UK.

Bennett, J. and Jayes, S., 1998, *The Seven Pillars of Partnering: A Guide to Second Generation Partnering*, Thomas Telford Publishing, UK.

Burtonshaw-Gunn, S.A., 2001, *Strategic Supply Chain Management: Critical Success Factors to Partnering Relationships in the UK Construction Industry.* PhD Thesis, Manchester Metropolitan University.

Burtonshaw-Gunn, S.A., 2004, *Management of Risk in Private Funded International Infrastructure Projects.* Proceedings of 2nd International Conference on Business & Technology Transfer (ICBTT 2004). University of Central Lancashire, UK.

Confederation of British Industry, 1998, *Partnering Arrangements Between the Ministry of Defence and Its Suppliers*, Crown Copyright 1998, www.mod.uk/commercial/pe/dgcom/partnering/index.html.

Construction Industry Board, 1997a, *Partnering in the Team*, Construction Industry Board Working Group 12, Thomas Telford Publishing, UK.

Construction Industry Board, 1997b, *Selecting Consultants for the Team*, Construction Industry Board Working Group 4, Thomas Telford Publishing, UK.

Construction Industry Council, 2002, *A Guide to Project Team Partnering*, 2nd ed, CIC, London.

Construction Industry Development Agency, 1993, *Partnering – A Strategy for Excellence: A Guide for the Building and Construction Industry*, Commonwealth of Australia and Master Builders Inc., Paragon Printers, Australia.

Cousins, P., 1992, Choosing the Right Partner, *Purchasing and Supply Management*, March 1992, pp. 21–23.

Cousins, P., 1995, *Strategic Procurement Management in the 1990's: Concepts and Cases*, Chartered Institute of Purchasing and Supply, Lamming, R. and Cox, A. (eds), UK.

Critchlow, J., 1998, *Making Partnering Work in the Construction Industry*, Chandos Publishing Limited, Oxford.

Davidson, P. and Trinick, A., 1997, Partnering and collaboration, *Facilities Management*, Vol. 4, No. 5, pp. 6–8.

Department of Trade and Industry 2003 *Construction Industry Statistics*. DTI, London.

Egan, Sir John,1998, *Rethinking Construction*, The Report of the Construction Task Force to the Deputy Prime Minister, on the scope for improving the quality and efficiency of UK construction, Department of the Environment, Transport and Regions, UK, July 1998.

Emmerson, H., 1962, *Survey of Problems Before the Construction Industries*, HMSO, London.

Galliford 1998, *Construction Partnering Survey Report*, Galliford (UK) Limited.

Gerhart, B., Wright, P.M. and McMahan, G., 2000, Measurement error in research on the human resources and firm performance relationship: further evidence and analysis, *Personnel Psychology*, Vol. 53, No. 4, pp. 855–872.

Green, S.D., Newcombe, R., Fernie, S. and Weller, S., 2004, *Learning Across Business Sectors: Knowledge Sharing Between Aerospace and Construction*. University of Reading, EPSRC Report.

Guest, D.E., 1997, Human resource management and performance: A review and research agenda, *International Journal of Human Resource Management*, Vol. 8, No. 3, pp. 263–276.

Hofstede, G., 1998, Attitudes, values and organizational cultures: disentangling the concepts, *Organization Studies*, Vol. 19, pp. 477–492.

Korczynski, M., 2000, The political economy of trust, *Journal of Management Studies* Vol. 37, pp. 1–21.

Latham, Sir Michael, 1994, *Construction the Team: Joint Review of Procurement and Contractual Arrangements in the United Kingdom Construction Industry*, Final Report, HMSO, London.

National Economic Development Council, 1991, *Partnering – Contracting Without Conflict* NEDC, London.

O'Keeffe, T., 2002, Organisational Learning: A New Perspective, *Journal of European Industrial Training*, Vol. 26, No. 2–4, pp. 465–486.

Purcell, J., 1999, Best practice and best fit: chimera or cul-de-sac?, *Human Resource Management Journal*, Vol. 9, No. 3, pp. 26–41.

Ritchie, B. and Brindley, C., 2004, Risk Characteristics of the Supply Chain – A Contingency Framework, in *Supply Chain Risk* (ed. C. Brindley), Ashgate, UK, pp. 28–42.

Sako, M., 1992, *Prices, Quality and Trust: Interfirm Relations in Britain and Japan*, Cambridge, Cambridge University Press.

Simon, Sir Ernest, 1944, *The Placing and Management of Building Contracts*, HMSO, London.

14 The impacts of workforce integration on productivity

Paul Chan and Ammar Kaka

Abstract

The construction industry remains one of the most labour-intensive, project-based industries in the UK that contributes significantly to the economy. This contribution justifies the need for on-site labour productivity improvements. However, productivity improvements have conventionally focused on improving the work content and environment from a managerial perspective, with comparatively little regard for workforce issues across different levels of the construction workforce. It is therefore essential to re-examine the factors affecting construction labour productivity, especially those that are pertinent to on-site labour. In the context of the growing importance of employee involvement, this chapter presents observational research findings from two case studies that illustrate practical examples of employee involvement initiatives supporting the notion that embracing differences of perspectives between white-collar managers and blue-collar operatives could lead to the attainment of high productivity levels on-site. The study also reaffirms the significance of benign paternalism and qualifies the need to supplement such a paternalistic approach with the management of controllable aspects of productivity. Furthermore, the discovery of the existence of social cliques offers a potential insight into how a transient workforce could be managed at the project level.

Keywords: blue-collar, construction labour productivity, employee involvement, participant observation and white-collar.

14.1 Introduction

The concept of employee involvement is not new and has been around in the UK in various guises over the last century. According to Marchington and Wilkinson (2002), the concept evolved from the days of collective bargaining at the end of the First World War to the rising interest in industrial democracy in the 1970s to management-driven employee involvement initiatives that stressed direct communication with individual employees. This development followed the declining role of trade unions in organising labour (see Clarke and Wall, 1996 and *Labour market trends*, 2001, both estimating around 30 per cent union density). At the same time, the emphasis and purpose of employee involvement also shifted from the principles of collective bargaining in determining wages and working conditions to the statutory emphasis of worker rights in the 1970s (Ackers *et al.*, 1992) to the contemporary focus on economic performance. For Marchington and Wilkinson (2002), this movement implied 'a neo-unitarist win-win approach, which is moralist in tone' (p. 438) and which illuminates Claydon and Doyle's (1996) reconciliation of high performance and employee autonomy.

Today, there is renewed interest in employee involvement with the Labour government's promotion of *partnership*. Furthermore, the introduction of the European Communities Information and Consultation Directive in April 2005 mandates firms of a certain size to work together with employees to establish appropriate information and consultation arrangements (Blyton and Turnbull, 2004). Yet, Sisson and Storey (2000) painted a rather grim picture of the state of employee involvement in the UK, as they cited statistics (European Foundation for the Improvement of Living and Working Conditions, 1997) to show that the UK, and particularly UK construction, was below the ten-country average in consultative arrangements for employee involvement.

Similarly, the desire to improve construction labour productivity has been reinforced over the last two decades. Despite much research on identifying the factors that influence productivity, the problem of low productivity levels persists in UK construction (see Latham, 1994 and Egan, 1998). A review of previous productivity research (see Section 14.2) found a significant lack of studies that investigated how employee involvement could enable productivity gains. Instead, past research relied on strong quantitative traditions mainly from a managerial perspective, which were inadequate to tackle a complex phenomenon like productivity. This chapter therefore focuses on construction labour productivity. It attempts to re-examine the factors that affect productivity at the project level from a practical perspective through participant observation on two project sites. In so doing, the study illustrated here not only goes beyond simply looking at the managerial perspective, but also emphasises the views of blue-collar operatives as well.

The reasons for concentrating on construction labour productivity at the project level are twofold. First, construction is deemed to be highly dependent on people's efforts (Movement for Innovation, 2000); second, the measure of construction labour productivity is a crucial performance measure of construction's core activity, that is the on-site production of the built facility. This view is borrowed from Groák's (1994) critique of the existence of a construction industry, in which he proposed that the construction project symbolises the kinds of uncertainty that makes the characteristics of construction unique. Groák (1994) argued that 'the notion of the dominance of the project changes ideas [...] on what we focus for productivity improvements' (p. 290).

This chapter highlights one such focus by reporting on findings which suggest that the integration of any differences in the perspectives of white-collar managers and blue-collar operatives could lead to productivity improvements. The first section briefly reviews past productivity research and demonstrates the need to investigate the views of blue-collar operatives. Next, the methodology employed in this study, that is participant observation, is explained, before a discussion of the key findings.

14.2   Review of construction labour productivity research

Past research into the factors affecting construction labour productivity could be divided into two main camps: *work content* factors involving issues pertaining to the management of work methods and *work environment* factors primarily involving studies dealing with the motivation of workers.

Work content factors

The use of delay surveys (Tucker *et al.*, 1982) to establish causal factors of labour productivity has been common hitherto. Borcherding and Garner (1981), for instance, reported results of a longitudinal study employing the craftsmen questionnaire survey technique on over a thousand carpenters, electricians and pipefitters. Olomolaiye *et al.* (1988), Zakeri *et al.* (1996) and Kaming *et al.* (1997) also employed such surveys to investigate the work content factors influencing construction labour productivity in Nigeria, Iran and Indonesia respectively. Throughout these studies, it is notable that material and tool availability, rework due to design changes, weather or poor workmanship, crew interference due to scheduling problems, craftsmen turnover and absenteeism were all recurrent problems that curtailed productivity.

Apart from the problems highlighted above, building design also emerged as a critical factor. Gray and Flanagan (1984) inferred that designers who took account of the use of construction technology tended to improve on-site productivity. Hinze and Parker (1988) also discovered the influence of design and technology on the productivity of curbing operations. Furthermore, Herbsman and Ellis (1990) explicitly identified two strains of what they called 'construction productivity influence factors' and broadly classified them into technical and administrative, the former defined as design related and deterministic and the latter as management related and stochastic.

The role of management is vital in influencing the work content element of construction labour productivity. Thomas *et al.* (1983) discussed how different organisational structures for construction project management could, in theory, affect the decision-making process, which would impact on productivity levels. In the same spirit, Tavakoli (1985) studied road construction projects and developed a quantitative analytical system to enable the timeliness and accuracy of the decision-making process. Similarly, Herbsman and Ellis (1990) noted the importance of management in their two-tier classification of construction productivity influence factors, mentioned above.

However, there is often a desire to seek a causal link between factors through quantifying them and measuring their impacts on productivity (Lemna *et al.*, 1986; Herbsman and Ellis, 1990; Thomas *et al.*, 1990). This is arguably one of the pitfalls of the dominant quantitative paradigm in previous studies. The quantitative paradigm results in a reductionist approach in the identification and isolation of factors, usually based on the researcher's interpretation (Calvert *et al.*, 1995; Seymour and Rooke, 1995). This interpretation often relates, in turn, to a managerial perspective (Macarov, 1982). Another consequence of the reductionist approach is demonstrated by the narrow focus of past productivity studies on particular construction operations (such as concrete works), thus failing to take into account the entire construction process (Chan and Kaka, 2004). In this respect, the conclusions of past studies are limited in their scope to offer plausible recommendations for implementation for the improvement of construction labour productivity.

Work environment factors

It is widely accepted that motivation plays a part in enhancing construction labour productivity (Smithers and Walker, 2000). For instance, through a review of studies on motivation and productivity in the 1970s (see Borcherding and Oglesby, 1975; Borcherding *et al.*, 1979), Parker (1980) identified communication as the underpinning cause of motivation that led to productivity improvements. Laufer and Moore (1983) advocated the use of financial incentive programmes to improve construction labour productivity. Autonomy and comradeship (Edwards and Eckblad, 1984) were also found to be important contributors to construction workers' self-motivation in their work.

However, much research depends on Herzberg's two-factor theory of motivation (Smithers and Walker, 2000) and this seems inadequate for understanding the blue-collar operatives' perspective. Mullins (2005), for example, criticised the use of Herzberg's sample and suggested that the results might not apply to manual labourers. Furthermore, Hofstede (1980) decried such motivational theories as merely offering the description of a specific value system, that of the American middle class. Hofstede's objections reinforce the misgivings expressed at the start of this chapter about the *ad nauseam* emphasis on the managerial perspective in the quest to improve productivity.

Apart from Herzberg's theory of motivation, past productivity research also employed the expectancy motivational model (see Laufer and Jenkins, 1982) to study the motivation of construction workers. Unlike the Herzberg model, where the link to productivity is only inferred, the expectancy model explicitly links productivity to motivation quantitatively. Motivation, according to this theory, is 'a multiplicative function of the expectancies that individuals have concerning future outcomes and the value (as perceived by the workers) placed on those outcomes' (Laufer and Jenkins, 1982: 535). Maloney and McFillen (1985), for example, used the model to study the motivational impact of work crews on labour productivity.

The expectancy model was also regarded highly by Thomas and Yiakoumis (1987) who tried to incorporate it within their factor model of labour productivity. However, it appears that the desire to combine expectancy and factor models came to nothing, perhaps because the latter requires factors to be quantified

(Thomas *et al.*, 1990). Apart from invoking the points made earlier in the discussion of the quantitative nature of the work surrounding work-content factors, such incorporation does not take heed of warnings by Laufer and Jenkins (1982). In their conclusions, they argued that the use of quantification in the expectancy model should be mainly for illustrative purposes as they explained that the complexities of human behaviour transcend that which a model can predict.

A further limitation is the reliability of statistics. Radosavljevic and Horner (2002) revisited formwork and masonry data sets compiled by Professors Thomas and Horner across 11 sites in the USA and the UK, only to confirm their suspicion that productivity is not normally distributed. This consistent finding bears significant implications in that 'some basic statistical diagnostics [...] may give misleading results' (p. 3). The authors therefore concluded that 'test statistics that rely on normality usually have been taken for granted, and consequently not much could have been done to achieve a better understanding of the ubiquitous complexity' of construction labour productivity (p. 11).

Employee involvement and construction labour productivity

The construction labour productivity research reviewed above appears primitive where employee involvement is concerned. Researchers have hitherto promulgated the managerial perspective in their efforts to seek the eradication of productivity variations. Whereas statistical inferences may appear to be impressive on the surface, it remains questionable whether such egalitarian approaches are plausible given that complexities of human behaviour can sometimes be beyond the control of management. Arguably, the constant fixation on the quantifiable aspects of productivity results in a regrettable dearth of attention to the *labour* element of construction labour productivity.

Indeed, there is often an underlying assumption that the goals of management and workers converge in the drive towards stabilising and improving productivity levels. This assumption echoes the unitarist view (Fox, 1966) of human resource management where 'differing views cannot exist within the organisation because [...] management and employees are working to the same goal of the organisation's success' (Beardwell *et al.*, 2004: 19). Such a perspective appears to validate the focus of past construction labour productivity research on merely management's agenda of controlling productivity. Arguably, the pursuit of productivity has downplayed workers' interests (Kelly, 1998) in the desire to maintain the cause of performance.

By contrast, the pluralistic approach acknowledges and deals with conflicting interests within organisations (Blyton and Turnbull, 2004). According to Beardwell *et al.* (2004), relatively stronger union structures as compared to those in the United States could explain UK academics' preference for pluralism. Indeed, the presence of strong union traditions in the UK is believed to give rise to what Blyton and Turnbull (2004) term 'institutional pluralism', where separate interest groups within an organisation negotiate a sense of balance through the institutions representing them, such as trade unions. However, unionisation is on the decline.

Coupled with the trend of employees becoming less collectivistic and more individualistic (Kelly, 1998; Blyton and Turnbull, 2004) – epitomised in the construction industry by the high level of self-employment – the adequacy of institutional pluralism is questionable. Moreover, for institutional pluralism to work, effective representation of interests is necessary (Brown, 1988). Therefore, low union density should cast further doubt on the efficacy of representation of workers' interests.

The notion of managing stakeholders' interests has been seen to pervade present day managerial discourse (Poole *et al.*, 2001). In a sense, 'stake-holding' presents an alternative to pluralism; managers are assumed to be accountable to the organisation's stakeholders, including the workforce (Blyton and Turnbull, 2004: 31). However, critics have questioned whether it was ever feasible to meet stakeholders' interests equally (Beardwell *et al.*, 2004) and whether such an approach simply toyed with the notion of pluralism while still maintaining a degree of unitarism (Boxall and Purcell, 2003). This is perhaps what Marchington and Wilkinson (2002) meant by 'neo-unitarism', where a win-win situation between managers and workers is desirable. Given the lack of consideration for employee involvement in past construction labour productivity research, this study aims to shed some light as to whether a win-win situation can be achieved.

14.3 Methodology

Access was granted on two projects for the purpose of site observations (see Jorgensen, 1989) in order to understand what in practice affects construction labour productivity. Table 14.1 below provides a descriptive profile of the two projects.

The researchers pursued an iterative, sense-making process of observing productivity levels of the two projects and identifying reasons for, and thereby ascribing particular factors to, variations that occurred. In project A, productivity was measured using the operatives' time sheets to derive the output produced per work-hour. Because the project was fairly straightforward – a rectangular building divided into 12 standard plot sizes – it was relatively easy to work out the aggregate productivity of square metre per work-hour by dividing the respective plot area by the total number of work-hours utilised to complete a particular plot.

Table 14.1 Profile of the two projects observed

	Project A	*Project B*
Project type	Multi-storey car park	HQ of a commercial bank
Contract value	Not given	£335 million
Contract type	Framework agreement	Construction management
Project duration	June 2001–May 2002	June 2002–August 2005
Project location	Glasgow	Edinburgh
Site visit	108 hours over 18 days between October 2001 and February 2002	16 hours over 2 days in August 2003

Due to the limited access provided in project B, productivity for project B was obtained by the project team using the Last Planner System. The Last Planner system was developed by Ballard (2000) to track what could be done (the plan) against the actual progress in a project. The key performance indicator of per cent-plan complete (PPC) (i.e. how much of the planned work was actually achieved) provided a useful measure for gauging productivity.

Key events that affected productivity were noted. Such questions as 'What's going on here?' 'How does this work?' and 'How do people do this?' (Massey, 1998; see also Jorgensen, 1989) were resolved mainly through visual observations, as well as through casual conversations with workers and managers, available evidential documents (e.g. programmes, minutes of site meetings etc.) and other means. The visual observational data collection was enabled by maintaining a research diary to keep track of occurrences.

14.4   Two case studies

This section reports on key issues that emerged from the site observations undertaken, with a particular emphasis on how integrating the workforce led to productivity improvements.

Case study A: Saving the site engineer who slipped up

Project A involved the construction of a multi-storey car park that was planned for completion at the beginning of May 2002. Despite its simplicity, a major event threatened to delay the entire project delivery. This observed event related to a site engineer's setting-out error and accounted for a dip in productivity recorded in the middle of November 2001. See Figure 14.1 below for an illustration of the productivity levels measured in project A, with a dramatic fall in productivity exhibited by plot 3–1.

The engineer concerned was one of a team of four site engineers. Because she had graduated only four months prior to the time of observation, she should have been mentored by a more senior, qualified engineer. However, time and resource pressures at other sites located near the project meant that the site engineers were often shared between and spread across the various sites. As a result, the engineer was left alone with another junior engineer to perform the setting out of the service voids for the third storey. The senior engineer had assumed that since the two junior engineers had been involved with the project from the start, they were sufficiently far up the learning curve. The reality was otherwise. The service voids were set out inaccurately and the steelfixers and concrete labourers consequently built to the wrong layout. The resultant dip in productivity was therefore caused by lack of experience on the part of the junior engineers and a lack of supervision.

It was only when the project team moved on to the preparation of level 4 that the voids were found to be off the mark. The identified error necessitated rework, which construction operatives normally frown upon. The rework involved

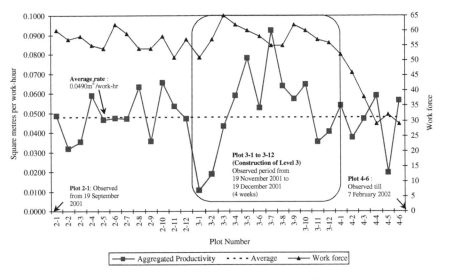

Figure 14.1 Aggregated productivity graph for project A collected during the time of observation (19 September 2001 to 7 February 2002).

strengthening and placement of reinforcement in the erroneous voids before patching up with concrete, and the hacking of concrete to create the service voids in the proper position. However, although this error affected the entire construction of level 3, the dip in productivity levels seemed to manifest only in plots 3–1 and 3–2 (see boxed section of Figure 14.1). This limited impact was due to the quick initial reaction of the project team to the unexpected crisis, which necessitated emergency planning. At that point, the planner had estimated the rework to take a total of four weeks to complete on the basis of full-time allocation of the manpower to enact this rectification. This in turn would potentially cause a three-week delay to the overall programme. However, the overall project was delivered on time in May 2002. The schedule acceleration in this instance was made possible through the mobilisation, and more importantly the motivation, of the workforce. It must be emphasised that the number of allocated operatives remained consistent; it is remarkable that they managed to perform the rework whilst concurrently moving on to level 4 (see the manpower curve depicted in Figure 14.1).

The effective motivation of the workforce was observed to have emanated from the general foreman, who rallied the operatives together to explain the situation. The fact that 'the engineers screwed up' appeared to have provided the impetus for the workers to prove that they were better. Furthermore, instead of defending their position and appear confrontational, the engineers accepted responsibility for the error and this seemed to have fuelled the operatives' desire to set things right. Despite this apparent 'them and us' situation between the engineers and the operatives, we are reminded here of the Phelps-Brown (1968) report, which insisted that 'the criticisms ranged at the fragmented nature of the industry arise from a lack of understanding of its function' (pp. 170–171). Arguably, the attainment of the operatives' high productivity in resolving the problem here is a testament of understanding and engaging with the age-old divide between the white-collar (i.e. the engineers) and the blue-collar (i.e. the operatives) workers.

Case study B: Initiating into the family

Project B involved the construction of the headquarters of a commercial bank in Edinburgh. As mentioned previously, the project team utilised the Last Planner system to keep track of productivity levels. Figure 14.2 below shows a graphical representation of project B's Last Planner activity completion for the four active work packages on site for the first 36 weeks of the project up to the point of observation in August 2003. Due to the limited access obtained for project B, it was not possible to observe and ascribe particular events to movements in productivity levels. Nonetheless, the documents from the progress meetings highlighted a number of common issues that were found by the project team to cause the variations in PPC levels. These included design changes, rework, materials, sequencing and interference, resource allocation and weather (i.e. *work content* factors mentioned in the literature review above).

Nevertheless, the overall productivity of the project could be considered as high. Indeed, on close inspection, the average PPC levels achieved by each of the four work packages were 73 per cent, 84 per cent, 79 per cent and 71 per cent respectively. The overall average of 76 per cent was therefore above the average suggested by Ballard (2000) of around 50 per cent. Undoubtedly, the tight

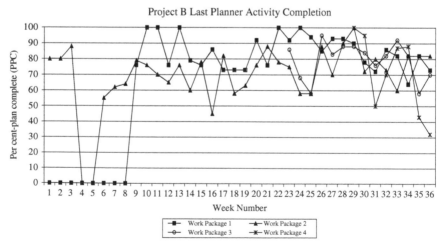

Figure 14.2 Project B progress for all four work packages for the first 36 weeks collected during the time of observation.

monitoring of PPC levels and the proactive attitude in analysing reasons for non-performance contributed to the attainment of such high levels. Apart from the pre-determined list of reasons, the management of the workforce also held the key to success. It is here that interesting observations were made about the site-induction sessions at this project.

While most induction sessions cover primarily the statutory health and safety aspects on construction sites, project B's director went a step further to set up the induction session as a means of initiating the worker into the family and rein-forcing the company's commitment to invest in its people and facilities. Two characteristics of the induction session emerged as pertinent in the current context.

First, the joint induction session was useful because it brought together the white-collar managers and the blue-collar operatives together. What was crucial about the joint induction was the underlying intention of making the participants aware of other personnel working around them, that is of establishing the place of each individual in the entire construction process. This approach was valuable since it emphasised the importance of sequencing and interference and combated the problem of cocooning within individual trade boundaries. Furthermore, the facilitator asked each individual to name one thing that would help improve their personal productivity. Fascinatingly, it was observed that managers seemed to focus on the timeliness of information, particularly cost information (see e.g. Clarke and Herrmann, 2004); on the other hand, the operatives, with one excep-tion, talked unanimously of the availability of tools and equipment. The one anomalous operative had suggested the need for an on-site cash machine so that they did not have to do a twenty-minute walk to and from the nearest village to get cash for lunch (the project was located in a remote greenfield site). Whilst this set many of the operatives laughing, the facilitator reiterated the significance and validity of this particular comment to point out the need to get feedback from everybody to seek improvements. This experience highlighted the existence of a divide between managers and operatives where the former emphasised on planning issues whilst the latter were mainly more concerned with operational detail.

The second significant characteristic of the induction session involved embracing the different views of all workers through tear-out sheets found in the *aide-memoire* issued to workers at the session. The feedback sheets allowed workers to inform their superiors of positive instances of on-site practices and to suggest possible improvements. Information from the completed forms was collated and discussed at daily foremen's meetings, weekly toolbox talks and, where necessary, at directors' meetings. In order to encourage contribution, an incentive in the form of a gift voucher was offered to the project team for every 10 implemented improvement suggestions. The induction sessions were essential for ensuring enthusiasm and commitment from the workers from the outset so as to create an exemplary project for the industry (Walker, 2002). This was indeed a major first step that brought about high productivity signified by the high average PPC levels obtained.

Cross-case reflection: Paternalistic treatment of workers

Throughout the observations, it was interesting to note the way in which the 'one in charge' (the general foreman in project A and the project director of project B) appeared to uphold a benign paternal role in relation to their confraternity of construction operatives. For example, the general foreman on project A prided himself on knowing all his 'lads' on a first-name basis, while the director of project B talked passionately about the career progression of his workforce. (There was also an on-site careers support adviser for the operatives, a rare phenomenon in the industry.) As project B's director stated, 'All of that is born from the desire to get those guys to understand that they are more than just workers, and that's a difference. You get to understand what these guys really want as individuals in terms of their understanding of respect.'

Of course, paternalism is not unproblematic. Blyton and Turnbull (2004), for instance, questioned the cliché of 'one big happy family' and suggested that 'like family relations, feelings of "good" and "ill" tend to be more intensely held' (p. 301). They argued further that 'such relations may achieve a degree of integration and a veneer of industrial harmony, but [...] their continuance is threatened by change' (ibid.). None the less, these problems were not apparent from the two projects observed. Credit should be given to the managers and operatives of the two projects for sustaining a harmonious, fraternal relationship, albeit working within an ever-changing construction project environment, characterised by Groák (1994) as 'temporary coalitions in a turbulent environment' (p. 291).

Arguably, the development of such a bond between the white-collar managers and blue-collar operatives built on respect and understanding is crucial to a construction industry that works well. After all, as project B's director remarked, 'It's quite a small, incestuous kind of industry.' His observation was borne out in relation to project A when it was found out later that 11 out of the 12 gangs moved *en masse* to a new project in Newcastle towards the end of the project. In fact, the general foreman claimed to have worked with most of the operatives for around 10 years as he reminisced about the first time they were involved on a project in Southampton. Similarly, it was also observed that groups of workers on project B tended to form cliques on the basis that they had also moved around together over a length of time.

Whilst the cliques were fascinating to observe, a limitation of this study is its inability to observe the dynamics associated with the social cliques formed over time. It was not clear, for example, whether the social cliques were formed over the duration of a project or perhaps over a more considerable period of time. Also, did the operatives move from project to project continuously or intermittently, and did they stay within their specific gangs throughout? It is suspected that the make-up of gangs did change over time, as Groák (1994) suggested that the nature of construction projects necessitates 'unpredictable [...] configurations of supply industries and technical skills'. In other words, whereas Blyton and Turnbull (2004) may be right to suggest that change might threaten the state of 'family relations', constant change in configurations in construction projects implies that 'families' are shaped from project to project. It can be argued therefore that 'families' in construction do not get rooted sufficiently to allow the intensification of 'good' and 'ill' feelings, to use Blyton's and Turnbull's (2004) analogy. In a sense, the observations of Wild (2005) are preferable, as he suggested that 'within a mobile, masculine confraternity creativity is re/de-constructed as yarns; that projects are completed is a profound tribute to those who do the work both because and in spite of the situations within which they labour'. Indeed, with regard to the phenomenon of social cliques and the implications of such affinity (see Dainty *et al.*, 2005; also Nicolini, 2002), a deeper ethnographic approach would shed more light and could usefully form the basis for further research.

Nevertheless, the *know-who* appeared to count much more than the *know-how*, as much of the deployment of human resources was performed on an informal basis. On both projects it was found that site managers did not rely on formal records to keep track of the workers' abilities and experience. The general foreman of project A depended on his personal knowledge of his workers based on working with them over a considerable length of time. Despite sub-contracting all the work packages on project B, the project director there claimed to make an effort to maintain a daily rapport with his workers to understand their strengths. Interestingly, Haas *et al.* (2001) also observed that while HR databases were often set up to help keep track of skills base, this was never consulted in practice due to the fact that decision-making on resource allocation on the ground was left to the foremen on site, who knew who possessed what skills and hence were able to make sensible deployment decisions. Besides, there was a tendency to perceive the HR function as external to the construction process (see Renwick, 2003).

The observations here have placed a greater emphasis on promoting the integration of the workforce by involving employees at all levels in seeking productivity improvements. Needless to say, managers often juggle with resources. Understandably, controlling resources and variations in the system becomes a preferable route for management. As we have seen, this preference has given rise to numerous studies on construction labour productivity with the inherent desire to stabilise levels rather than to accept that human behaviour can sometimes be erratic. Of course, past efforts should be applauded in advancing our understanding of what influences productivity. However, as Rosabeth Moss Kanter commented in an interview with Merriden (1997), 'management means managing an entire context; if you strip out one element and apply one methodology, it won't work'. Undeniably, previous studies have been too prescriptive in their approach to dealing with work content and environment variables of productivity. As a result, past research has often lost sight of embracing the equally important labour element. Indeed, it can be argued that without the appreciation of this element, project B's director would not have genuinely believed in the value of the induction sessions that he managed to convince the client to finance. Similarly, the general foreman in project A would not have been able to mobilise his operatives to effectively resolve the setting-out error described above. These events *inter alia* brought about high productivity levels during the projects.

14.5   Implications

Two important findings emerged from the site observations. First, the interpretation of paternalistic organisation of labour showed, to some extent, how paternalistic human resource practices based on mutual understanding and respect could stamp out labour resistance, thereby improving business performance. Given the continuing trend of declining unionism in UK construction, 'the new rhetoric surrounding these changes within the employee relations field concerns the need to build *partnerships* with employees in order to enhance business performance' (Loosemore *et al.*, 2003: 141; original *emphasis*). This new rhetoric relates to the rising interest in employee involvement. Examples of good practice in this respect are reported here.

A word of caution, however: good practice has to be subject to a level of contingency! As Marchington and Wilkinson (2002) and Boxall and Purcell (2003), amongst others, have observed, the danger with proposing best practice as a universal panacea to the productivity challenge is the fact that most practices are inimitable. Hence, good practice is the preferred term here and the case studies portrayed in this chapter are intended to offer practical examples for construction managers to see how employee involvement initiatives may be implemented with a view to improving labour productivity. Indeed, identification of such good practice should complement, and even supplement, the scientific analysis of the controllable factors influencing labour productivity as the existing literature illustrates.

None the less, it was evident that the two case studies observed went beyond Ramsay's (1977) superficial 'tea, towels and toilets' syndrome. While the operatives did not maintain discretion over the choice of employee involvement mechanisms – project A was an example of crisis management whereas the induction sessions in project B resulted from management's decision – benefits were observed to extend beyond productivity gains to include improved employee attitudes and commitment (Wagner, 1994). This improvement was manifested in a sense of pride either through raising the status of operatives in problem-solving in project A or working cooperatively towards delivering an exemplary project in project B. Thus, this study echoes similar studies into the effects of embracing employee voice (Marchington *et al.*, 2001). Arguably, by participating in employee involvement initiatives that yielded improved performance, the operatives were also party to shaping their individual job prospects as shown by the move *en masse* to Newcastle with the general foreman at the end of project A and the careers advisory support provided within project B.

The second key finding from the observations concerns the organisation of blue-collar operatives. It was found within the context of the two projects observed that despite being mobile, operatives tended to stick within their social cliques. This finding has a significant implication for the way academics and practitioners view the longer-term workforce issues that impact on construction labour productivity. Chan and Kaka (2003) demonstrated that practitioners placed more emphasis on shorter-term, controllable work content and work environment factors when thinking about productivity improvements. Longer-term workforce issues were often assumed to be naturally in place in organisations and so managers run the risk of ignoring them. The existence of social cliques and the extent to which these relationships are embedded in the organisation of the construction project means that this can be a potential (albeit latent) 'resource' to mobilise. However, the management of blue-collar operatives tends to be left out of human resource managers' scope of focus as evident, for example, in the lack of reliance on formal HR databases and the 'delegation' of human resource allocation decisions to site managers.

It is suggested that such behaviour stems from the normative view of construction organisations as flexible firms (Atkinson, 1984). According to Atkinson, firms increasingly rely on three groups of workers, namely a core group of primary labour, a first periphery group of secondary labour and a second periphery group of temporary labour (see also Handy, 1989; Langford *et al.*, 1995). Such distinction develops a pecking order that justifies human resource practices which concentrate efforts first on the professionals, then on support staff and lastly on the site operatives. Yet, the phenomenon of social cliques highlights that, despite the temporary nature of projects, relationships formed amongst site operatives actually run deeper and are considerably more permanent than is currently thought. This implies a need, therefore, to place an emphasis and effort on longer-term workforce issues. This could create a vicious circle that, in turn, could reap overall improvements in the industry's productivity. Indeed, efforts on the part of project B's management to create an exemplary project with an emphasis on employee involvement at all levels indicate the desire to reach the altruistic goal of achieving the spill-over effects of productivity improvements across the industry, starting with their current partners.

14.6 Conclusions

This chapter has re-examined the literature on construction labour productivity and found that previous research has been too prescriptive in advocating the need to control the work content and environment elements of productivity. Consequently, past efforts have often lost sight of the labour element of construction labour productivity in their desire to eradicate productivity variations. At the same time, the notion of employee involvement has been given scant attention by construction labour productivity researchers. Indeed, the cross-cutting theme that emerged from this study is that interaction and integration between management and workers is pertinent to the cause of achieving improvements. The participant observations made sense of how differences between white-collar and blue-collar operatives were embraced, for instance, through project B's induction sessions. A further example was extracted from project A, where blue-collar operatives were mobilised effectively to rectify an engineer's error resulting from a lack of experience and lapses in supervision. Of course, these good practices supplemented the analysis of the controllable element of productivity, as represented by the productivity measurements in project A and the documentation obtained from progress meetings in project B. However, it is imperative that managers are genuine and sincere in their approach towards accepting that blue-collar operatives have a significant level of (tacit) knowledge that could unlock productivity improvements. Besides, managers' ability to take on a benign paternal role in developing a personal understanding and respect for their workers could go some way to align workers' interests with the goal of achieving higher productivity. As de Silva (1997) commented, 'efficiency [...] and equity are not antithetic concepts'. A balance between the controllable aspects of productivity as identified in past research and the less or even uncontrollable aspects of labour should provide a more rounded approach towards our understanding of construction labour productivity.

References

Ackers, P., Marchington, M., Wilkinson, A. and Goodman, J. (1992) The use of cycles: explaining employee involvement in the 1990s. *Industrial Relations Journal*, **23**(4), 268–283.

Atkinson, J. (1984) Emerging UK work patterns. *In: Flexible Manning: The Way Ahead.* IMS report no. 88. Brighton: Institute of Manpower Studies.

Ballard, G. (2000) *The Last Planner System of Production Control*. PhD thesis, University of Birmingham, UK.

Beardwell, I., Holden, L. and Claydon, T. (2004) *Human Resource Management: A Contemporary Approach*, 4th edn, Essex: Pearson Education.

Blyton, P. and Turnbull, P. (2004) *The Dynamics of Employee Relations*, 3rd edn, Hampshire: Palgrave Macmillan.

Borcherding, J.D. and Garner, D.F. (1981) Workforce motivation and productivity on large jobs. *Journal of the Construction Division*, ASCE, **107**(CO3), 443–453.

Borcherding, J.D. and Oglesby C.H. (1975) Job dissatisfaction in construction work. *Journal of the Construction Division*, ASCE, **102**(CO2), 415–434.

Borcherding, J.D., Sebastian, S.J. and Samelson, N.M. (1979) Improving motivation and productivity on large projects. *In: Proceedings of the ASCE Annual Convention and Exposition and Continuing Education Programme*, 22–26 October.

Boxall, P. and Purcell, J. (2003) *Strategy and Human Resource Management*. New York: Palgrave Macmillan.

Brown, R.K. (1988) The employment relationship in sociological theory. *In*: D. Gallie (ed.) *Employment in Britain*. Oxford: Blackwell, 33–36.

Calvert, R.E., Bailey, G. and Coles, D. (1995) *Building Management*, 6th edn, Oxford: Butterworth-Heinemann.

Chan, P. and Kaka, A. (2003) Construction labour productivity improvements. *In*: G. Auoad and L. Ruddock (eds), *Proceedings of the Third International Postgraduate Research Conference in the Built and Human Environment*, 3–4 April 2003, ESAI Lisbon, 583–598.

Chan, P. and Kaka, A. (2004) Construction productivity measurement: A comparison of two case studies. *In: Proceedings of the Twentieth Annual ARCOM Conference*, 1–3 September 2004, Heriot-Watt University, Association of Researchers in Construction Management, **1**, 3–12.

Clarke, L. and Herrmann, G. (2004) Cost vs. production: Disparities in social housing construction in Britain and Germany. *Construction Management and Economics*, **22**, 521–532.

Clarke, L. and Wall, C. (1996) *Skills and the Construction Process: A Comparative Study of Vocational Training and Quality in Social Housebuilding*. The Policy Press, Housing and Construction Industry Research Programme.

Claydon, T. and Doyle, M. (1996) Trusting me, trusting you? The ethics of employee empowerment. *Personnel Review*, **25**(6), 13–25.

Dainty, A.R.J., Bryman, A., Price, A.D.F., Greasley, K., Soetanto, R. and King, N. (2005) Project affinity: The role of emotional attachment in construction projects. *Construction Management and Economics*, **23**, 241–244.

de Silva, S. (1997) The changing focus of industrial relations and human resource management. *In: The International Labour Organisation Workshop on Employers' Organisations in Asia-Pacific in the Twenty-first Century*. Turin, Italy, 5–13 May.

Edwards, B. and Eckblad, J. (1984) Motivating the British construction industry. *Construction Management and Economics*, **2**, 145–156.

Egan, J. (1998) *Rethinking construction*. Report from the construction task force, UK: DETR.

European Foundation for the Improvement of Living and Working Conditions (1997) *New Forms of Work Organisation: Can Europe Realise Its Potential? Results of a Survey of Direct Employee Participation in Europe*. Luxemburg: Office for the Official Publications of the European Communities.

Fox, A. (1966) *Industrial Sociology and Industrial Relations*. Royal Commission research paper no. 3. London: HMSO.

Grant, R.M. (2002) *Contemporary Strategy Analysis: Concepts, Techniques, Applications*, Oxford: Basil Blackwell.

Gray, C. and Flanagan, R. (1984) US productivity and fast tracking starts on the drawing board. *Construction Management and Economics*, **2**, 133–144.

Groák, S. (1994) Is construction an industry? Notes towards a greater analytic emphasis on external linkages. *Construction Management and Economics*, **12**, 287–293.

Haas, C.T., Rodriguez, A.M., Glover, R. and Goodrum, P.M. (2001) Implementing a multiskilled workforce. *Construction Management and Economics*, **19**, 633–641.

Handy, C. (1989) *The Age of Unreason*. London: Business Books.

Herbsman, Z. and Ellis, R. (1990) Research of factors influencing construction productivity. *Construction Management and Economics*, **8**, 49–61.

Hinze, J. and Parker, R. (1988) Productivity study of extruded concrete curbing operations. *Journal of Construction Engineering and Management*, ASCE, **114**(2), 256–262.

Hofstede, G. (1980) Motivation, leadership and organisation: Do American theories apply abroad? *Organisational Dynamics*, 42–63.

Jorgensen, D.L. (1989) *Participant Observation: A Methodology for Human Studies*. Thousand Oaks: Sage Publications Inc.

Kaming, P.F., Olomolaiye, P.O., Holt, G.D. and Harris, F.C. (1997) Factors influencing craftsmen's productivity in Indonesia. *International Journal of Project Management*, **15**(1), 21–30.

Kelly, J. (1998) *Rethinking Industrial Relations: Mobilization, Collectivism and Long Waves*. London: Routledge.

Labour Market Trends (2001) Central statistical office. London: HMSO, September.

Langford, D., Hancock, M.R., Fellows, R. and Gale, A.W. (1995) *Human Resources Management in Construction*. Englemere: The Chartered Institute of Building (CIOB).

Latham, M. (1994) *Constructing the Team*. London: HMSO.

Laufer, A. and Jenkins Jr, D.G. (1982) Motivating construction workers. *Journal of the Construction Division*, ASCE, **108**(CO4), 531–545.

Laufer, A. and Moore, B.E. (1983) Attitudes toward productivity pay programmes. *Journal of Construction Engineering and Management*, **109**(1), 89–101.

Lemna, G.J., Borcherding, J.D. and Tucker, R.L. (1986) Productive foremen in industrial construction. *Journal of Construction Engineering and Management*, ASCE, **112**(2), 192–210.

Loosemore, M., Dainty, A. and Lingard, H. (2003) *Human Resource Management in Construction Projects: Strategic and Operational Approaches*. London: Spon Press.

Macarov, D. (1982) *Worker Productivity: Myths and Reality*. California: Sage Publications.

Maloney, W.F. and McFillen, J.M. (1985) Motivational implication of construction work. *Journal of Construction Engineering and Management*, ASCE, **112**(1), 137–151.

Marchington, M. and Wilkinson, A. (2002) *People Management and Development: Human Resource Management at Work*, 2 edn, London: CIPD.

Marchington, M., Wilkinson, A., Ackers, P. and Dundon, A. (2001) *Management Choice and Employee Voice*. London: CIPD.

Massey, A. (1998) 'The way we do things around here': The culture of ethnography. *In: Proceedings of Ethnography and Education Conference*, Oxford University, 7–8 September.

Merriden, T. (1997) Rosabeth Moss Kanter. *Management Today*. February. 56.

Movement for Innovation (M⁴I) (2000) *A Commitment to People 'Our Biggest Asset'*. Respect for People (RfP) working group report, Rethinking Construction.

Mullins, L.J. (2005) *Management and Organisational Behaviour*, 7th edn, Essex: Pearson.

Nicolini, D. (2002) In search of 'project chemistry'. *Construction Management and Economics*, **20**, 167–177.

Olomolaiye, P.O. (1988) *An Evaluation of Bricklayers' Motivation and Productivity*. PhD thesis, Loughborough University of Technology, UK.

Parker, H.W. (1980) Communication: Key to productive construction. *Journal of Professional Activities*, ASCE, **106**(EI3), 173–180.

Phelps-Brown, E.H. (1968) *Report of the Committee of Inquiry Under Professor E.H. Phelps Brown into Certain Matters Concerning Labour in Building and Civil Engineering*. London: HMSO.

Poole, M., Mansfield, R. and Mendes, P. (2001) *Two Decades of Management: A Survey of the Attitudes and Behaviour of Managers Over a 20 Year Period*. London: Institute of Management.

Radosavljevic, M. and Horner, R.M.W. (2002) The evidence of complex variability in construction labour productivity. *Construction Management and Economics*, **20**, 3–12.

Ramsay, H. (1977) Cycles of control: Worker participation in sociological and historical perspective. *Sociology*, **11**(3), 481–506.

Renwick, D. (2003) Line manager involvement in HRM: An inside view. *Employee Relations*, **25**(3), 262–280.

Seymour, D. and Rooke, J. (1995) The culture of the industry and the culture of research. *Construction Management and Economics*, **13**, 511–523.

Sisson, K. and Storey, J. (2000) *The Realities of Human Resource Management: Managing the Employment Relationship*. Buckingham: Open University Press.

Smithers, G.L. and Walkers, D.H.T. (2000) The effect of the workplace on motivation and demotivation of construction professions. *Construction Management and Economics*, **18**, 833–841.

Tavakoli, A. (1985) Productivity analysis of construction operations. *Journal of Construction Engineering and Management*, ASCE, **111**(1), 31–39.

Thomas, H.R. and Yiakoumis, I (1987) Factor model of construction productivity. *Journal of Construction Engineering and Management*, **113**(4), 623–639.

Thomas, H.R., Maloney, W.F., Horner, R.M.W., Smith, G.R. and Handa, V.K. (1990) Modelling construction labour productivity. *Journal of Construction Engineering and Management*, ASCE, **116**(4), 705–726.

Thomas, R., Keating, J.M. and Bluedorn, A.C. (1983) Authority structures for construction project management. *Journal of Construction Engineering and Management*, ASCE, **109**(4), 406–422.

Tucker, R.L., Rogge, D.F., Hayes, W.R. and Hendrickson, F.P. (1982) Implementation of foreman-delay surveys. *Journal of the Construction Division*, ASCE, **108**(CO4), 577–591.

Wagner, J. (1994) Participation's effects on performance and satisfaction: A reconsideration of research evidence. *Academy of Management Review*, **19**(2), 312–330.

Walker, D.H.T. (2002) Enthusiasm, commitment and project alliancing: An Australian experience. *Construction Innovation*, **2**, 15–31.

Wild, A. (2005) Uncertainty and information in construction: From the socio-technical perspective 1962–66 to knowledge management. What have we learnt?. In: *Knowledge Management in the Construction Industry: A Socio-technical Perspective*. Hershey: Idea Group Inc.

Zakeri, M., Olomolaiye, P.O., Holt, G.D. and Harris, F.C. (1996) A survey of constraints on Iranian construction operatives' productivity. *Construction Management and Economics*, **14**, 417–426.

15 The quest for productivity revisited, or just 'derecognition by the back door'?

Neil Ritson

Abstract

While the oil and chemicals plants' productivity objectives are a distinct, well-documented feature in the various oil refinery studies, any mention of contractors is only in passing and without any coherent theme (Flanders, 1964; Oldfield 1966; Gallie, 1978; Young, 1986; Ahlstrand, 1990), the problems of management by the clients of the contracted work are seldom mentioned in the literature despite the prevalence of contracting out of maintenance in UK industry. This chapter uncovers the employee relations implications for contractors in the engineering construction industry engaged in the repair and maintenance operations of a large UK oil refinery.

This process of contracting-out was controlled by means of a change in the industrial relations arrangements. The contracting firms benefited from their client's strategy of increasing productivity through 'numerical flexibility', which culminated in the transfer of more work and of many of the client's maintenance workers to the contractor firms.

To increase managerial control of this work, a 'single union agreement' (SUA) was devised for the contractors' workforces, by which route the other unions were derecognised. This agreement was imported from other industries, re-written by the client management and then imposed on the contractors in an attempt to change the endemic culture on the refinery site. The National Agreement for Engineering Construction (NAECI) was considered and rejected, as it did not serve this purpose. The client refinery thus created a controversial industry 'first', which was copied by two other refiners, and which indirectly led to changes in the NAECI.

This change is analysed under the frameworks of transaction cost economics (TCE), and institutional theory, with its notions of 'isomorphism' or 'industry recipes'.

The critical discourse of efficiency, it will be argued, is based in the first instance on managerial ideology, followed by, where possible, measurable data. In its absence, under conditions of uncertainty and complexity, firms seek to copy the strategies of others, creating 'isomorphism' or 'industry recipes'.

Implications for the wider construction industry are then discussed.

Keywords: engineering, contracting, efficiency and ideology.

15.1   Introduction

The chapter is organised under seven sections. First, the engineering construction industry is introduced and explained in relation to the broader construction industry. The low-productivity culture of contracting is then discussed. The next section addresses the shortcomings of the literature on the industry by reference to two major theoretical approaches: transaction cost economics and institutional theory. The following section combines the earlier discussions to discuss the nature of ideology and its role in creating isomorphism or recipes for productivity. The research study is then described, followed by discussion and analysis of the contribution of economic and institutional factors to its results. The chapter concludes with some general issues for construction.

15.2   The Engineering Construction Industry

The construction industry contains a wide spread of task and skills, and in its wider definition it incorporates the engineering construction industry sector (ECI). Work in this sector is based on the use and manipulation of metal, usually on the sites of large clients. The ECI is different from the wider construction industry in this important respect: it is based on narrow, capital-intensive industry markets, where specialist clients buy from specialist contractors. There is no wholesale or retail segment as there is in house building, for example. The ECI's clients are very large enterprises: owners of power stations, offshore platforms and rigs and, of particular importance for our discussion, oil and chemical installations.

The Engineering Construction Industry Training Board's sector development plan states:

> The engineering construction industry has a big impact on the UK economy, directly contributing around 1.5 per cent of GDP. The importance goes beyond direct GDP contribution, because engineering construction underpins the UK production of crucial products such as oil, gas, chemicals and power generation. Around 50,000 people are employed in the industry in both site and offsite (design and procurement) work. The type of site work has changed in the last decade as fewer new build projects have been undertaken. Repair and maintenance work now makes up the majority of work in the sector.
>
> (ECITB, 2005: 1)

The ECI itself is therefore split into two distinct spheres: construction of plant – 'new builds', clearly forming part of the construction sector – and the repair and maintenance (R&M) of existing plant. In R&M, the work relates to construction in two ways. First, and most importantly, the short-term nature of contracting work and its 'atmosphere' or culture is identical. Indeed, large maintenance projects – 'shutdowns' or 'turnarounds' – closely resemble construction projects. Secondly, the ECI firms involved in repair and maintenance also operate in wider construction activity. Hence this chapter has implications beyond its initial context, as will become apparent.

The industry originated in construction works undertaken by a number of separate firms from widely differing backgrounds. These firms had different employer associations, and each had its own set of relationships and its particular employee relations policies and agreements. Tasks involved relatively simple sheet steelwork (using the 'Outside Steelwork Erection Agreement', for example) and more complex projects involving boiler-making (using the 'Water Tube Agreement'). These and other agreements, though detailed in terms of 'works rules' (as Salamon, 2000 describes them), were not necessarily about efficiency. They were more about systematically separating out spheres of influence, predominantly not only between client and contractor workforces, but also between the different groups within the ECI firms, which were usually single trade (e.g. scaffolding, lagging, and welding). This separation related to a discourse about the various employer power-bases and their ability to get work and, given the insecurity of the contracting environment, this in turn gave rise to multi-unionism.

This diversity was problematic as it led to huge delays in the UK's major construction projects, particularly power stations. Accordingly, the government commissioned the National Economic Development Office to investigate and make recommendations. Its report 'Large Industrial Sites' was finally published in 1970 (NEDO, 1970). The separation of the firms' competences and single-trade working was not highlighted; instead, the labour issues were considered significant. The report suggested that 50 per cent of hours in the ECI were unproductive. Three quarters of the labour force was composed of a temporary, drifting mass, often paid 'on the lump'. By this means – popular with employees – income tax and National Insurance was not paid and employees often used aliases to avoid detection by the authorities (Austrin, 1980). Such arrangements contributed to a low-quality/low-pay culture which featured inter-craft 'demarcation disputes' and 'leapfrogging claims'. This latter occurred where unions on one site copied the negotiated agreement of another – and this then in turn was copied by others.

The Report recommended a National Agreement (NAECI) and a National Joint Council (NJC) to oversee it. This Agreement was not finally signed-off until 1981, ten years later, when its importance was politically recognised: the Labour government actively pushed for its establishment (Korczynski, 1996).

According to Garfit (1989), with this new structure and its processes the ECI then became an entity. The NAECI, he suggested, initiated a new era and led to a more coherent industry. Negotiated by disparate and competing employers and unions, the NAECI formed some kind of normative consensus, though very long and complex with 120 pages (National Joint Council, 1994). Williams (1989) the NJC chair at the time, wrote an article for *Personnel Management* magazine in which he described the subsequent increase in engineering construction industry productivity over the period 1980–89 as 'Construction's IR miracle'.

However, the ECI's clients, who all had their own parochial site agreements, rejected the NAECI for Repair and Maintenance work. There is little in the literature concerning these agreements despite the fact that maintenance costs have a considerable indirect effect on overall costs: Whiston (1988) presented evidence that even minor maintenance failures, consolidated over the year, ensured that no profits for the year were declared by two plants.

There is a widespread use of contracting to lower costs in the UK generally (Atkinson, 1984), particularly in the oil and chemicals sector (Cross, 1989). Keenoy (1979) in his 'Lagger's Tale', describes how contracted-in employees in the repair and maintenance area are paid considerably less than client employees, creating conflict.

The general lack of capital investment in the UK has also featured in the literature (Blyton and Turnbull, 1998), and the use of contractors avoids extensive capital costs by their clients. This avoidance reinforces the strategy of using contractors, with their low labour costs. Cibin and Grant (1996) argue that in the oil and chemicals sector, over a considerable period of time, there had developed a focus on internal 'static' efficiency, which was in the process of being replaced by its 'dynamic' cousin. But there is a relative dearth of information in the literature on 'dynamic efficiency' – on the crucial link between contractors, craft industrial relations and maintenance on the one hand, and plant productivity on the other. Contracting and supply-chain issues are generally ignored, and vertical-integration is taken as a given (e.g. Sampson, 1975; Yergin, 1991). Yet the key issue in maintenance is not budget (unlike the wider construction activities in the ECI) but schedule. Plants are run at capacity and 'downtime' cannot be made up later through overtime.

Ahlstrand's intensive study (1990) demonstrated how at Esso's Fawley refinery, as early as 1975, 'static' efficiency needs resulted in a change to the structure of the internal craft workforce by the removal of craft demarcations through derecognition of the relevant unions. This strategy has extended across the client firms. Table 15.1 shows the results of more widespread craft derecognitions within the internal labour forces of the client firms in recent times.

Table 15.1 arguably shows the effect of the strategy of increasing 'static efficiency' by removing craft demarcations of the internal maintenance work-force through derecognitions. With a new concern for 'dynamic efficiency' the question is, how were the client's plants to increase the efficiency of contractors in R&M? The clients already used construction firms because they were cheaper than in-house provision, but these ECI firms were apparently only 50 per cent efficient according to NEDO. Was further action necessary? Were the craft unions to be derecognised as they had been in the client firms? What were the alternatives? Could the NAECI act as a coordinating device?

Table 15.1 Trade Union derecognition in major UK oil companies

BP Chemicals (Baglan Bay)	Craft and process operators,
BP Oil (Llandarcy)	Craft and process operators
Esso (Fawley)	Craft and process operators
Mobil (Coryton)	Craft and administrative
Mobil (Birkenhead)	Craft and administrative
Shell UK (Shellhaven)	Craft and process operators
Shell UK (Stanlow)	Craft and administrative

Source: Adapted from the House of Commons' Select Committee report 'The future of Trade Unions' evidence from the TGWU (House of Commons, 1994).

15.3   Contracting and its low-productivity culture

There is an employer discourse around the idea, if not the exact nature, of productivity and efficiency, which is often based (wrongly, according to Goldratt and Cox, 1993) on input (labour) and output ratios. This is a key concept in unravelling the culture of the work of contractors. The survey evidence of Cross (1989) showed increasing subcontracting and decreasing direct labour, caused by the economics of the 'production function' where, in the short run, only labour can be cut and where efficiency equates to productivity. Where this is set as 'output per employee', the internal loss of jobs does not have to mean loss of tasks or expertise, as these can be made up for by using contractors. Though such a crude, neo-classical idea of efficiency has been extensively criticised (e.g. by Marginson, 1993), it may still be a driving force for managements.

The ECI needed to address this problem, and it was a large one. It required a change in the culture of the construction industry as it affected the clients – from under-utilisation of labour to the short-term goals of the ECI firms themselves as well as their employees. This culture had resulted in the infamous inter-trade 'demarcation disputes' mentioned above. There was no NAECI in R&M, the predominant branch of the industry: no coordination mechanism except the local site agreements. Thus, the sites had none of the advantages ascribed to the ECI in construction of 'new build' projects. In contrast to Williams' assertion that the NAECI played an important role, Korczynski (1993) argued that only the emergence of the coordinating managing contractor solved the inherent weaknesses of the employment situation. Of course, it could be a joint achievement.

Nevertheless, this culture of low pay, insecurity and low productivity had led to a situation where Lloyd (1990) in his *History of The EETPU* described the contracting culture or atmosphere as follows:

> Like much of the rest of the construction industry, contracting was a rough tough young man's game, with more than its fair share of sharks and tear-aways among both employers and workpeople. With the opening and closing

of sites, it was in perpetual turmoil. The industry had no stability at local level. The quality of supervision was a lottery. Frequently the sites were not properly organised, with shop stewards difficult to elect. Often, known shop stewards found it hard to find work due to a fairly effective blacklisting of 'militants' by employers.

(Lloyd, 1990: 500)

There is, however, a lack of analysis of the practices in the industry and how its problems might be resolved. Any resolution might have implications for the rest of the construction industry, which has a similar culture and structure. Will the client firms find a formula, for managing contractors, which results in increased 'dynamic efficiency' and so can be exported to the wider construction industry, or is the oil and chemical sector bound by its own ideology or recipe? In other words, do clients act rationally and strategically in economising and seeking productivity for construction firms they employ, or do they simply copy each other's existing recipe? If the former is true then there may be an extension of the economics to the wider industry, if the latter then not.

The lack of analysis may result in part from the lack of an appropriate theoretical framework. In the debates about flexibility (e.g. Dore, 1973, 1986; Pollert, 1988; Wood, 1989; Legge, 1996), little theorising has been attempted. Neither Atkinson (1984) in his 'flexible firm' model nor his critics provide a theory of efficiency and derecognition, nor the reasons for outsourcing work to contractors in what he termed 'numerical flexibility'.

The next section seeks to apply appropriate theory, and to evaluate its application to the case in question.

15.4 Theoretical frameworks

This section attempts to address the lack of discussion of productivity in the literature by identifying appropriate theory. The strategic management literature has little or nothing to say about human resources (HR) activity and its influence on productivity, or on strategy (see Johnson and Scholes, 2002; Grant, 2002). There is little evidence of the efficacy of union deals. It is appropriate to consider the rational theory of economical necessity in describing the changes in the ECI and this is attempted through an analysis of transaction cost economics (TCE). On the other hand, Table 15.1 illustrates the widespread use of derecognition in the oil and chemicals sector of clients' own craft workforces, and this may suggest an industry recipe which can be analysed using institutional theory.

15.5 Transaction cost economics

As discussed above, the issue of market supply and its 'dynamic efficiency' is being raised now that potential gains from 'static efficiency' under the internal hierarchies have been exhausted (Cross, 1989; Cibin and Grant, 1996). This theory of increased market contracting can be developed within the framework of TCE (Williamson, 1981). TCE seems appropriate because the out-sourcing of functions may be due to a resurgence of the issue of poor industrial management and TCE was described by Willman (1986) as the 'institutional failure framework'. In this way, economics can shed additional light on the problem of industrial relations interfaces (Douma and Schreuder, 1991; Rowlinson, 1997).

According to Coase (1937), the existence of the firm, and its boundary, is a balance between the costs of using the market price system – the transaction costs – and the costs of organising an extra transaction within the firm. Thus, the size of the firm is regulated by the number and complexity of the arrangements for production, including its subcontracting arrangements. The firm retains 'specific assets' or skills and uses the market for non-specific ones, which are readily available when required. If the costs of internalisation – of keeping the specific assets – are driven up, then externalisation – outsourcing to contractors – becomes more attractive, especially as contractor wages are lower.

However, there are costs in using the market: the transaction costs. Arrow (1969) classifies transaction costs into *ex ante* costs (negotiating and writing contracts etc.) and the *ex post* costs (executing, policing and remedying disputes). Transaction cost theory thus predicts that managers of organisations will attempt to reduce the complexity of internal and external transactions.

As external transactions have increased for clients using the ECI through the additional use of subcontracting, the clients' management has the challenge of controlling these additional transactions and thus incurring additional costs, unless increased efficiency can be achieved. Such costs would include ways of counteracting the uncertainty over management and industrial relations issues in the contracting industry documented above.

Williamson (1985) describes as 'atmosphere' the factors surrounding the process of decision-making under the constraints of environmental factors. Uncertainty and complexity exhibited in the environment lead to 'small-numbers exchange' – a reliance on only a few suppliers who could then act in concert to bid up the price. This can happen because the buyers have 'bounded rationality': this is 'behaviour which is *intendedly* rational but only *limitedly* so' (Simon, 1961: xxiv, emphasis in original).

In order to reduce the effect of these factors, the extent of subcontracting is therefore decreased and this leads to small-numbers exchange. The idea of small-numbers exchange can include the Korczynski argument for creating a single managing contractor to oversee the project, and in addition the saving of time writing contracts under the NAECI and adjudicating on issues using the National Joint Council. Even more savings on bounded rationality would ensue from dealing with only one party, and reducing the length of the 120-page NAECI – for example, the notion of the 'single union deal' which, as many authors have noted, is used extensively by Japanese car and electronic plants.

This reduction of complexity reduces bounded rationality but of itself this leads to the threat of opportunism, defined by Williamson as 'self-interest-seeking with guile'. This happens because, with less redundancy and with smaller numbers of agents, the contractor and his employees now have more power (Porter, 1985). Thus, any change to small-numbers exchange should be managed carefully within constraints – in the case of single-union deals, this has often been accompanied by a 'no-strike clause' (Beaumont, 1990).

Hence, TCE can provide a framework for analysis as it can encompass the process of management of contracts via industrial relations agreements.

15.6   Institutional theory and isomorphism

As an alternative to TCE, and given 'bounded rationality', an institutional theory framework might suggest (as did Grinyer and Spender (1979a, b, 1985) in their research into 'business recipes') that institutions might simply copy each other. Spender (1989) extended the idea of the business recipe (such as efficiency and productivity, as mentioned above) to a wider discussion of 'industry recipes'. These might result, for example, from adopting concepts such as 'world class manufacturing' or 'strategic management' resulting in the so-called 'benchmarking' surveys among a group of firms. This 'industry recipe' would need to be shown to have originated as a method of avoiding blame and uncertainty, according to the literature, and not by some back-to-basics analysis of the economics of contracting to ECI firms, as suggested under TCE.

This notion of recipes has been usefully analysed as 'isomorphism'. The latter word replaces the more mundane idea of a 'recipe' and derives from the Greek meaning 'similarity in unrelated forms' (Macdonald, 1977). Coined by DiMaggio and Powell (1983, 1991) within the framework of institutional theory, it extends the idea of recipes to identify three types of similarity based on causative factors – coercive, normative and mimetic isomorphism – which allude to the origin of the existing 'industry recipe'.

Coercive isomorphism is exercised formally or informally by other organisations or by society via rules, laws or sanctions, including the use of contract and power within buyer–supplier and other inter-institutional relationships.

Normative isomorphism is a process by which organisations adopt certain procedures, policies and structures advocated by professional bodies such as trade associations, universities or consultancies, and as such is more deeply hidden.

Mimetic isomorphism takes place at a cognitive level through the adoption of similar practices by organisations within the same field. These may be deliberate and objective – by use of benchmarking exercises, for example – or they may be intuitive and taken for granted. More often than not they need to be culturally supported norms of behaviour.

Table 15.1 showed an industry recipe of internal craft derecognition, but the evidence of causative factor(s) is not there: the similarity between the actions of the different firms may be a result of economic factors or of ideology. This is discussed below.

15.7   Ideology and recipes for productivity

Origins of ideology

Ideology can be defined as a set of inter-related belief systems which underpin behaviour. Salamon (2000) describes three paradigms which represent points of managerial and employee ideology within the employee relations cognate area: unitarism, pluralism and Marxism or conflict theory. Unitarism relates to the original paternalism of the capitalist employer, whether nurturing or not. Pluralism accounted for the rise to power of different interest groups where conflict was inevitable and reduced by a system of rules, such as those within Trade Union Agreements. Marxism or critical theory only accepts the idea of class conflict on a larger scale, and thereby rejects the conflict of pluralism within the rules established by capitalism. Pluralism is accused of dividing the general power of workers into separate interest groups (i.e. the trade unions).

It has been suggested that the current managerial ideology across most sectors in the UK should be seen as a move from pluralism back to unitarism. This move is heralded by the arrival of the idea of a 'new industrial relations' (Grant D, 1993, 1994, 1996) and of human resource management (HRM), a set of supposedly new policies and process (Keenoy, 1979; Guest, 1987, 1989; Legge, 1996; Storey, 1992). HRM has also been termed 'macho management' (Purcell, 1982; Mackay, 1986). Thus, ideology may underlie the attempt by firms to provide an 'appropriate' recipe – managements may only look in certain favourable directions when seeking corroboration of their position by other firms.

Whereas the client firms formerly recognised unions at a level of almost 100 per cent, and so had a pluralistic ideology, the derecognitions seen in Table 15.1 suggest a dramatic change to unitarism. However, as all of the firms are subject to similar environmental pressures, a degree of similarity is to be expected. As the change has been dramatic – set as it is against almost 100 per cent unionisation of manual workers in the petrochemicals industries (Kelly, 1998) – and recent, this change would point to an economic motive.

On the other hand, the clients are different from the UK industry norms. Many are American and have been through what Kochan *et al.* (1989) call 'the transformation of American industrial relations'. These authors point to the relocation of factories in non-union areas as part of a union-avoidance strategy. Mobil was singled out among such firms for its new, explicitly non-union Joliet refinery in Chicago. Furthermore, in the UK, Wolfson *et al.* (1996) demonstrate the resistance of all these major firms to collectivisation in their 'upstream' functions, perhaps also deriving from a transatlantic influence.

Thus, the question which must be addressed by TCE and institutional theory is whether the cause of the isomorphism seen in the craft derecognitions in Table 15.1 is indeed a recipe, perhaps based on a managerial ideology or, as identified by Cibin and Grant, an economic rationality in support of the focus on efficiency.

This evidence provided by the House of Commons' Select Committee in Table 15.1 illustrates that the derecognition of the clients' craft unions is a unifying feature. However, neither in the Select Committee's separate reports nor elsewhere is there evidence given of the industrial relations issues of the large contractors' workforces on oil and chemical plants, nor are the arrangements for their collective bargaining described.

There is an important link here between the internal and the external workforces. Gallie (1978) noted the fear of BP's internal maintenance employees that contractors might take their jobs in the future. This concern turned out to be well-founded: the crucial mechanism for the preponderance of internal craft derecognition was the substitution of available contractor labour, which provided leverage and subsequent exploitation of the weakness of the internal crafts trade unions (Claydon, 1989, 1996). Thus, the ECI firms substituted labour rather than adding to it when demand arose, as they had in the past. The question is why? This question was addressed in the research study on which this chapter is based.

15.8 The research study

The research set out to explore the basic propositions identified above regarding management motivations in the organisation of construction work at various plants. The evidence of a move to unitarism was hypothesised as the cause of the derecognitions in Table 15.1, but equally these might have been economically based due to the similar environment of the firms and the need for static efficiency by way of removal of craft demarcations. In this study, the client firm attempted to improve its dynamic efficiency – that is, the contractors' performance – and in so doing it had to manage the economic exchanges as described under TCE.

The study began with a survey of the R&M function in the oil and chemical industry in the UK. This study revealed that one refinery had rejected the NAECI in favour of a single union agreement (SUA). This agreement was taken up by a neighbouring plant and subsequently by one some hundreds of miles away. The question raised by this isomorphism is whether the spread of the SUA idea would endanger the role of the NAECI in other plants by its substitution by this much simpler arrangement. Should the trend of SUAs continue, the Joint Council would lose both status and the money charged for the NAECI administration when it is adopted on sites.

As discussed earlier, a deeper analysis of possible causative factors should give a better picture of the way in which ECI firms might be managed in other locations – would a 'recipe' be adopted, or would an economic calculus be rigorously applied instead? Data is needed at the level of the plant to assess the relative merits of these two alternatives: in particular is there evidence of extensive calculations of the costs and benefits of alternative strategies? Or, alternatively, is there evidence of 'benchmarking' against the end results of others?

The following case study is based on interviews with line managers and HR specialists, triangulated with the officers of the relevant trade unions and the industry bodies within the construction industry.

15.9 The refinery's internal maintenance craftwork

Productivity deals were continually introduced in the refinery during the 1960s (Oldfield, 1966) and 1970s and into the 1980s (Young, 1986, 1992). Management negotiated hundreds of inter-craft flexibilities for maintenance tasks (and some operator-craft flexibilities). It was difficult, however, for both managers and trade unions to manage all these transactions effectively via 78 pages of individual 'task-sheets', and gradually demarcations began to reappear in the late 1970s through different interpretations of the wording and context (Young, 1992).

This development led to a new craft agreement in 1984 which swept away all the detailed inter-craft flexibilities described in the task-sheets, and rationalised the craft workforce by outsourcing the lower level trades to contractors (Young, 1986). This did not work: demarcations began to reappear and the refinery managers felt able to ask the unions to honour the agreement or to devise a new 'flexible' approach. The local craft union's full-time officers (FTOs) were rigidly opposed to this proposition. As a result, management put more and more work out to contract, and advertisements for 'fully flexible craft technicians' appeared anonymously in the local press. This action led to the derecognition of the craft unions on the instigation of the craft workforce.

However, given the success of the various tactical moves to outsource work to contractors, management deemed that the days of static efficiency gains were over, and that continual wars of attrition with the craft maintenance workforce, as in the recent past, were no longer an option. Economic performance problems, and government initiatives (McInnes, 1987) illustrated by BP's substantial de-layering and downsizing (Cibin and Grant, 1996), led them to consider dynamic efficiency.

Subcontractors

Previously, the refinery's own site agreement, like most others in the sector, pre-dated the NAECI and followed the general principles of the existing National Agreement (the 'Black Book') between the Engineering Employers Federation and the 16 Confederation of Shipbuilding and Engineering Unions. However, its detail was written and controlled locally by the refinery. This agreement was not subject to the transforming productivity agreements in the refinery. While production operators were already flexible, as they all performed a broadly similar function, maintenance craftwork was managed under inter-trade demarcation arrangements (described as 'professionalism' by Stinchcombe, 1959) This arrangement was due partly to the difficulty of coordinating many contractors and the 16 trade union signatories, and the lack of economic incentive given the lower cost of contractor craftwork.

Not only did inter-craft demarcations still exist, but also the contractors were typically single-craft, single-union, offering a single service, such as scaffolding. This exacerbated attempts to raise overall productivity on the site for maintenance generally, as synergy between contractors and internal crafts was difficult to achieve. However, in 1991, the economic recession and government initiatives in reining back the power of the trade unions gave managers a 'window of opportunity' resulting in the successful changes to the internal crafts who were derecognised and whose lower trades were transferred out to contractors.

Buoyed by these successes, the refinery management attempted to renegotiate the site agreement.

The management set about a benchmarking exercise, looking for an acceptable external model. However, the survey results (Table 15.2) showed that there was no existing appropriate agreement elsewhere: there was nothing to copy.

The HR manager then surveyed the refinery's own engineers and assessed their criticisms of the way contractors operated. These criticisms were then collated and codified to create a set of principles in order to begin negotiations to change the agreement (which Walton and McKersie, 1965 would term 'integrated bargaining') and so to satisfy engineering efficiency. Following various mergers over time, there were three unions signatory to the existing site agreement: the General Municipal and Boilermakers Union (GMB), the Manufacturing Science and Finance Union (MSF) and the Amalgamated Union of Engineering Workers (AUEW). Their FTOs were asked to draft a new agreement covering the requirements for increased flexibility. As these craft FTOs were still opposed to radical change, the refinery gave them three months' notice of

Table 15.2 Arrangements for controlling maintenance contractors' work at selected oil and chemical plants in the UK, 1991

Plant	Contractors' agreements
1 GULF	No fixed approach
2 TOTAL	In-house agreement
3 SHELL	In-house agreement
4 LINDSEY	Contractors local agreements
5 CONOCO	Contractors local agreements
6 ICI Teesside	Own personnel
7 BP Hull	In-house agreement
8 BP Grangemouth	In-house agreement
9 SHELL Stanlow	Site agreement
10 ESSO Fawley	Site agreement
11 TEXACO	NAECI – non-nominated
12 ELF	Whatever is cheapest

Source: Company documents

termination of the site agreement. At the expiration of the notice, no proposals had been received from any of the FTOs.

The personalities of the FTOs had certainly influenced some of the options, such as derecognising the union with the most difficult FTO under Option 3 below. On the other hand, although over time the unions had appointed different FTOs, the current incumbents were still seen in the refinery management documents as '60s men'. This evaluation reflects the general 'atmosphere' and culture of construction, referred to above, and the role of shop stewards who progress to FTOs. Indeed one of the FTOs had allegedly been blacklisted by the refinery in the mid-1970s and was stridently critical of the firm and the oil industry, whose members he saw as colluding in the general derecognition of the internal unions.

The aim of the changes was to alter the culture of the construction site to attain more of a factory atmosphere, linked to the refinery infrastructure, as opposed to a traditional temporary construction site, separated from the refinery temporally, culturally and physically. Efficiency would be the main objective. Management therefore drafted a new agreement with the following principal features:

1   There was to be full flexibility of trades and employers, depending only on safety and competence. A training scheme was to be set up under the National Skills Development Scheme of the ECITB to permit additional payments to craftsmen for additional skills, giving them a 'passport' useable within the locality.
2   The procedural agreement was to comprise only three stages, the final one requiring the National Officer of the Union to attend in person (thus making it an unlikely third step except for major collective issues). Unlike other similar agreements in other sectors of the economy, there was to be a no 'no-strike clause'.
3   The selected union was to represent all tradesmen (who were allowed to retain their original union membership).
4   The overtime rates were changed to reduce the incentive payment for weekend working (allegedly abused to increase earnings).
5   Start and finish times were to be the same as the refinery's, and washing and changing were to be carried out in employees' own time; the collective morning tea-break was replaced by an individual one taken depending on the needs of the task.
6   Canteen, toilet, smoking and other facilities, including transport, were to be duplicated round the site, permits to work in dangerous areas were to be simplified and stores access enabled.

There remained the difficulty of how to progress these ideas. Because the same unions and FTOs would again have to be involved in negotiations, the need for a clean break with the past was becoming more evident in the various documents produced during managerial 'away days'. Alternative options, some drastic, had been discussed:

1   retaining the status quo
2   the use of the NAECI
3   selective derecognition of one union
4   complete deunionisation
5   staff status and even
6   self-employed status for some trades.

In the meantime, high-level talks had taken place with senior officers of the AUEW who seemed sympathetic to the idea of a single-union deal. Single-union deals were current in the media – especially in the new Japanese transplants – and signalled the 'new industrial relations'. For the managers at the refinery, they seemed to emphasise a policy that said that any change was simply to increase efficiency and if the AUEW national officers were keen on negotiating, then a new deal would secure the refinery's future viability, because so much work had been contracted out. Documents stated that a single-union deal 'would be pro-change but not anti-union'.

The key issue for the refinery was that the AUEW could represent all the trades credibly in a single-union deal. For example, welders were represented by the AUEW as well as the General Municipal and Boilermakers Union, the GMB. Similarly, the sheet metal and the heating and ventilating trades were not a problem: though represented by the Manufacturing Science and Finance union, their skills were neither so specific nor so frequently used to deserve more specific representation.

The idea of a credible single-union deal was discussed at some length informally with AUEW senior National Officers and finally agreed and implemented after employee consultations. The agreement, unlike many other single-union deals, does not contain a 'no-strike' clause, and the documents seen in the research study concentrate on changes to methods of working and economic cost savings (overhead savings were estimated at £1 million per year). There is reason to suppose that any derecognition of the other two trade unions, the GMB and the MSF, had more to do with economic rationality than ideology, that is with efficiency rather than power.

Thus the motive force was economic, and the resulting agreement efficient. At that level, it was not based on copying other plants which, as Table 15.2 showed, had not initiated any such changes. It was 'first' initiated by the refinery itself. However, it was isomorphic with those single-union deals outside the sector, and imposed on the contractors.

15.10   Discussion

As we have seen, uncertainty and complexity exhibited in the environment lead to small-numbers exchange, because buyers have bounded rationality. Of course, there are costs in internalisation of craftwork: so craft derecognition which ended the demarcations between the different crafts was cost-efficient as it saved time and effort. Work proceeded faster due to flexible inter-trade team working, and management did not have to discuss work assignments in detail or consider various trade-based interests, nor did the unions have to make mutual adjustments to incorporate craft demarcations in a 'balance of trades' as under the task-sheets arrangement.

The extension of this economic motive, however, creates the isomorphism seen in the industry.

Although the contractors were economically cheap, they still created expensive transaction costs due to a lack of flexibility between trades. As industrial relations can be defined as a 'system of rules' (Dunlop, 1958), the TCE framework transforms these rules into explicit costs: designing and negotiating the agreements can be regarded as the *ex ante* costs of contracting, and the procedural agreements within them as part of the *ex post* costs of monitoring and policing the contract.

So, the refinery management achieved a reduction in the transaction costs by the use of a SUA for contractors of £1.7 million per year. There was no direct economic saving: contractors were not asked or required to reduce their tenders for refinery work. This was not the ideology of unitarism, though it could be said to have had the same end result.

Ideology and 'mimetic' isomorphism

We can surmise, however, that a kind of managerial ideology prevailed in the pursuit of labour-based savings as opposed to capital investment. To change the nature of maintenance work (by using high-lift hydraulic platforms instead of scaffolding, for example), extreme demands were placed upon the workforce to exhibit flexibility. Though the client firms can well afford to pay high salary rates (an anti-union strategy in the USA, according to Kochan *et al.*, 1989), they cannot report to shareholders' evidence of inefficient working practices, especially in the ratio of employees to output (barrels of oil per man per day) still used in the industry. Though this measure may seem outdated, it is a common means of assessing productivity throughout UK industry.

In these circumstances, there was perhaps only a general feeling among the managements of the legitimacy of change due to the economic downturn and the encouragement by the government of the day for a kind of 'macho management'. This was not a deal in the same way that the SUAs in Japanese plants were: they had, as a core value, the incorporation of employees in a kind of Works Council (Beaumont, 1990). It has more in common with the 'scrabbing' or labour intensification described by Dunn (1993).

The seeking of legitimacy through isomorphic practices was a key initial feature of the case. When it was clear that there was no isomorphic solution, the management decided on a far more risky, individual, course. This might have implications for the wider construction industry, which is also not efficiently organised at the site level, and suffers from casualisation, multi-unionism and single-trade suppliers resulting in a low productivity culture. Lack of unions – deunionisation – is not transaction-cost saving as there remains the problem of those inter-group relativities that the unions usually manage (Walton and McKersie, 1965). In construction, an SUA solves this problem. It exists in Germany, for example, with Industry Group unions, and in Japanese enterprises in the UK. It is effectively a management device, which gives some legitimacy to change and yet reduces transaction costs.

15.11   Conclusions

Given the fact that in this case the refiner created an industry 'first', there are implications for other oil and chemical sites. The case could be used as a basis for 'mimetic' isomorphism. It is important to remember that the starting point for the refinery management was a search for an agreement that it could emulate, as seen in Tables 15.1 and 15.2. We might therefore see further SUAs and derecognitions on other sites; after the conclusion of the research study, two new SUAs were reported to the NJC. Thus, isomorphism is a present and continuing rationale for management.

If the role of economic necessity is paramount, instead of legitimacy through isomorphism, it may have implications for the wider engineering construction industry. The aims of the new agreement were basic and related to efficiency. The alignment with refinery working times, and the various tactical changes made for working in teams – a more efficient method than single-trade working – and the loss of the washing-up and tea-break times were an example of this approach despite the risk of upsetting previous established practices like the afternoon tea-break. This latter item was highlighted in the documents as it 'carried an IR warning'.

We already know from Garfit (1989) that the NAECI contained inefficient practices and its procedure was long and complex. Subsequent to the research study, according to the NJC, the economism of the SUAs led to a move to change the NAECI by using a local site Addendum to make the conditions and practices more efficient. The savings on transaction costs would then match those of the SUAs. This is 'stage 1' of the change, based on economics.

This new 'product' is now available to R&M operations which had previously rejected the full NAECI terms; using it to progress economic change is not as risky as an SUA as it is part of a set of legitimate National negotiations. On this basis, a new survey would reveal that, according to the NJC, this NAECI-SA (Supplementary Agreement) has been adopted in a mimetic fashion in a number of client firms – that is, without extensive internal economic calculations or justification. This is 'stage 2' of the change process.

These analyses indicate a complex inter-relationship between economic needs and ideology, resulting in isomorphism, or common practices. In terms of its isomorphic extension to the wider construction industry, which has a similar 'atmosphere' and efficiency problems, there are significant barriers. There is an important difference between the ECI's clients and those of the wider industry, as discussed earlier: contractors in R&M have a relatively stable existence on the site, and a single local solution to local issues is therefore possible. The same is not the case for construction in the wider sense, as sites are temporary and geographically dispersed, making the contractors and their labour force far more mobile. Accordingly, in coordinating these differences, a National Agreement is more efficient than a series of single site agreements.

Acknowledgements

This chapter is based on fieldwork and associated reports to ESRC under Award No. R000-22-1670.

References

Ahlstrand, B.W. (1990) *The Quest for Productivity: A Case Study of Fawley After Flanders*, Cambridge: Cambridge University Press.

Arrow, K. (1969) The organisation of economic activity: issues pertinent to the choice of market versus nonmarket allocation, in *The Analysis and Evaluation of Public Expenditure – the PPB System*, Vol. 1 US Joint Economic Committee 91st Congress, 1st session, US Government Printing Office, pp. 54–73.

Atkinson, J. (1984) Manpower strategies for flexible organisations, *Personnel Management*, 16, 8 August, 28–31.

Austrin, T. (1980) The lump in the UK construction industry, in T. Nichols (ed.), *Capital and Labour: A Marxist Primer*, London: Fontana, pp. 302–315.

Beaumont, P.B. (1990), *Change in Industrial Relations*, London: Routledge.

Blyton, P. and Turnbull, P. (1998) *The Dynamics of Employee Relations*, London: Macmillan.

Cibin, R. and Grant, R.M. (1996) Restructuring among the Worlds Leading Oil Companies 1980–1992, *British Journal of Management*, 7 December, 283–307.

Claydon, T. (1989) Union derecognition in Britain during the 1980s, *British Journal of Industrial Relations*, 27: 214–223.

Claydon, T. (1996) Union derecognition: A re-examination, in I.J. Beardwell (ed.), *Contemporary Industrial Relations: A Critical Analysis*, Oxford: Oxford University Press, pp. 151–174.

Coase, R.H. (1937) The nature of the firm, *Economica N S*, 4: 386–405.

Cross, M. (1989) *A Study of Maintenance Subcontracting in UK Industry*, City University, unpublished report.

DiMaggio, P.J. and Powell, W.W. (1983) The iron case revisited: Institutional isomorphism and collective rationality in organizational fields, *American Sociological Review*, 48: 147–160.

DiMaggio, P.J. and Powell, W.W. (1991) Introduction, in W.W. Powell and P.J. DiMaggio (eds.), *The New Institutionalism in Organisational Analysis*, Chicago: University of Chicago Press.

Dore, R. (1973) *British Factory, Japanese Factory – The Origins of National Diversity in Industrial Relations*, Cambridge: University of California Press.

Dore, R. (1986) *Flexible Rigidities*, London: Athlone.

Douma, S. and Schreuder, H. (1991) *Economic Approaches to Organisations*, London: Prentice Hall.

Dunlop, J.T. (1958) *Industrial Relations Systems*, New York: Holt.

Dunn, S. (1993) From Donovan to wherever, *British Journal of Industrial Relations*, 31: 2 June 1169–1187.

Engineering Construction Industry Training Board (ECITB) 2005 *Sector Work-force Development Plan*, at http://www.ecitb.org.uk/_db/_documents/ECITB_Sector_ Workforce_Development_Pla %5B20031203045332 %5D.pdf.

Flanders, A. (1964), *The Fawley Productivity Agreements – A Case Study of Management and Collective Bargaining*, London: Faber & Faber.

Gallie, D. (1978) *In Search of the New Working Class – Automation and Social Integration within the Capitalist Enterprise*, Cambridge: Cambridge University Press.

Garfit, T.C.N. (1989) *The Making of the Engineering Construction Industry: A Case Study of Multi-Employer and Multi-Union Bargaining*, Warwick Papers in Industrial Relations, 26 September.

Goldratt, E. and Cox, J. (1993) *The Goal – A Process of Ongoing Improvement*, London: Gower.

Grant, D. (1993) *Japanese Manufacturers in the UK Electronics Sector: The Impact of Production Systems on Employee Attitudes and Behaviour*, unpublished PhD thesis, LSE.

Grant, D. (1994) New-style agreements at Japanese transplants in the UK: The implications for trade union decline, *Employee Relations*, 16(2): 65–83.

Grant, D. (1996) Japanisation and new industrial relations, in I.J. Beardwell (ed.) *Contemporary Industrial Relations: A Critical Analysis*, Oxford: Oxford University Press, pp. 203–233.

Grant, R.M. (2002) *Contemporary Strategy Analysis: Concepts, Techniques, Applications*, Oxford: Basil Blackwell.

Grinyer, P.H. and Spender, J.-C. (1979a) Recipes crises and adaptation in mature businesses, *International Studies of Management and Organisation*, 9(3): pp. 113–133.

Grinyer, P.H. and Spender, J.C. (1979b) *Turnaround: The Fall and Rise of the Newton Chambers Group*, London: Associated Business Press.

Grinyer, P.H. and Spender, J.C. (1985) *Turnaround: Managerial Recipes for Strategic Success*, London: Associated Business Press.

Guest, D.E. (1987) HRM: And industrial relations, *Journal of Management Studies*, 24(5): 503–521.

Guest, D.E. (1989) Personnel and HRM: Can you tell the difference? *Personnel Management*, 2: 48–51.

House of Commons (1994), *Select Committee on Employment Report: The Future of Trade Unions*, Appendix 3, pp. 195–199.

Johnson, J. and Scholes, K. (2002) *Exploring Corporate Strategy*, FT: Prentice Hall.

Keenoy, T. (1979) *Invitation to Industrial Relations*, London: Macmillan.

Kelly, J. (1998) *Rethinking Industrial Relations Mobilisation Collectivism and Long Waves*, London: Routledge.

Kochan, T. Katz, H. and McKersie, R.B. (1989) *The Transformation of American Industrial Relations*, New York: Basic Books.

Korczynski, M. (1993) *Capital Labour and Economic Performance in the Engineering Construction Industry 1960–1990*, Unpublished PhD Thesis: University Of Warwick.

Korczynski, M. (1996) Centralisation of collective bargaining in a decade of decentralisation: the case of the engineering construction industry, *Industrial Relations Journal*, 28(1): 14–26.

Legge, K. (1996) *Human Resource Management – Rhetoric and Realities*, London: Macmillan.

Lloyd, J. (1990) *Light and Liberty: The History of the Electrical Electronic Telecommunications and Plumbing Union*, London: Weidenfeld and Nicholson.

Macdonald, A.M. (ed.) (1977) *Chambers Twentieth Century Dictionary*, Edinburgh: Chambers.

Mackay, L. (1986) The Macho Manager: It's No Myth, *Personnel Management*, 18: 1.

Marginson, P. (1993) Power and Efficiency in the Firm: Understanding the Employment Relationship, in Pitelis, C. (ed.), *Transaction Costs, Markets and Hierarchies*, Oxford: Blackwell, pp. 133–166.

McInnes, J. (1987) *Thatcherism at Work – Industrial Relations and Economic Change, Milton*, Keynes: Open University Press.

National Joint Council for the ECI (1994) *National Agreement for the ECI.*

NEDO (1970) *Large Industrial Sites*, Report London: HMSO.

Oldfield, F.E. (1966) *New-look Industrial Relations*, London: Mason Reed.

Pollert, A. (1988) The flexible firm – fixation or fact? *Work, Economics and Society*, 2(3): 281–316.

Porter, M.E. (1985) *Competitive Advantage: Creating and Sustaining Superior Performance*, New York: Free Press.

Purcell, J. (1982) Macho management and the new industrial relations, *Employee Relations*, 4(1): 3–5.

Rowlinson, M. (1997) *Organisations and Institutions*, London: Macmillan.

Salamon, M. (2000) *Industrial Relations Theory and Practice*, London: FT Prentice Hall.

Sampson, A. (1975) *The Seven Sisters: The Major Oil Companies and the World They Made*, London: Hodder and Stoughton.

Simon, H.A. (1961) *Administrative Behaviour*, 2nd edn, New York: Macmillan (original 1947).

Spender, J.C. (1989) *Industry Recipes: The Nature and Sources of Managerial Judgement*, Oxford: Basil Blackwell.

Stinchcombe, A.L. (1959) Bureaucratic and craft administration of production: a comparative study, *Administrative Science Quarterly*, 4(168):187.

Storey, J. (1992) *Developments in the Management of Human Resources*, Oxford: Blackwell.

Walton, R.E. and McKersie, R.B. (1965) *A Behavioural Theory of Labour Negotiations*, New York: McGraw Hill.

Whiston, J. (1988) Quality improvement process. A view from the chemical industry, in *Zero Defects A New British Quality Standard?* Conference Report CBI/Rooster Books Ltd., Stanstead Abbots, Hertfordshire, pp. 74–85.

Williams, I. (1989) Engineering construction's IR miracle, *Personnel Management* March.

Williamson, O.E. (1981) The modern corporation: origins, evolution, attributes, *Journal of Economic Literature*, 19 December, 1537–1568.

Williamson, O.E. (1985) *The Economic Institutions of Capitalism: Firms, Markets, Relational Contracting*, London: Free Press.

Willman, P. (1986) *Technological Change Collective Bargaining and Industrial Efficiency*, Oxford: Clarendon Press.

Wolfson, C., Foster, J. and Beck, M. (1996) *Paying for the Piper: Capital and Labour in Britain's Offshore Oil Industry*, London: Mansell.

Wood, S. (ed.) (1989) *The Transformation of Work*, London: Unwin Hamlyn.

Yergin, D. (1991) *The Prize: The Epic Quest for Oil Money and Power*, New York: Simon & Schuster.

Young, K.L. (1986) The management of craft work – A case study of an oil refinery, *British Journal of Industrial Relations*, 24(3): 363–380.

Young, K.L. (1992) *Flexibility and Craft Work – A Case Study of An Oil Refinery*, Unpublished PhD thesis: University of East London.

16 The culture of the construction industry

Emergence, recognition and nature?

Paul W. Fox

Abstract

In the period 1994–2001, several countries have produced studies of their construction industries with a view to further development.[1] Almost all of these have identified the need for the industry culture to change. The nature of culture within construction projects and construction firms has been the subject of study with growing interest, especially since the CIB Task Group 23 was set up in 1996.[2] Yet, surprisingly, *industry culture*, as distinct from construction project culture or construction company culture, is something which has hitherto been neglected both in terms of conceptualising it and in terms of empirical study. Almost all the research on culture in the construction industry relates to project and company culture, even though the term *industry culture* is frequently used in those same studies. Thus the boundary lines between these separate concepts are blurred. Such fuzziness is not surprising, as we realise that the people who work in construction companies and the projects they engage in do share common experiences within the construction business environment. Yet, as this chapter will argue, this environment possesses its own distinct nature in cultural terms. People who belong to the construction industry, whether they are employed, self-employed, unemployed, or under education and training, may share the same cultural characteristics.

16.1 Introduction

As stakeholders in the industry, both scholars and practitioners alike acknowledge the *existence* of a construction industry culture. Yet its *emergence* (i.e. Can we see it and describe it?) and its *recognition* (i.e. Do we agree on its definition and its full nature/character?) are steps not fully completed in the process of industry development. This is important, since to understand the nature of the industry culture is to better prepare and inform all stakeholders of the industry. How can we improve the construction industry culture if we don't understand its basic conceptualisation and character?

The chapter first uses mainstream management literature to explore the conceptual foundations and nature of a *construction industry culture* as distinct from the culture found in a construction firm or construction project. It adopts a holistic definition of the construction industry and then maps the mainstream conceptual foundations onto the construction industry characteristics. Then it uses construction industry literature to trace the transition from emergence to recognition of an industry culture. It does so in two ways. First, the chapter reviews the literature going back 50 years, tracing the path from where aspects of a construction industry culture may have first emerged. Secondly, it draws upon a recent international study of construction industries (Fox, 2003), where the nature of a construction industry culture has been articulated through a generic model of construction industry development. Through these two strands, the author explores the construction industry culture, developing the argument that, through various descriptions of artefacts and underlying values, the concept has found acceptance and hence *recognition* in the international construction community generally. In particular, this acceptance is established in the minds of construction industry stakeholders in various countries, both developed and developing. Both of these strands are used to support the assertion that the construction industry culture exists, has emerged and become visible to an extent, but in terms of recognition its nature has not yet been fully agreed upon between stakeholders. The second strand elaborates the possible nature of this culture in its various guises. Finally, some suggestions are made as to the extent to which this culture may be managed and fostered, so as to bring benefits to all industry stakeholders, both within the industry and those which it serves.

16.2 The emergence of an industry culture in mainstream management

In mainstream management literature, the concept of culture has grown in importance and especially concerning the nature of national culture (Hofstede, 1980, 2001) and organisation culture (Trompenaars, 2001). Various levels of culture have been the subject of studies, the main focus being those of corporate culture, project culture and team culture. These three concepts have also been explored and research into them has also become well established in the construction management field (Hall, 1999; Ankrah and Langford, 2005), such that the nature of culture within construction projects and construction firms has been the subject of study for over a decade. Yet the concept of an *industry culture* is something which has hitherto been neglected. Not only is there a dearth of studies within the construction domain about industry culture, but also in mainstream management. For example, Margaret Phillips found theoretical support for the existence of industry culture as reviewed from the domains of institutional theory, industrial economics, marketing, organisational behaviour and strategy (Phillips, 1994). However, she asserted that empirical support was weak, such that

> *it has not yet been determined if a shared culture exists among the wider set of participants in a well-defined industry...*

Hence, her study set out to identify and map industry *mindsets* – a term which she used interchangeably with culture. Given the need to start with a well-defined industry, the next section addresses this concern.

16.3   The definition of a construction industry

The culture for a construction industry must inevitably rest on the concept of a construction industry and how that industry is defined. If we cannot define the industry, we cannot define its culture. The definition of the industry is by no means a 'given', since various scholars have presented arguments that either support or negate a definition of a holistic industry. At the supportive end of the spectrum, conceptions of a whole industry go back at least 50 years to the earliest reports in the UK, such as Simon (GB MPBW, 1944) and Phillips (GB MoW, 1950). Alternatives to a holistic view have continued to surface periodically (Powell, 1979; Groák, 1994). It is likely that this dialectic will continue for some time. Recent support for the holistic idea is reinforced in several studies, both industry/practitioner orientated (Latham, 1994; Egan, 1998) and academic (London and Kenley, 2001; Fox, 2003).

The dialectic between a fragmented/aggregated view and a holistic view, as outlined above, has been apparent within mainstream construction literature. However, there is a growing body of knowledge from mainstream management scholars which is gaining ground in support of a holistic view. Notably, the work of Phillips has been seminal in not only arguing in support of an industry conceptualisation, but also exploring the nature of industry-based cultural groupings (Phillips 1994). The argument which follows is based on a holistic definition of the construction industry.

16.4   Linking mainstream management studies to a construction industry culture

The mainstream management developments referred to in the previous paragraph confirm not only the existence of a holistic industry concept, but also that it has cultural dimensions. Phillips (1994) explored the existence of *industry* cultures, as distinct from national culture or the culture of a firm. She drew upon the work of anthropological and sociological theory in which culture is viewed as '...*ideas, beliefs, and knowledge*...' (Spradley, 1972: 6), describing cultural knowledge as '... *a set of assumptions shared by the group of people*' (Phillips, 1994). Phillips was not alone in her conceptualisation, since, at the time of writing, almost 20 scholars have cited her work, none of whom have challenged it. Indeed, several have found support from within their own empirical work (Christensen and Gordon, 1999; Bierly *et al.*, 2000), including from the construction industry (Riley and Clare-Brown, 2001; Root, 2002). Phillips' work is seminal in the sense that her findings not only confirmed the distinction between national and organisational cultures, but additionally asserted the existence of *industry cultures*. Her empirical work did not involve the construction industry: she explored the concept of an industry culture through data collected from the wine-producing industry and that of fine-arts museums. She described the industry cultures as '... *broad-based assumption sets comprising the cultural knowledge widely shared among organizational participants within two industries (fine arts museums and California wineries)*...'. Later, she continued with the statements

> *The surfacing of distinct industry mindsets reinforces the emerging belief that a multiplicity of dynamic, shared mindsets exist within an organization's environment. A new cognitive lens – that of industry – is offered, through which scholars and managers alike can view behavior in organizational settings.*

She posited that management scholars would, through adopting an industry-based focus, be able to explore the source of existing cultural assumptions, as well as to enquire about what effect these have on the evolution of the industry. We have a particular interest in the latter question relating to the construction industry, as the later sections will elaborate. The significance of her contribution is that she invites further investigation into culture at the level of the industry as a single level of cultural analysis. This analysis, together in combination with analyses at other levels is necessary in order to sort out overlaid industry-based influences, prior to asserting that a culture is uniquely organisational or sub-organisationally based.

Recognition of these underlying values influencing behaviour at a sub-industry level was made by Mearns *et al.* (2004) in their study of psychosocial and organisational factors in offshore safety.

The significance of the work of Phillips was recognised by Lyle (1998) in her own study of culture, also at the level of an industry. Her framework and data collection were based on modifications to Phillips' work. Her findings were that industry cultures underlie corporate cultures, and they also provided empirical support for the existence of an 'industry culture'.

The multi-stranded nature of culture is important to the construction industry. Here is used a simple definition of culture, which is shared values, the attitudes and perceptions arising from them and the behavioural patterns and artefacts which manifest them. This broadly matches the definition by Schein (2004), who argued that a single culture could be conceptualised with different levels. The deepest level is those *unconscious basic values* and assumptions about life, reality, human nature, human activity and human relationships. Above that lie the *espoused values*, beliefs and attitudes. Above the second level lie the *artefacts*, the more visible manifestations of the first and second levels of concepts such as organisation structures and processes (Schein 2004: 26).

Whilst Schein's definition relates to any particular level of culture, different levels of culture can be interpreted in another way. This is more akin to a systems hierarchy where, for example, sub-systems form part of a larger system. Baker (2000: 220), asserted that '... *within any one distinct culture, one would expect to find a number of sub-cultures each with its own distinctive features and different from other sub-cultures'*. This multi-level conceptualisation is echoed by Alvesson (2002: 155), which reinforces the idea of a hierarchy of meta-cultures, cultures and sub-cultures. The conceptualisation is consistent with the idea of a national culture, with various different industry-level subcultures, containing, in turn, their own sub-cultures at the level of firms and industry associations/institutions. Alvesson specifically elaborates on the conceptualisation of an entire industry in terms of it being a sub-culture of the whole society, *'...Here the entire industry is conceptualised as a subculture. Individual organizations then may appear as sub-sub-cultures'* (Alvesson, 2004:155). For the construction industry this is not difficult to imagine, as we can readily identify examples at these levels, as well as at even lower levels (the multi-level subcontracting system, for example). The most obvious example, below the level of the firm, is the level of the construction project.

A systems-thinking approach leads on to another aspect of the conceptualisation of the construction industry with its associated culture. Many proponents of systems thinking state that every system has a purpose or function. It is this purpose which distinguishes a system from merely an aggregate of parts. Going along with this purpose is an associated set of values which underlie it. Checkland refers to this as the *Weltanschauung* (Checkland, 1981). Given the system's stated purpose, one can measure and evaluate how well it meets this purpose. Such evaluation would include an evaluation of the various component parts which it needs. However, an evaluation cannot avoid reference to the underlying values. Whether consciously or sub-consciously, the person undertaking the evaluation will have a reference framework of values which they draw upon in selecting the criteria used. It is this set of values that constitutes the foundations of culture. Thus, a conceptualisation of a construction industry culture would need to have its own set of values, and there may be several systems that can be proposed, each with their own set of values, each matching the root definition of the system defined. Some of these may be more relevant than others in helping stakeholders of the industry to agree upon and hence commit themselves to in terms of collective action (Fox, 1989).

Similarities with this mode of thinking can be found in the work of Spender (1989), who focussed on managerial judgement and the uncertainty that incomplete information creates. Spender articulates this through his explanation of 'industry recipes' which he describes as *'...a shared pattern of beliefs that the individual can choose to apply to his experience in order to make sense of it...'* (Spender: 61). He goes on to state that a recipe is not a theory or formula, but a complete framework of heuristics. As such it needs to be understood in a specific context which gives it meaning. Concerning governance, the making of managerial judgements requires a holistic understanding of the organisation. Having worked in several industries, Spender recognised a pattern of how managers within each industry make these managerial judgements which were *appropriate to that industry and its characteristics* (the author's emphasis added). He labelled this *'...professional common sense...'* or body of knowledge as the *'industry recipe*'. He further went on to explain the process of socialisation which newcomers to the industry/organisation would need to go through before they too could make good judgements. Although Spender recognises the term 'organisational culture', he does not use it, but prefers a looser, more ambiguous concept (Spender, 1989: 7). However, it is clear that his 'industry recipes' contain the set of assumptions and values that constitute an industry culture.

Turning this line of thought towards the construction industry, one might posit its purpose, state its definition, and then attempt to measure and evaluate how well it meets this purpose. How narrowly or how broadly its boundaries and the purpose, respectively, are defined will be a matter of debate for all stakeholders. For example, Gordon (1991: 401) argued '... *that if an organization is to survive, it will be built on certain assumptions required by the industry, and it is from these assumptions that certain values emerge, which, in turn, help define useful forms*'. We might regard this as a 'process' view of culture. Such an approach emphasises a process of enquiry into the idea of emergence of the values and their subsequent recognition by the stakeholders. Thus, in following a process approach to culture, in order to find evidence for the emergence of a construction industry culture, one might adopt such a process of enquiry. The next section attempts to address this process.

16.5 Tracing the emergence and recognition of construction industry culture

In the argument developed so far, the evidence for the emergence of a construction industry culture rests on the premise that if a generic industry-level culture exists, as Alvesson, Phillips and Spender indicate, such a phenomenon should exist in all industries, including construction. At the time of writing, the author knows of no empirical work that has set out to demonstrate this. Given that a culture is based on a set of values held in common and recognised as such by a group, the search within the case of the construction industry for the emergence of such concepts must surely start with studies which focus on the construction industry as a whole. The idea of emergence is consistent with Phillips' expression of the '... *surfacing of distinct industry mindsets* ...', as well as it being one of four key characteristics of systems, according to Checkland (1981). The value of tracing the emergence is to not only confirm the existence of a construction industry culture, but also give some clues to its recognition and its nature. Emergence will not be a sudden event. Some aspects are more visible than others at the early stages of the emergent process. Culture will be first seen through its artefacts, then ultimately through the underlying shared values.

Five holistic studies are presented here, each one providing some evidence of industry-level cultural characteristics sought. The first two are linked studies, carried out by the Tavistock Institute of Human Relations (Higgin and Jessop, 1965; Crichton, 1966). The second of these two coincided with the publication in the same year of the third one, namely Bowley's study of the British building industry (Bowley, 1966). The fourth study tried to develop the Tavistock work further (Napier, 1970). The fifth study by Ball (1988), continues on from Bowley's work. These five early studies which demonstrate the emergence of the industry culture may be regarded as seminal, having arisen more by chance than by research design. They are followed by over a dozen studies dating from Latham (1994), and covering the most recent decade, all of which confirm the emergent theme through their mostly explicit reference to culture.

The period 1960s–1980s showing implicit evidence of culture (emergence phase)

The Tavistock Studies 1966

The first two studies that introduced concepts related to an industry culture came from the work of the Building Industry Communications Research Project (BIRCP) at the Tavistock Institute of Human Relations (Higgin and Jessop, 1965; Crichton, 1966). Both of them refer to a system of social and personal relations. The first explored the social status and attitudes of stakeholders, whereas the second study incorporated these cultural artefacts in systems and operations research models. It can be argued that the existence of an industry culture can be shown through these artefacts.

Bowley 1966

The third study presented here to identify a cultural characteristic of the industry was by Bowley (1966) in her book of the British building industry covering an 80-year period prior to publication. Given her background in political science, she introduced a unique perspective. It is not known to what extent she was aware of the work by Tavistock. However, there is a remarkable similarity to Tavistock in the way she describes the socio-political characteristics and their strong influence at that time. Bowley claimed that the main obstacle to efficiency in the building industry was the '*outworn pattern of organisation*' (Bowley, 1966: 441), or simply '*the system*' whereby the roles of the main stakeholders (building owners, building professions and the builders) and relationships between them were based on 'social class-distinctions'. Preservation of these relationships was more important than economics, with the consequence that throughout the whole of the twentieth century, the building industry failed to satisfy a large proportion of its clients, despite a series of reports and studies which clearly spelt out the nature of the problems (Murray and Langford, 2003: 1–84). It should be emphasised that *The system* referred to the building sector rather than civil engineering, although Bowley included descriptions of both sectors and compared them. Most dominant in her thinking was the role of architects as an elite, at the top of a social hierarchy. Below them were engineers who were closely associated with trade and industry, and, as a consequence, of lower social status. Surveyors were even more socially inferior, whilst 'builders' were forming the lowest ranks along with the workmen:

Builders, however wealthy or successful, were in trade and small builders were little more than glorified craftsmen, 'cap-in-hand builders' as they are sometimes termed.

(Bowley, 1966: 350)

Within *The system*, Bowley also identified *The establishment*, which was simply the version of *The system* approved by the architects, in which the architect was recognised as head of the hierarchy. As Bowley describes,

This was the only form of the system that was really respectable, and which building owners concerned with culture or social prestige could use.

(Bowley, 1966: 350)

Although it can be seen that there were numerous consequences of the system, [for example, failure in communication between specialists, absence of competition in design, and divergence of interests of designers and clients], the main causes of these characteristics were the values and attitudes of the stakeholders, particularly those in positions of power and influence, such as the architects. The contribution by Bowley is clearly significant, particularly in highlighting the social relations, and the values and attitudes important in the culture of the industry and its stakeholders. It is these factors which seem to have been overlooked until more recently. In short, this was a cultural phenomenon.

The fourth study by Napier (1970) was a further development from the work of the Building Industry Communications Research Project (BICRP) at the Tavistock Institute of Human Relations. He was strongly influenced by the then emerging use of systems theory, particularly the work on the UK building industry by (Higgin and Jessop, 1965; Crichton, 1966) at the Tavistock Institute. Napier attempted to model the Swedish construction industry and conceptualised a '*model of reality*'. There was no acknowledgement of key stakeholders including government (national, regional and local), educational and training bodies, or professional institutions, to name a few. However, he mentioned the importance of various stakeholder functions, their values and interests. He asserted that these stakeholders are key factors, and in their behaviour he advocated several additional concepts of (1) power, (2) status, (3) learning, (4) boundaries, (5) goal evaluation, (6) innovation and (7) sentient group values as being important (Napier, 1970: 36). Of the four main stakeholders, namely clients, producers [consultants, contractors, subcontractors and suppliers], users and society, he considered that the latter two groups were especially important.

Of all the concepts which Napier considered as being able to explain important current problems of the industry, he included these seven numbered above (Napier, 1970: 136), added to the following four sections of his theoretical model of the formal system and system of resource controllers, namely

1   The value system
2   The communication system
3   The informal system
4   Change.

His conceptual framework is difficult to follow as he used several interlinking models to present his thinking. Nevertheless, his emphasis on social and political concepts, especially the sentient values of all stakeholders, is highly relevant to the idea of an industry culture.

Ball 1988

Ball considered a number of theoretical positions postulated by various authors, for example, Hillebrandt's 1974 analysis using models of oligopoly, Winch's use of transaction costs (1985), and contributions using mainstream Marxism relationship between capital and labour. However, these were all rejected by Ball on the grounds that they did not fit, or did not adequately explain, and did not take into account the various stakeholders, and political, market and historical contextual forces that he argued were important.

Ball recognised the unique contribution of Bowley, concerning the importance of social factors in developing the industry (Ball, 1988: 33). At the same time he was critical of her analysis for it being one-sided, and contended that it was not supported adequately by evidence and explanation. In his own analysis, he argued that social relationships are indeed important. To begin with, Ball made a surprising claim about previous studies attempting to look at the industry as a whole, and in particular a definition of the industry:

> *There has been little previous work looking at the economic operation of the industry as a whole. One reason could be that the industry is assumed to be essentially no different from any other. The next chapter, however, will argue that it is different in a number of crucial respects, ones that derive essentially from the ways in which distinct social agents combine in the physical act of construction. It is the combination of the social and physical nature of the construction process which simultaneously defines the boundaries of the industry and highlights its relative uniqueness, as later chapters will argue.*

(Ball: 18–19)

Ball then developed his analysis over several chapters. The main themes of his argument are explained in Chapter 11 '*A New Social Balance in the Construction Industry?*'

Similar to Bowley, he believed that the Contracting System embodied the most important set of social relations between architects, other design professionals, surveyors, building contractors and organised labour. Ball's inclusion of the social relations as an essential part of the system further supports the notion of an industry culture, although he did not use this term explicitly.

These five reports each show evidence of an industry-level culture operating in the construction domain. Together they provide clues as to when this culture became emergent, mainly through artefacts and implicit references.

The period 1990s–2005 showing mostly explicit evidence of culture (the recognition phase)

Since the publication of these early reports/studies, much more is known about culture through research in mainstream management, as well as within the construction domain, for example through CIB TG23.[3] In the period 1994–2006, there have been many reports/studies about the construction industry, not only in the UK context, but internationally. A sample of these is briefly reviewed here in chronological order, starting with Latham. The studies were all framed from within the industry, even if some of the originators were not (such as Latham and Egan), strictly speaking, 'construction people'. These collectively reinforce the notion that the construction industry does indeed have its own culture. That these studies speak of an industry culture is itself confirmation that the concept is gaining acceptance amongst the industry stakeholders. These are covered in chronological order and description of each is kept deliberately brief, merely to demonstrate their linkage.

Although Ofori (1993) does not write in terms of a construction industry culture, he does explore the industry's *purpose* (pp. 35–41) and refers to a socio-cultural context (UNCHS, 1984 and Wells, 1986) in which the various stakeholder roles are played out. This is followed up by discussion of the measures for improving the industry, where it needs to be considered in holistic terms.

Latham (1994: 39) was quite explicit in his use of the term *culture* as applied to the industry as a whole, although he did not use this term frequently throughout his report.

Seymour and Rooke (1995) made strong opening arguments in their paper that there was a strong need to review the industry culture, but this could only occur if the research culture changed first. Here we see reference to not only the industry culture, but also a subset of it, the research culture.

Although Lenard's (1996) focus was at the level of construction firms (in a comparison with their counterparts in manufacturing), in carrying out his study, he was the first to explicitly focus on cultural aspects of the industry. The main factors from an industry-wide view were

- the need for cross industry information systems and knowledge bases
- the need for co-operative learning between industry participants (Lenard, 1996: 201).

His advice was not only about the nature of construction industry culture, but also how to deal with it.

Egan (1998) followed up with Latham's report, and confirmed the framework that it used, including its at least six specific references to an industry culture.

Singapore Construction 21 (1999) picked up on the Latham and Egan reviews, and further confirmed the cultural concept.

South Africa (1999) also made reference to Latham, but had independently conceptualised a need to introduce cultural change, through its use of the industry transformation framework. The emphasis on culture is not explicit, but implied through the strong use of vision, empowerment through participation and training, and affirmative action.

Akintoye (2000: 166) refer to the industry culture as having a significant influence on the introduction of supply chain management to the UK construction industry, it being a barrier.

Similar references to industry culture had become accepted when considering the Swedish and Hong Kong construction industries (Flanagan *et al.*, 2001; Hong Kong, 2001).

Bröchner *et al.*'s (2002) paper included this concept as a given, and included it in the title. The paper reviewed a century of construction literature in order to conceptualise and define the nature of the Swedish construction industry culture.

Rooke *et al.* (2003) open the introduction to their paper with a statement about the industry culture, and link to the Tavistock studies.

The final example of evidence of a construction industry culture is from Fox's 2003 study. His survey and conceptualisation of the construction industry development model further consolidated the evidence for existence and nature of the industry culture. The grounded theory approach adopted in the research study meant that the concept of industry culture was not a given factor at the outset. It only emerged and its nature recognised as the data was analysed. The study collected data relating to two separate analyses/frameworks. The first used data about the current strength of factors which contribute to develop the construction industry. The second set used data about the future importance of factors. Each data set was analysed in a similar approach using statistical factor analysis. The analysis reduces a large number of known and measured variables (the visible artefacts) into a small number of factors. The technique is well established and the data processing arranges similar variables into groups such that several within a group can be represented by an underlying construct, described as a Factor, with its own name. Since the Factor has not been measured directly in the data collection process, it only emerges after the variables have been statistically grouped together. Consequently, as well as the statistical linkage between variables and the Factor, there should be a logical linkage between the Factor and the variables which are grouped under it.

Of the eight factors in his model contributing to current industry development, four of them were cultural factors. These were

1   *Human skills and a culture of transparency*
2   Financial resources and investor confidence
3   Government policies and strategies supporting construction business

4 *A self-reliant construction culture*
5 Institutional support
6 *Supportive attitudes from aid agencies*
7 Research and development for construction
8 *Industry-based better practice and culture.*

In the above list, the four highlighted in bold were new factors and were cultur-ally based. The others had been identified from previous studies. The ranking of them in this list corresponds only approximately in relation to their strength. Item 6 is not a cultural influence from within the industry, so this will not be further considered here.

Of the six factors in his model contributing to future industry development, he included three traditional factors and three new factors. These were

1 *Thinking the best and behaving the best (a better practice culture)*
2 *Long-term vision and policy for the industry*
3 Financial and human resources
4 *A learning culture*
5 Techniques and technologies supporting high production performance
6 Basic resources and infrastructure (physical and institutional).

All three new factors (highlighted above in bold) were culturally based. Each of the three was articulated as necessary and distinct dimensions. The nature of these three are explained in more detail in the section which follows. Most of the commentary is based on interpretations of the statistical data, but reference is also made to interview data collected as well as the literature.

16.6 The nature of a construction industry culture

Given that the earlier sections have traced the emergence of a construction industry culture, this might imply that the industry culture is mono-dimensional. However, this section will elaborate on the concept and show that there are several dimensions to construction industry culture. These dimensions have two different time perspectives. One is current, and the other relates to the future.

In the previous section, the conceptual frameworks of Fox (2003) identified four currently active factors and a further three dimensions of future importance to construction industry development. Discounting the attitudes of aid agencies, the remaining six factors were concerning industry culture. Between the two sets, currently active and future importance, there is one common factor, the better practice culture. This is explained separately in the discussion sections which follow.

Dealing with the three currently active factors first, the most dominant is *Human skills and culture of transparency*. There is no doubt there is wide-ranging human skills and human resources that are required for a successful construction business. Construction is a highly complex process involving a large number of specialists, an enormous range of materials, specialist plant and equipment, organisations, and a huge variety of unique products. It would not be difficult to argue that of all the sectors that a person could choose as a working occupation the construction sector is the most complex. In modern society we take for granted the need for a wide range of specialists each with their own skills and areas of special knowledge. As Lawrence and Lorsch (1967) have shown us in their study of differentiation and integration within organisations, for a higher level of performance the various specialists (highly differentiated) need to be effectively co-ordinated (integrated). We have learnt that highly educated and trained people as specialists need also to have generalist skills, so that they may communicate effectively with each other and co-operate (Goleman, 1996; Fox, 1996, 2002). A series of research studies, of government reports and more general articles over the last 50 years, have stressed the need to integrate people in the construction industry more effectively (Banwell, 1964; Crichton, 1966; Wood, 1975; Latham, 1994). Yet human capital cannot fully realise its potential unless there are appropriate co-operative attitudes hand in hand with the specialist knowledge (Rwelamila and Hall, 1994; Walker, 1995). At the heart of every construction project is teamwork, and it is no accident that Latham's reports about the UK construction industry were called 'Trust and Money' and 'Building the Team'.

The second strand of this factor is labelled a *culture of transparency*. For people to co-operate with each other there needs to be an attitude of willingness to co-operate. Over a period of time, successful partnership can develop, and can be defined as a good level of trust between the parties. In turn, trust depends on honesty and integrity and openness between the parties. As well as several research studies or reports which advocate a greater levels of trust, co-operation and more formal mechanisms such as partnering (Latham, 1994; HKHA, 2000), a number of interview respondents (Fox, 2003) mentioned values and attitudes of significance.[4] Two of them even mentioned this as being THE most important factor. To a certain extent the strand of transparency is a political dimension.

A government's concern for its image was put forward by one of the respondents to explain how governments sometimes make decisions incorrectly in order to present a 'correct' image to the public. Thus, instead of making decisions in the best interests of the industry, a more expedient choice is preferred because it has a more attractive appearance. Two examples of this were in developing countries, Thailand and the Philippines, where government would prefer to use mechanisation in construction instead of labour-based methods, even though this was more expensive and even though there was ample labour supply for the latter. Image was more important than economics. Making the correct decision requires more effort in explaining to the key stakeholders, and hence it demands, in turn, a higher level of transparency as well as greater availability of information.

This factor therefore is concerned with both the construction industry environment as well as the general business environment and the wider social environment in society generally.

The second cultural factor currently active was a *Self-reliant construction culture*.

In the developed countries, the construction industry has well-established skills specialists, education and training, all of which contribute to a self-reliant construction culture. Some studies have argued that the major contractors have been too strong and powerful for the good of the industry as a whole (Ball, 1988; Cockerill, 1993). For example, Ball argued that many of the industrial relations problems in the 1960s and 1970s stemmed from a manipulation by main contractors in order to extract more money from their clients through claims.

Evidence from Japan indicates that the main contractors have been very influential on political leaders and political parties, such that the government has awarded construction contracts when they are not really justified. It is only in recent years that this latent influence has become more widely known to the public at large. However, it does illustrate well a strong influence of contractors and construction culture. Such self-reliant strength explains how Japanese contractors not only have understood the need for government support, but have actively taken steps to ensure it to their own advantage. This behaviour is of course not accepted by many members of society in Japan who regard it as covert and not in the best interests of the society as a whole. However, through these methods, the Japanese construction industry has become the world leader in construction Research and Development (Seaden and Manseau, 2001).

A strong construction culture is not unique to Japan. For example, in Hong Kong the main contractors of the construction industry pushed government for the setting up of their own training organisation, the Construction Industry Training Authority, before government had even thought about it for other industry sectors. This authority had become established, funded by a levy on all large construction projects, and established by statute, years before the government decided to introduce training institutions for a range of occupations more generally through the Vocational Training Council.

Thirdly, the factor namely *Industry-led better practice and culture* is considered. This was not necessarily the strongest factor in terms of the score given by the respondents. Thus we have to conceive of this factor as being more latent than apparent. Although it linked several variables together, the individual variables had been scored at a lower strength level collectively. From a developed country point of view it is clear that there is already a strong push to encourage better practice. Examples include the recent reports from Latham and Egan in the UK, supported by the output from construction industry institutes in the USA, Europe and Australia. Benchmarking, partnering and the use of construction IT are all characteristic of publications from these institutes. A strong emphasis on fostering an industry culture is also readily apparent from industry reports from the UK (Latham, 1994; Egan, 1998; Flanagan *et al.*, 1999: 7), Singapore (Construction C21, 1999), Hong Kong (HK CIRC, 2001) and Australia (AUS ISR, 1999).

From a developing country point of view, there is also a concern for the use of IT, and certainly there is an aspiration to do things better than has been normal practice in the past. The attitudes and desire in each case, whether developing or developed country, is the same. The main difference between these two types of country is the extent to which good practice or best practice has been achieved.

Turning now to the cultural factors which are important in the future development of the industry, there are three strands to the industry culture. The first of these is *Thinking the best and behaving the best (a better practice culture)*. If the construction industry wants to achieve better practice, it must also think the best. Fox (1999) had already indicated the strong influence of values upon industry development as captured through a series of interviews with experts in construction industry development.[5]

Following this statistical combination, four variables were grouped under the factor of *Thinking the best and behaving the best (a better practice culture)*. In sequence, these were

1   Ethical behaviour
2   Communication between government and contractors
3   Attention to organisation culture
4   Government's understanding of the construction industry.

The first of these four, 'Ethical Behaviour', is self explanatory. The third, the variable of 'Attention to organisation culture', has a common concept of values which connects culture with ethical behaviour. After all, culture is all about shared values. The linking of individual values in order to establish a culture is clearly important.

Of the two remaining variables loading on this factor, one is almost as strong, namely *Communication between government and contractors*. This is important to encourage good ethical behaviour and can assist in identifying good practice. Some studies have put forward strong arguments that since government is a major client of the industry that it can (Wells, 1998; Hindle, 2000: 245; Milford, 2000: 242) and should use its influence as a client in order to promote and to encourage better practice (Egan, 1998: 39; HK CIRC, 2001: 2). Examples from the UK and Hong Kong both support this point. Not loading quite so strongly but also consistent with this factor is the variable *Government's understanding of the construction industry*.

The factor *Thinking the best and behaving the best (a better practice culture)* has a strong ethical and communications flavour as shown in the analysis. Furthermore, the government can take a prominent role in promoting best practice within the industry, through its role as a client and its good understanding of the industry as a whole.

The second dimension is the one labelled *Long-term vision and policy for the industry*. Fox argued that this factor would need a champion for change, and that proposing a vision was something that needed to be formulated consensually amongst all stakeholders if it were to be successful. A vision is strong only if it impacts on the people needing to receive it. A strong vision and a coherent strategy need strong leadership, and this is an important implication of his findings, since seldom does this need come through in earlier research studies. Most of all, it would require that any force for change should come from within the industry itself, from its own vision, from its proactive role towards excellence and from its own participants taking responsibility for personal and organisational development. The implication of this characteristic is that collectively the industry needs to change its culture from a reactive one to a proactive one. This is a very distinctive difference between the current situation and that of the future. Readers are reminded that Fox's model was developed from an international survey, and thus represents a generic conceptualisation, rather than one founded on only one particular country's industry.

Turning now to the third factor, A *learning culture*, there are three variables which load on this factor and they are to do with help, self-help and self-development combined with the organisation culture. The first two variables originally came from experience in South Africa, namely *The mentor system* and *Encouragement for contractor's self-development through ladder of opportunity*. It is interesting to note that the respondents in the wider survey generally have considered this an important value. The third variable, the *Use of computing skills*, initially appears not to fit with the other two variables. It can be argued that the use of computing skills is something which encourages individual development and self-learning, particularly with access to the Internet. Thus this factor is labelled 'a learning culture' that is important for the development of the industry as a whole. From a developed country perspective, this emphasis comes through in a number of countries as identified in national reports of their respective construction industries, for example, UK Latham report (1994) Hong Kong CIRC (2001) and HKHA (2000), and Singapore Construct 21 (1999). Evidence from developing countries of the need for a learning culture is not so common. A learning culture is something that is necessary and important within the construction industry, and so a large number of stakeholders within the industry need to be involved. In addition, stakeholders outside the industry would also need to play their part, especially the national government through any appropriate changes in the education and training processes generally.

Apart from the findings by Fox, other industry cultural aspects do not yet appear to have been systematically identified. However, we should not overlook assertions that the industry has a claims culture (Chan and Tse, 2003; Rooke *et al.*, 2003), although this may apply more to the countries that have adopted the British way of doing things. Thus, in terms of recognition of the nature of industry culture, further studies need to be done, perhaps on a country-by-country basis, to tease out points of difference and points of similarity to a generic model presented here.

In conclusion to this section, it may be seen that the culture of the industry is not a simple concept. Rather it is complex and multi-dimensional. It varies somewhat between developed and developing countries, but these differences are not large. The differences should be seen more as a matter of degree, or emphasis, rather than totally different dimensions.

16.7   Managing the construction industry culture

Having described the various strands of a construction industry culture, this final section makes some suggestions about the extent to which this may be managed. The material presented here does not rely on new empirical data, but instead draws upon that already presented in the earlier sections, and may be seen as a logical extension, perhaps by implication, in some respects.

Management of the construction industry culture at the national level may be through central government direction through government departments and ministries. This will hardly ever apply in practice in pure form. Rather, the government will likely take a lead and work together with stakeholders such as industry development boards, research institutions, educational institutions, training institutions, contractor federations, professional associations, trade associations, trade union associations, client federations, pressure groups, and the national press. The extent to which each of these institutions plays its role will depend on the maturity of the country concerned and the extent to which the industry has developed the depth of institutional support characteristic of those countries. The most mature, such as the USA, will have strong private-sector-led support through agencies such as the Construction Industry Institute (CII) that provide the leadership necessary to research and bring about change in the industry. For developing countries, all of which lack the richness and variety of comparable institutions, will need stronger direction and leadership from central government sources.

Exactly how this change may be brought about is still uncertain. Despite the efforts and exhortations of governments and industry leaders, the UK industry was highly resistant to change over a period of 40 years from the Second World War era. Perhaps broader economic and political influences have a greater impact than focussed construction-orientated bodies such as the Construction Industry Board or similar bodies. Perhaps leaders delude themselves if they believe that they can initiate change in the direction they desire. Such actions may be counterproductive. Perhaps the best we can do is to monitor some key aspects, and publish the results, and let the industry culture evolve naturally and organically.

In presenting the evidence for the nature of a construction industry culture, the nature of culture at the level of construction firms or construction project has not been acknowledged. At these other levels, the nature of culture needs no justification, since many researchers, including members of CIB TG23, are already in the process of exploring them comprehensively (Fellows and Seymour, 2002). However, insufficient understanding exists about the boundary lines between the three levels of culture: industry, firm and project. Some aspects have already been considered, if not systematically or sufficiently. For example, some have considered the boundary line between industry and national culture through the UK experience (Morrell, 1987), Singaporean experience (Ofori, 1994), South African experience (South Africa, 1999), Hong Kong experience (HK CIRC, 2001), and the mainland Chinese experience (Lu and Fox, 2001).

16.8 Conclusions

The chapter has accepted the existence of a construction industry culture, then attempted to argue the case for its emergence and recognition through a review of selected holistic construction industry studies, by reference to studies from mainstream management. Emergence is demonstrated first implicitly through the appearance of cultural artefacts in the period 1960s–1980s. Explicit emergence followed thereafter. The recognition of an industry culture has indeed been shown through wide acceptance amongst stakeholders of the industry, particularly through the work of Fox (2003).

It has then explored the nature of a construction industry culture, by presenting evidence from a variety of sources and various countries. The nature of this culture has also been explored through the study by Fox 2003. This exploration considered not only the current strength of three cultural factors in the industry, but also the future importance through a further three dimensions. Having established them, we should not merely accept these six dimensions. Certainly, they help to give direction to any vision which may seek to change the culture, but this is not sufficient on its own. The culture, having a dynamic of its own, through the eyes of its stakeholders, needs to be further explored and articulated. The vision needs to be embraced by all stakeholders, and those who would seek to lead the change need to champion their cause. They will need the political skills to influence those around them, in whatever institutional setting they find themselves. The benefits, if any, will be for those stakeholders to judge. Such an analysis may simply help such a vision to be established and explored, if not fulfilled.

Notes

1 For example, Australia (Australia ISR, 1999), Hong Kong (Hong Kong (China), CIRC, 2001), Singapore (Singapore Construction C21 Committee, 1999), South Africa (South Africa, Ministry of Public Works, 1999), Sweden (Flanagan *et al.*, 2001) and UK (Latham 1994).
2 International Council for Research and Innovation in Building and Construction (http://www.cibworld.nl/website/).
3 Ibid. Now working commission W112 "Culture in Construction".
4 Fox's study incorporated 25 interviews of construction industry development experts. These were used to compile the two questionnaires used to collect data on CURRENT STRENGTH and FUTURE IMPORTANCE of variables affecting construction industry development.
5 The 25 interviews represented different country experiences.

References

Ankrah, N.A. and Langford, D.A. (2005) Architects and contractors: A comparative study of organizational cultures, *Construction Management and Economics*, 23, 595–607.
Akintoye, A. (2000) A survey of supply chain collaboration and management in the UK construction industry, *European Journal of Purchasing & Supply Management*, 6, 159–168.
Alvesson, M. (2002) *Understanding Organizational Culture*, London: Sage Publications.

Australia, Industry Science Resources (1999) *Building for Growth: An Analysis of the Australian Building and Construction Industries*, Commonwealth of Australia.

Baker, M.J. (2000) *Marketing Strategy and Management*, Basingstoke: Macmillan Press Ltd.

Ball, M. (1988) *Rebuilding Construction: Economic Change and the British Construction Industry*. London: Routledge.

Banwell, H. (1964) see Great Britain, Ministry of Public Building and Works.

Bowley, M. (1966) *The British Building Industry*, London: Cambridge University Press.

Bierly, P.E. III, Kessler, E.H. and Christensen, E.N. (2000) Organizational learning, knowledge and wisdom, *Journal of Organizational Change Management*, 13 (6) 595–618, Emerald: Bradford, ISSN 0953–4814.

Bröchner, J., Josephson, P. and Kadefors, A. (2002) Swedish construction culture, quality management and collaborative practice, *Building Research & Information*, 30 (6) 392–400.

Chan, E.H. and Tse, R. (2003) Cultural Considerations in International Construction Contracts, *Journal of Construction Engineering and Management ASCE*, 129, 4, 375–381.

Checkland, P.B. (1981) *Systems Thinking, Systems Practice*, London: John Wiley & Sons.

Christensen, E.W. and Gordon, G.G. (1999) An exploration of industry, culture and revenue growth, *Organization Studies*, 20 (3) 397–422, W. de Gruyter: Berlin, ISSN 0170-8406.

Cockerill, J.E. (1993) *The Construction Industry in Belfast 1800–1914*, unpublished PhD thesis, Department of Economic & Social History, The Queen's University in Belfast.

Crichton, C. (ed.) (1966) *Interdependence and Uncertainty: A Study of the Building Industry*, London: Tavistock Publications Ltd. [see also The Tavistock Institute of Human Relations].

Egan, Sir J. (1998) see Great Britain. Dept. of the Environment, Transport and the Regions (DETR).

Fellows, R. and Seymour, D. (eds) (2002) *Perspectives on Culture in Construction*, Task Group 23 – Culture in Construction, CIB Publication No. 275, Rotterdam: International Council for Research and Innovation in Building and Construction, ISBN 9063630301 (pbk).

Flanagan, R., Jewell, C., Larsson, B. and Sfeir, C. (2001) *Vision 2020 – Building Sweden's Future*. Dept of Building Economics and Management, Chalmers University of Technology, Göteborg.

Fox, P.W. (1989) *A Study of the Hong Kong Construction Industry Using a Systems Approach*, Unpublished MSc Thesis, University of Salford.

Fox, P.W. (1996) Training for quality in the Hong Kong construction industry, *Proceedings of 1996 CIB Beijing International Conference*, Beijing, PRC, 21st–24th October 1996 CD ROM-[no page numbers].

Fox, P.W. (1999) Construction industry development: Exploring values and other factors from a grounded theory approach, *Proceedings of CIB W55 & W65 Joint Triennial Symposium*, Cape Town, September 1999. CIB Publication 234, ISBN 0-620-23944-1, V1 121–129.

Fox, P.W. (2002) Training for quality in the construction industry: Lessons from Hong Kong, In Ogunlana (ed.), *Training for Construction Industry Development*, CIB W107 Report, CIB Publication No. 282, ISBN 974-8208-52-4, pp. 41–60.

Fox, P.W. (2003) *Construction Industry Development: Analysis and Synthesis of Contributing Factors*, unpublished PhD thesis, Queensland University of Technology, Brisbane, Australia.

Goleman, D. (1996) *Emotional Intelligence*, London: Bloomsbury.

Gordon, G.G. (1991) Industry determinants of Organizational culture, *Academy of Management Review*, 16 (2) 296–415, Academy of Management: Ada, Ohio, ISSN 0363-7425.

Groák, S. (1994) Is construction an industry? *Construction Management and Economics*, 12, 287–293.

GB DoE. (1998) Great Britain. Dept. of the Environment, Transport and the Regions (DETR)(1998) *Rethinking construction*, The Report of the Construction Task Force to the Deputy Prime Minister, John Prescott, on the scope for improving the quality and efficiency of UK construction, chaired by Sir John Egan (The Egan Report).

GB, EDC. (1975) Great Britain, EDCs for Building and Civil Engineering (1975) *The Public Client and the Construction Industries* (Chairman: Sir Kenneth Wood), NEDO, HMSO: London (The Wood Report).

GB, MoW. (1950) Great Britain, Ministry of Works (1950) *Survey of Problems Before the Construction Industries*, A report prepared for the Ministry of Works by Sir Harold Emmerson, HMSO: London (The Phillips Report).

GB, MPBW. (1944) Great Britain, Ministry of Public Buildings and Works (1944) *The Placing and Management of Building Contracts: Report of the Central Council for Works and Buildings* (Chairman: Lord Simon), HMSO: London (The Simon Report).

GB, MPBW. (1964) Great Britain, Ministry of Public Building and Works (1964) *The Placing and Management of Contracts for Building and Civil Engineering Work*, A report of the Committee under the chairmanship of Sir Harold Banwell, HMSO: London (The Banwell Report).

Hall, M.A. (1999) *International Construction Management: The Cultural Dimension*, Unpublished PhD thesis, Liverpool University, 403pp. [pagination error: page 281 to 295 are incorrectly numbered].

Higgin, G. and Jessop, N. (1965, 2001) *Communications in the Building Industry*, London: Tavistock Publications. ISBN 0415264405.

Hillebrandt, P.M. (1974) *Economic Theory and the Construction Industry*, London: MacMillan Publishers Ltd.

Hindle, R. (2000) Construction industry development through intervention – A right and a wrong way?, *Proceedings of the 2nd International Conference on Construction in Developing Countries, 'Challenges Facing the Construction Industry in Developing Countries'*, 15–17 November 2000 Gaborone, Botswana, ISBN 999 12-2-156-5.

Hofstede, G. (1980) *Culture's Consequences: International Differences in Work-related Values*, Beverly Hills, Calif.: Sage Publications.

Hofstede, G. (2001) *Culture's Consequences: Comparing Values, Behaviors, Institutions, and Organizations Across Nations*, 2nd edn, Thousand Oaks, Calif.: Sage Publications.

Hong Kong (China), Construction Industry Review Committee (2001) *Construct for Excellence: Report of the Construction Industry Review Committee*, Report of the Construction Industry Report Committee under the chairmanship of Henry Tang, Hong Kong, China. Hong Kong SAR Government. (The Tang Report).

Hong Kong Housing Authority (China) (2000) *Quality Housing: Partnering for Change*, Consultative Document.

Latham, M., Sir. (1994) *Constructing the Team: Final Report*, July 1994: Joint review of procurement and contractual arrangements in the United Kingdom construction industry, HMSO: London.

Lawrence, P.R. and Lorsch, J.W. (1967) *Organization and Environment*, managing differentiation and integration, Harvard University, Division of Research, Graduate School of Business Administration: Boston, Mass.

Lenard, D.J. (1996) *Innovation and Industrial Culture in the Australian Construction Industry: A Comparative Benchmarking Analysis of the Critical Cultural Indices Underpinning Innovation*, unpublished PhD Thesis, The University of Newcastle, Australia.

London, K.A. and Kenley, R. (2001) An industrial organization economic supply chain approach for the construction industry: A review, *Construction Management & Economics*, 19, 777–788.

Lu, Y.J. and Fox, P.W. (2001) *The Construction Industry in the 21st Century: Its Image, Employment Prospects and Skill Requirements – A Case Study from China*, Sectoral Activities Programme, Working Paper WP180 for the ILO, Geneva. 50pp, ISBN 92-2-112858-X.

Lyle, L.G. (1998) *The Performance of Industry Culture: Assumptions, Sources and Evolutionary Patterns as Revealed in the Paradigmatic Interplay of Reporting Structures and Communicative Processes*. Published PhD thesis. The University of Tennessee, Knoxville.

Mearns, K. *et al.* (2004) Evaluation of psychological and organizational factors in offshore safety: A comparative study, *Journal of Risk Research*, 7 (5) 545–561, Carfax Publishing.

Milford, R.V. (2000) National systems of innovation with reference to construction in developing countries, *Proceedings of the 2nd International Conference on Construction in Developing Countries, 'Challenges Facing the Construction Industry in Developing Countries'*, 15–17 November 2000, Gaborone, Botswana, ISBN 999 12-2-156-5.

Morrell, D. (1987) *Indictment: Power and Politics in the Construction Industry*, Faber & Faber: London.

Murray, M. and Langford, D. (2003) *Construction Reports 1944–98*, Blackwell Science: Oxford.

Napier, I.A. (1970) *A Systems Approach to the Swedish Building Industry*, Stockholm: National Swedish Institute for Building Research, Document No. D9:1970.

Ofori, G. (1993) *Managing Construction Industry Development*, Singapore: Singapore University Press.

Ofori, G. (1994) Construction Technology Development: Role of An Appropriate Policy, *Engineering Construction and Architectural Management*, 1, 147–168.

Phillips, M.E. (1994) Industry mindsets: Exploring the cultures of two macro-organizational settings, *Organization Science*, 5 (3) 384–402.

Powell, E. (1979) The short and long term role of the building industry *Building Technology & Management*, January 1979, Institute of Building: Ascot, UK, 4–8.

Riley, M.J. and Clare-Brown, D. (2001) Comparison of cultures in construction and manufacturing industries. *Journal of Management in Engineering*, 17(3), 149–158.

Rooke, J., Seymour, D. and Fellows, R. (2003) The claims culture: A taxonomy of attitudes in the industry, *Construction Management and Economics*, 21, 167–174.

Root, D. (2002) Validating occupational imagery in construction: Applying Hofstede's VSM to occupations and roles in the UK construction industry, in R. Fellows, and D. Seymour (eds), *Perspectives on Culture in Construction*, Task Group 23 – Culture in Construction, CIB Publication No. 275, Rotterdam: International Council for Research and Innovation in Building and Construction, ISBN 9063630301, 151–171.

Rwelamila, P.D. and Hall, K.A. (1994) An inadequate traditional procurement system? Where do we go from here?, In R.G. Taylor (ed.), *CIB W92 'North meets South' Procurement Systems Symposium Proceedings*, Durban, 473–482.

Schein, E. (2004) *Organizational Culture and Leadership*, 3rd edn, San Francisco: Jossey-Bass.

Seaden, G. and Manseau, A. (2001) Public policy and construction innovation, *Building Research and Information*, 29 (3) 182–196.

Seymour, D. and Rooke, J. (1995) The culture of the industry and the culture of research, *Construction Management & Economics*, 13, 511–523.

Singapore, Construction 21 Committee (1999) *Construction 21: Re-inventing Construction*, Ministry of Manpower and Ministry of National Development, ISBN 9971-88–709–6.

South Africa, Ministry of Public Works (1999) *Creating an Enabling Environment for Reconstruction, Growth and Development in the Construction Industry*, White Paper, Government Printer: Pretoria.

Spender, J.-C. (1989) *Industry Recipes: An Enquiry into the Nature and Sources of Managerial Judgement*, Oxford: Basil Blackwell.

Spradley, J.P. (1972) Foundations of cultural knowledge, in J.P. Spradley (ed.) *Culture and Cognition: Rules, Maps and Plans*, San Francisco: Chandler Publishing Company, 3–38, in Phillips (1994).

Trompenaars, A. (2001) *Riding the Waves of Culture: Understanding Cultural Diversity in Global Business*, 2nd edn, New York, NY: McGraw Hill.

UNCHS (Habitat). (1984) *The Construction Industry in Developing Countries: Vol. 1 Contributions to Socio-Economic Growth*, HS/32/84/E Nairobi.

Walker, D. (1995) The influence of client and project team relationships upon construction time performance, *Journal of Construction Procurement*, 1 (1) 4–20.

Wells. (1986) *The Construction Industry in Developing Countries: A strategy for Development*, unpublished PhD thesis, University of Swansea.

Wells, J. (1998) *The Construction Industry in Developing Countries, and in countries that are not developing* in R. Hjerppe (ed.), Urbanisation: Its Global Trends, Economics and Governance, United Nations University/World Institute for Development Economics Research (UNU/WIDER) Helinski, 1998, 137–157.

Winch, G.M. (1985) The Construction Process and the Contracting System: A Transaction Cost Approach, *Proceedings of the 7th Bartlett International Summer School*, Vaulx-en-Velin, UCL, London.

Wood. (1975) *see* Great Britain, EDCs for Building and Civil Engineering, London: NEDO, HMSO.

17 Respect for people – The dawn of a new era or mere rhetoric?

An historical analysis of labour relations in construction

Steven McCabe

Abstract

Rethinking Construction (Construction Industry Task Force, 1998) has been a catalyst for renewed debate about what British construction must do to improve itself. The report has led to a number of initiatives, one of which is 'respect for people', intended to address the need to make people a key component ('driver') of improvement. This chapter explores the importance of people in construction. It argues that even though the recognition of the importance of people is welcome, it must be seen in the context of industrial relations in construction over the last two hundred years. An analysis is carried out which suggests that an initiative such as 'respect for people' can be viewed as an attempt by employers to deal with the threat of increased power by workers. The current lack of skilled workers may be viewed as a direct consequence of the preceding twenty-five years. Construction firms, in an attempt to reduce cost – still a major component of the 'Egan agenda'[1] – vigorously engaged in 'contracting' as a means to shift risk to individuals who were encouraged to become self-employed. As a result, long-term improvement through training, education and recruitment of apprentices was severely curtailed. Construction, especially when compared to other industries, was seen as 'backward' (Ball, 1988). However, some of the other industries most notably motor manufacturing and retailing have recently been criticised for their treatment of people. It is against such a background that 'respect for people' has emerged. The key question that this chapter seeks to answer is whether 'respect for people' is part of sustainable cultural change in construction or simply what will be a short-lived attempt to control wage levels.

Keywords: Workers, contracting, unions, markets, improvement and respect for people.

17.1 Introduction: Analysing the role of people in construction

All societies require some construction to create an environment that allows survival and development. As societies advance, so too does the requirement for the built environment in terms of the buildings in which people carry out their day-to-day activities. Advanced societies need an elaborate infrastructure that allows travel, and provides continuous power and clean water. All of this must be created using construction, which itself needs an intention by a provider to secure funding and procurement of those with appropriate and sufficient skills and expertise. In small-scale works, these will be individuals with particular trades. Larger-scale work, though, usually means employing a contractor: an organisation capable of employing individuals or firms known as subcontractors. This chapter explores the way that the contracting system evolved and how it has been used to organise construction work and, in particular, to regulate the pay and conditions of workers.

The focus of analysis is historical and socio-economical. British construction has traditionally operated on the basis of a free market in which price is governed by supply and demand. If work is required, those with the most sought-after skills usually obtain the highest price. Medieval stonemasons became an elite group of workers because their skills were highly prized by the rich and powerful. To protect their interests, they formed guilds that enabled them to ensure there was regulation of entry and control over the training of apprentices, something that ensured they could maintain high rates of pay and a mystique that still exists in the form of freemasonry. Other construction trades have sought to follow the example of the stonemasons. Significantly, the creation of elite groups within construction – a phenomenon Postgate (1923) termed 'labour aristocracies' – has explicitly served only the interest of that trade. An analysis of the history of such relationships shows that such groups have not been averse to undermining the interests of other workers, particularly if they believe they can be used as a cheaper alternative. Those who procure construction (clients or their representatives) maintain their right to do so at the lowest cost. In doing this, they frequently find that others involved in employing construction workers are only too willing to collude in the diminution of pay and conditions. It is argued that reliance on the 'power' of the market has caused disputes in the past and led directly to the current shortage of skilled labour.

The analysis indicates that labour relations in construction are notable for the way that competing groups attempt to defend and improve their interests. Importantly, much depends on the power that a particular group is believed to possess. In the case of clients, this will be the financial ability to initiate work. In the case of workers it is based on their knowledge and expertise. As will be described, the emergence of contractors and subcontractors in the eighteenth century became the key link between clients and workers and, in particular, the mechanism through which regulation of pay and conditions occurred. As supply and demand for construction shifts, employers and workers seek to defend their interests. The 'respect for people' initiative may be viewed as an attempt to redress what employers see as the shift in power to construction workers whose skills are currently in short supply.

The following are the main objectives of this chapter:

- Analyse the historical evolution of organised labour in construction.
- Describe why 'contracting' became a favoured method of procuring and carrying out construction in Britain.
- Explore the major changes in construction since the Second World War.
- Consider the potential impact of 'respect for people'.

17.2   The origins of organised labour in construction

Guilds and local disputes

As buildings became more regularised in the Middle Ages, their construction required particular skills learned through apprenticeship. The need to ensure that standards of work were consistent became important in that it meant wage levels could be kept as high as possible, depending, of course, on the market. To do this required some form of organisation, the earliest example of which were the guilds. Guilds, it is believed, date back to the Norman Conquest (Reid, 2004:8). Their purpose was to protect the interests of the members who lived and worked in towns or villages. The small guilds were controlled by 'masters', who could become privileged members of the local society. In order to increase their influence, masters often established 'liveries' which created alliances through the invitation to merchants to join. Such elevation in the status of masters meant that less powerful members of the guild began to sense inequity in their treatment. These 'journeymen', members who had completed their apprenticeships but were not yet able to become masters, frequently believed that breaking away to form their own independent group would increase their advantage. Agitation by such groups for what they believed to be a fairer distribution of wealth and influence brought them into conflict with those who would previously have been their allies.

Even though the advent of unions organised on a more widespread basis was not to occur until the nineteenth century, it was not uncommon for such breakaway groups to engage in industrial action in support of fellow workers. Masters increasingly sought legal remedy to attempt to control the behaviour of journeymen. The eighteenth century was punctuated by increasingly bitter disputes concerning wages. According to Reid, the 'building trades' in particular were involved in disputes in the period between 1757 and 1761 (ibid.: 13). This was not to be the last time groups within construction would take on others to assert power. At this point, however, such disputes were localised and much depended on the willingness of either side to engage in prolonged dispute. The nineteenth century would be the period when Britain changed and the consequences for those involved in construction would be profound.

The great transportation boom and contracting

The Industrial Revolution brought tremendous change to Britain. Workers migrated to the factories, which in turn created markets for goods in the rapidly developing towns and cities. The need for efficient means of transporting coal from mine to factory provided the impetus for canal construction. Canal-building required a Bill of Parliament, which was sought by whoever would operate the canal, often the factory owners. An engineer would be appointed to oversee initial design and construction. Creating canals needed a huge number of men and horses, which the engineer looked to others to procure. As such, he *contracted* this work and created the need for organisations known as 'contractors'.

Given the scale of the work required, the contractor appointed agents to be responsible for particular stretches of canal who, in turn, hired subcontractors. The subcontractor employed a number of 'gangers' who actually hired the men and superintended the work carried out on a day-to-day basis. Gangers were, according to Coleman, the 'corporals of the enterprise' (1965: 51). Because speed and cost were crucial, gangers were the key to ensuring that the subcontractor could achieve profit. (Many went bankrupt trying to do so.) Even though pay was good – sometimes three or four times what could be earned elsewhere – there was little concern for the safety or living conditions of the workers. Creating the 'Inland Navigation System' (from which the nickname 'navvies' is derived) was both dangerous and extremely arduous (Cowley, 2001). Workers were frequently housed in hovels and huts created from whatever material came to hand. However, speed of construction overrode any concerns. Even though the next phase of the transportation boom, the building of railways, was little better, it did cause some people to question the treatment of navvies.

Railways quickly superseded canals. Those with expertise in digging and earthmoving were highly prized. Contractors, their agents, subcontractors and navvies adapted accordingly. As with canals, however, making profit on railway construction was uncertain and bankruptcy was common. Some subcontractors, realising that their payment for work would not cover wages and materials, simply disappeared. And, as before, profit came before concerns about workers' conditions or safety. Intense competition between rival railway companies to open lines early put pressure on contractors to complete faster. Subcontractors responded by demanding that gangers force their men to work faster and take risks. For railway navvies, the work was extremely hazardous, especially when explosives were used for the creation of tunnels. Given the estimation that there was one fatality per mile of track laid, this would mean over 20,000 deaths occurred. Injuries were common and those most seriously disabled frequently faced a life of destitution. Compensation, apart from any benevolence by fellow workers, was unknown (Coleman, ibid.: 111). The living conditions which accompanied railway building, the hovels and huts used by canal navvies, were appalling. The constant damp, lack of clean water and poor sanitation led to regular outbreaks of smallpox and cholera.

In their desire to make profit, however, subcontractors and gangers had no incentive to improve conditions. Indeed, their use of the odious system known as 'truck' only added to the toll of suffering. Truck was such a sure way to make profit that it was possible to tender a price so low that a loss would be made on the work carried out, but compensated for elsewhere. In order to attract men from subcontractors, it was often necessary to offer a higher rate of pay. Once the men were hired, they would need to eat and drink, but the rural location of much work made obtaining food and drink difficult. Truck therefore came into play, with subcontractors making provision but at an extortionate profit. Workers were not paid until the end of a period (usually a month), and made any 'purchase' using a ticket, which would then be used to make deductions from their wages. It was only on surrender of the ticket that workers discovered the real cost (for produce that was often rancid anyway). Moreover, while not the sole factor, the high accident rate on railway construction may be partly explained by those operating truck encouraging workers to drink as much as possible while working: the lack of clean water made ale a normal substitute.

Given such exploitation, it was inevitable that trouble occurred, particularly between groups from different parts of the British Isles after payday, when it was common to go on what was referred to as a 'randy' (in effect a drunken riot). In effect, navvies were a law unto themselves. Discipline, if it existed at all, occurred through payment which, of course, was manipulated to the employer's advantage. However, any temptation for navvies to encourage discipline amongst themselves by collective actions (proto unions) was unlikely to be welcomed by their employers. Rather, it was actively discouraged and those navvies attempting to organise in order to engage in any action risked both the sack from their current employment and being 'blacklisted' by all other railway contractors.

The call for a change in conditions eventually came from such individuals as Thomas Brassey and Samuel Peto, railway contractors who, having witnessed such ill treatment and unscrupulous behaviour by some subcontractors and gangers, believed that there must be a better way. They argued that short-term profiteering by subcontractors and their gangers was counterproductive. Brassey's experience taught him that those who had been treated well (respected) could be relied on to be loyal and work efficiently which, of course, increased the likelihood of secure financial return (Helps, 1872). Peto, who entered Parliament in 1847, made similar assertions. A parliamentary inquiry into railway building in 1846 provided first-hand accounts from both workers and their employers. This inquiry generally showed workers to be decent and honest and their employers anything but. The publication of the inquiry raised awareness of the need for the construction industry to change the way it treated workers.

Urbanisation and trade unionism in construction after the Industrial Revolution[2]

The Industrial Revolution changed many things about the landscape of Britain, particularly in towns and cities. Industrial progress and expansion required a ready supply of those with the ability and resources to provide the necessary buildings: the factories for production, the warehouses for distribution, the shops to sell goods and, of course, housing for an increased population, together with the necessary infrastructure of sewers, water and, later, power. Canal and, more especially, railway construction had shown the advantages of the contracting system (cf. Cooney, 1955). It allowed clients to deal with one organisation for all of the works rather than having to organise the various individual trades. Importantly, contractors promised to complete all the work for a predetermined price. The early contractors were those master craftsmen who used their knowledge of techniques and labour markets to ensure they could complete the work. Probably the best known of these was Thomas Cubitt whose development from carpenter to 'master builder' is described in exquisite detail by Hobhouse (1971).

Similar to those organisations involved in railway construction, contractors involved in urban building discovered that mobilising the resources required to carry out necessary work meant being able to react to market rates. Whereas in the railways the requirement was for physical ability to cope with tremendous effort and harsh conditions, being able to procure trades with sufficient skill meant paying the going rate. In towns and particularly cities where particular trades were well organised and sought after, contractors had little choice but to give in to demands for higher wages. The alternative, using migrant labour willing to work for less, was risky as it often resulted in disputes and disturbance.

Craft workers, who were frequently employers of labour themselves (having become subcontractors), viewed any increase in influence by general contractors as a direct threat to themselves. Most especially, they resisted attempts to dictate the rate at which they worked – through the use of incentive payments – and the intention to pay on an hourly basis as opposed to daily, allowing long days in summer and lay-offs in winter. Price (1980) explains that trades saw such alterations in practice as a diminution of their role as custodians of custom and practice. Many of those affected reacted to what they saw as attacks on their security of employment by organising resistance. In such a climate, localised strikes were common and, in order to avoid contractual claims, contractors were forced to give in. As a result, other groups who believed that their stature was thereby reduced attempted to redress the balance. Ball makes the point that though trades did not tend to cooperate with each other in challenging employers, gains made by one group (those willing, and able, to force employers to pay more) frequently led to benefit by all with 'recognised' skills (ibid.: 65). However, all trades were united in their opposition to any increase in the influence of those workers who were semi-skilled or unskilled. Importantly, such workers were frequently employed by trade subcontractors and any increase in wages paid to them raised cost (and lowered potential profit). Worse, though, trade subcontractors recognised the potential for such workers to be used in implementing new technology that would undermine the importance of those with recognised skills, precisely what subsequently happened.

The rise and fall of trade unionism

Postgate (1923) describes the situation in the late 1860s by which time recognised craft workers unions had accepted the contracting system and, despite their reservations about the changes it brought, realised that they could also benefit from it. In order to maximise their influence and interests, he explains, they accepted organising on a national basis would be useful. Accordingly, 'friendly societies' (Ball, 1988: 66) developed to represent collective interests of a wider range of trades. In such a group, however, a hierarchy of trades still existed in which the oldest were the most influential. One issue that did create solidarity was the belief that any attempts by the increasing numbers of semi-skilled and unskilled workers to organise on their own behalf should be resisted (including use of the Trades Union Congress). Such resistance proved futile and any hegemony enjoyed by particular craft unions proved to be short-lived, especially in the period between the two world wars when a sharp decline in economic activity caused clients to search for cheaper solutions to their building needs.

Within craft unions, tensions became apparent between local activists and national officers. The former argued that even though national agreement might benefit them in the short term, there was evidence of increased willingness by employers in the long term to implement new technology that would undermine the importance of skilled labour. Reinforced concrete, power-driven machinery and use of subcontractors employing non-union labour to operate such equipment – which required less training and skill to achieve output – reduced cost. Unsurprisingly, this reduction appealed to clients. At national level, though, unions accepted that subcontractors and their use of non-union labour were likely to increase. Members at local level disagreed with their national officers and frequently engaged in 'wild cat' action. Despite such protest, the tide of change was relentless and led members at branch level to believe that membership offered little or no protection. The loss of members in all craft unions further exacerbated this perception and undermined any ability to defend long-term interests. The consequence was amalgamations which, according to Ball, were often 'loose and fairly ineffectual' (ibid.). For example, the creation of the National Federation of Building Trade Operatives (NFBTO) in 1918 was a direct consequence. It was replaced in 1971 by UCATT (Union of Construction, Allied Trades and Technicians), which still represents crafts today.

The effects of the Second World War meant a vast increase in construction demand. First, there was the immediate need to repair bomb damage to major cities. Subsequently, there was a massive programme of urban destruction, reconstruction and development. In order to meet such demand, the use of techniques such as prefabrication – particularly in the use of standard timber components such as doors, windows and roofing – was widely adopted as standard practice. This, of course, reduced the need for carpenters, who had surpassed stonemasons as the pre-eminent trade in construction. Reinforced concrete on larger-scale building projects also became the norm, as a much cheaper alternative to steel, the use of which was strictly controlled. Accordingly, contractors were able to employ labour-only subcontractors whose workers usually did not need specific skills to carry out work. As in the earlier period when canals and railways were built, immigrants often became the favoured way to ensure a supply of cheap labour.

For those willing to forgo the immediate prospect of high wages for purely manual work, there was the possibility of training to become a tradesman. Many firms operating today were set up by individuals who started in the industry in this way. The majority, though, were attracted by the prospect of earning as much as they could. Because many were immigrants, and viewed themselves as temporary workers, there was a sense of accepting whatever work was offered. Efforts to organise such workers, by the likes of the TGWU (Transport and General Workers Union), were mostly unsuccessful. Because workers were willing to agree to incentive-based payment, a principle long resisted by craft unions, the loss of control over work by trades was exacerbated. It was precisely such principles that created the conditions that led to a conflict between employers and unions in the early 1970s.

The national building strike

The national building strike of 1972 was intended to improve pay and conditions for all construction workers through the aegis of the NJCBI (National Joint Council for the Building Industry). It subsequently proved to be a defining point for industrial relations in construction, particularly given that employers made explicit their intentions not to engage in meaningful negotiations. The inevitable collapse in relations led to bitter confrontation between some employers and local activists. At stake for the unions involved – UCATT, TGWU and NUGMW (National Union of General and Municipal Workers) – was the principle of national bargaining and, in particular, of bringing official wage rates into line with those being paid unofficially, which were higher. The central demand was £30 for a standard 35 hour week. Employers resisted such demands on the basis that they reserved the right to pay according to local markets and, more especially, to maximise production when necessary.

A major change that enabled employers both to pay according to output and to marginalise union influence was self-employment. In effect, workers became their own bosses. They were paid an amount – known as 'the lump' – which was determined by the quantity of work carried out and were responsible for payment of deductions such as tax, national insurance, holidays and any pension contribution they thought necessary. In the 1960s and early 1970s, when construction demand was high, those willing to work long hours could earn significant amounts, not unlike the navvies in earlier railway construction. For such workers, concern about the rights of others or securing agreements to cover pay and conditions of employment were not a priority. However, construction unions, many of whose members were employed in the public sector, believed that parity across all sectors was crucial.

Though the strike did end after thirteen weeks, the legacy of bitterness was to remain among those who had lost wages for such a protracted period and employers who lost money on contracts that had been picketed. The consequence for the latter was to become even more determined to use labour-only subcontractors and encourage self-employment. Among the workers who had been at the vanguard of picketing, usually local activists, there was disillusionment in the ability of their unions to negotiate at national level (Actor Ricky Tomlinson provides his reflections in his autobiography, Tomlinson, 2003.) Leslie Wood, as UCATT General Secretary, provides a very personal account of this period. Though he stresses some gains were made, he accepts construction unions' weakness in organisational terms, especially in an industry which operates on the basis of casualisation (1979: 150). As the period from the mid-1970s to 1980s demonstrated, if unions found difficulties in having a meaningful dialogue when construction was buoyant, they were ineffectual when demand reduced.

Contemporary developments in labour-relations

The period after the building strike was one in which union activism carried a high price. Some companies made clear that they would not employ those who openly admitted to membership. It later became well known that many major contractors subscribed to a right-wing organisation known as 'The Economic League' whose purpose was to vet workers for left-wing and trade union links. Unsurprisingly, union membership declined as self-employment became ever more widespread. The Labour government of 1974–79 had a somewhat conflicting approach to dealing with construction. In 1978 it introduced the P714 self-certification tax scheme for self-employed construction workers. However, the previous year, a policy background paper, 'Building Britain's Future', had proposed that, consistent with the nationalisation that had occurred in other sectors, a National Construction Corporation be formed. Such an organisation, it was explained, could be created by the acquisition of 'one or more major contractors, in consultation with trade unions in the industry' (Hillebrandt, 1984: 148). Major contractors responded by forming CABIN (Campaign Against Building Industry Nationalisation) with the intention of vigorously fighting this proposal.

The 1979 election was won by a Conservative administration avowedly believing in free-market principles and that too much was spent on the public sector. Allied to these beliefs was the intention to invite privately owned contractors to tender for work that had been previously carried out by public sector organisations. Construction unions, many of whose members worked for the 'contracting arm' of public authorities, known as Direct Labour Organisations (DLOs), feared that what had occurred in the private sector during the 1970s would be repeated in the public sector. The Conservative government of the 1980s made clear their intention to 'face-down' unions, regardless of the cost. The most notable example was the miners' strike of 1984–85, led by Arthur Scargill. The privatisation of public sector organisations was, undoubtedly, seen as part of the union struggle against 'Thatcherism'. However, like the miners, UCATT proved unable to resist change that caused reduction of membership and dilution of influence.

The mantra espoused by government and organisations in the 1980s and 1990s was 'efficiency'. Organisations, regardless of sector, attempted to reduce cost. The construction industry had demonstrated what was possible by the encouragement of subcontracting and self-employment. However, such a trend brought with it a corresponding reduction in 'avoidable costs' such as training and education. There was also a personal cost. Those who were self-employed (and who prospered during periods when workload was high) quickly discovered that when demand decreased they were laid-off without redundancy payment. Uncertainty, coupled with an industry-wide cut in rates paid to workers, caused many to question the point of being involved in an industry which uses casualisation as its favoured method in dealing with fluctuations in workload, a point precisely made in the Phelps Brown Report (1968). Warnings by unions such as UCATT and CITB (Construction Industry Training Board) that this approach was not conducive to long-term improvement and that safety was being severely compromised were mostly ignored (Buckley and Enderwick, 1989: 121). It would take a report commissioned by the incoming Labour government of 1997 to make clear just how bad things in the industry had become.

17.3 The provenance of the 'respect for people' initiative

The need to consider improvement in the way that people in construction are treated came as a result of the so-called 'Egan Report', *Rethinking Construction*, published in 1998 (Murray and Langford, 2003). The previous year, the new minister of construction (a post that no longer exists) had invited an influential group of people to become part of what would be known as the Construction Task Force (CTF). This group included representatives of such major clients as Tesco and Nissan, and an academic with expertise in 'lean manufacturing',[3] and was led by Sir John Egan, the chief executive of British Airports Authority.[4] A union representative was included but, significantly, from GMB (General and Municipal Boilermakers) not UCATT. The brief given to the CTF was to examine ways in which construction could radically improve what it produced so as to ensure greater success in meeting client expectations and underpin the growth of the British economy, believed to be approximately 10 per cent of gross domestic product.

Within ten months the CTF had published its findings as to how this might be achieved. In order to do so, the group drew from their knowledge and 'extensive experience' in other industries: car manufacturing, steel-making, grocery retailing and offshore engineering. As a consequence, a number of observations and recommendations – 'drivers of change' – were made. One observation was that improvement could occur only if people in construction were treated as an 'asset' (1998: 14). In 1999, in order to encourage and sustain the 'commitment to people' driver of change, the minister for construction established a working group comprising representatives from a number of bodies with interest and expertise in people in construction. Significantly, this group did include representatives from the two major unions involved in construction – UCATT and TGWU.

The objective of the working group was to consider and propose ways in which, through organisations of every size, the construction industry could implement measures that would ensure improvement in respect for people. In 2000 the group produced an interim report, 'A Commitment to People "Our Biggest Asset"', which recommended that supporting toolkits and performance indicators should be developed. The definitive report, *Respect for People* was published in October 2002.

17.4   What does 'respect for people' propose?

As part of implementing 'respect for people', the authors of the final report contend that there are four 'aspects of the change process' which create a framework (2002: 6). These are promotion of the business case, measurement of performance, the integration of reporting systems and provision of a network of support for managers. Following a trial study of various themes recommended in the 2000 report, six 'toolkits' were proposed that are intended to support improvement. These were

1   equality and diversity in the workplace
2   health and safety
3   working environment
4   training plan
5   workforce satisfaction
6   work in occupied premises.

Importantly, the authors of *Respect for People*: accept that even though managers accept that there might be a moral case for improvement through investment in people, they perceive it as 'altruistic', that is as having no 'immediate benefit to their business success' (ibid.: 10). In order to assuage potential concerns by those considering implementing the principles of 'respect for people', the authors present a 'business case'. This case contends that improvement in the treatment of people and increased return on investment are not mutually exclusive. Rather, it is argued, they are entirely complementary. Clients are more likely to be satisfied. People are paid more and enjoy better conditions which, it is believed, means they are 'happier, healthier and more productive' (ibid.). Finally, projects are more likely to be delivered on time, to cost and to a level of quality that will satisfy clients. This business case, it must be said, has strong resonance with the so-called 'Deming Cycle', named after the 'quality guru' Dr Deming, who helped Japanese manufacturers to achieve pre-eminence in the period after the Second World War. Deming was also known to pointedly ask of those who suggested the need for improvement, 'by what means' that improvement would be brought about. It is salutary to ask by what means respect for people will ensure that long-term change in construction occurs.

17.5   Will 'respect for people' change construction?

The philosopher George Santayna mused that 'those who do not remember the past are condemned to repeat it'. Since the Second World War, labour relations in construction have not been notable for close working between major employers and unions. On the contrary, employers have shown that they are only too happy to exclude unions, particularly if it allows them to be dynamic and cut costs during periods of decline. The fact that the working group for 'respect for people' includes two major unions with interest in construction is to be applauded. However, there are some – especially those with direct experience of the strike of 1972 – who believe that closer relations with employers should be approached with caution. As they would typically claim, the high wage levels that construction workers presently enjoy are a direct consequence of many years of under-investment by employers. Why should unions intervene to assist employers who believe wage costs have become prohibitive to ensuring profit? The counterargument is that unless unions are prepared to engage on behalf of all workers, they assist in the maintenance of the 'labour aristocracy' that was identified in the late nineteenth century. Moreover, if the benefit of high wages is to be enjoyed by all construction workers, it is incumbent on unions to negotiate the case accordingly.

While pay is always a concern for workers, their conditions of employment and security are often equally important. 'Respect for people' is assumed to address all of these issues through the duality of moral justification and economics. While better treatment of workers has a moral basis, there is also a strong business case in that organisations operate more efficiently. As such, respect for people is comparable to the belief of those who contended that Japanese manufacturers' achievement of pre-eminence in the automotive and electronics sectors was based upon their recognition of the importance of people. Those who advocate a 'quality' approach articulate this message and, it is to be noted, the Department of Trade and Industry (DTI) sponsored a trade mission to Japan in 1994 to attempt to learn how British construction might similarly improve (CIOB, 1995). Since the 1980s, successive governments have reinforced the message that efficiency, particularly in terms of productivity, is the key to economic prosperity. It might reasonably be assumed that the incoming Labour administration of 1997, wishing to continue economic strategies pursued by their predecessors, recognised that the apparent inefficiency of construction hampered growth. It should be remembered that election commitments were made to increase the building of hospitals and schools, albeit through Private Funding Inititiaves (PFIs). Might, therefore, 'respect for people' be viewed as an acknowledgement of the need to marry the interests of business with the concerns of people? But, as Sako (1992) stresses, this requires more than platitudes about improving relationships with suppliers and workers. The key question is whether construction is capable of the sort of change that is being called for.

The original membership of the CTF is notable for the strong representation by car manufacturers and supermarkets. Both of these sectors have, in recent years, been criticised for the treatment of workers, especially in what is known as the 'supply chain'. They justify their approach by stating their desire to give customers value. Accordingly, car purchasers can obtain products that are not only cheaper than predecessors (in real terms) but also superior in quality when specification and occurrence of faults are considered (Hiraoka, 2000). Large supermarket chains claim their customers benefit from lower cost and the ability to shop for all their needs 'under one roof'. Critics of large corporate business

(Klein, 2001; Lawrence, 2004; Blythman, 2004) argue that they exert ruthless control over suppliers – regardless of the consequences for workers. Ongoing and close business relationships (the expression 'partnership' is frequently used) can typically mean regular demands for price reduction, unreasonable expectations about delivery and stock levels and, in the case of supermarkets, payments towards advertising and product placement. All of this raises suppliers' costs while lowering the costs of those who sell the goods. Usually, there is the threat of losing business to suppliers with lower wage costs (often in developing countries), something that may happen anyway.

What differentiates car manufacture and retailing from construction is the way that power is concentrated. The development of self-employment and subcontracting in construction means that, apart from material producers, contractors are unable to exert the same degree of influence that large car manufacturers or supermarkets chains possess. It is a paradox that the fragmentation in construction that some critics condemn as making change difficult to implement, and which was originally intended to weaken workers' control over the production process, has undermined the dominance of large contractors. The trouble is self-employment and subcontracting, while intended to provide a dynamic and adaptable workforce, have led directly to a paucity of investment in training and the current skill shortage. 'Respect for people' must therefore function in an environment where no organisation (with the exception, it is suggested, of government, through direct intervention) can implement radical change such that others are forced to follow.

In carrying out analysis of labour relations in construction, it is all too easy to engage in a myopic stance that sees workers as good, employers as bad. To do so is simplistic. Construction has traditionally provided an industry open to people who, because of their educational background or ethnicity,[5] would have difficulty finding alternative employment. The fact that newer immigrants, especially from Eastern Europe, make up a large percentage of the current construction workforce should be seen as maintaining a tradition that dates back to the building of canals and railways. However, as in that period, these workers should not be exploited.

To suggest that all those who employ construction workers are wilfully manipulative and cynical is erroneous. Many employers believe their workers are a key resource that allows them to meet client expectations and achieve success. Like the enlightened railway contractors, they realise exploitation is not a good long-term strategy. Ultimately, though, 'respect for people' must be judged by the actions of the majority of organisations, particularly smaller firms. For example, will they still employ on a casual basis? And, should demand for construction work decrease, will they treat such workers as an avoidable cost and lay them off?

The experience of industries that have achieved culture change suggests that it is a long-term process of creating trust and mutual co-operation in securing gains for both employers and workers. Even in the Egan Report there was tacit recognition that if construction was incapable of change, other solutions, such as using manufacturing to increase the use of prefabrication and modularisation,[6] might be more widely adopted. Not for the first time, technology provides both the promise of advancement (for clients) and a threat to construction workers. Some critics proclaim that attitudes in construction are so embedded and intractable that, like the issue of health and safety, change can only be engendered by increased regulation and legislation. Experience of culture change elsewhere strongly suggests that an initiative such as 'respect for people' would not flourish if impositions such as these were used.

17.6 Conclusions

This chapter has provided the historical background to the 'respect for people' initiative. The construction industry's present problems – lack of skills, cynicism and poor image – are a legacy of two centuries of ill treatment of people. As such, the portents for success do not appear good. Any initiative intended to reverse institutional attitudes and create redress will always be ambitious. However, the treatment of construction workers must be improved. For too long it has been possible for those with malign intentions to exploit workers, something of which clients are only too aware but nevertheless choose to ignore. What seems crucial is that the culture and balance of power that exists in the industry should be aligned so that all involved in the 'labour process' (employers and workers, especially through trade unions) are dedicated to ensuring cooperation and mutual benefit. Should that happen, 'respect for people' is likely to be judged as a significant contribution both to improving the way in which construction workers are treated and to achieving desired efficiency. If such change does not occur, 'respect for people' will, like Samuel Peto's appeal for change in 1847, be viewed as failed rhetoric.

Dedication

This chapter is in memory of all those people who gave their lives to construction, some, in all too many tragic cases, quite literally. It is especially dedicated to the memory of my father James (1926–2004). Similar to many thousands of construction workers, though he spent the majority of his life in the industry and retired in 1991, he did not receive an occupational pension.

Notes

1 An eponymous reference to the chair of the Construction Task Force that published *Rethinking Construction*.
2 I am indebted to Michael Ball for his analysis and categorisation of the four phases in trade unionism in construction since the early nineteenth century.
3 An approach, originally developed by carmaker Toyota, that emphasises value-adding in production cycles, the elimination of waste, integration of sub-processes and dedication to 'end' customer/user satisfaction.

4 Murray (Murray and Langford, ibid.: 180) suggests that 'Egan' is now used both as a noun (as in the person himself) and as a verb (as a shorthand to indicate the desire to radically change construction).
5 The current lack of ethnic minorities in construction gives rise to concern, a point made by the 'respect for people' working group.
6 Prefabrication and modularisation are techniques based upon preassembly of elements of work that would normally be carried out on site. The intention is to reduce time on site, to decrease the reliance on particular trades and, usually, to be assured of 'factory standard' for finished components.

References

Ball, M. (1988), *Rebuilding Construction, Economic Change in the British Construction Industry*, Routledge, London.
Blythman, J. (2004), *Shopped: The Shocking Power of British Supermarkets*, Perennial Publishers, London.
Buckley, P.J. and Enderwick, P. (1989), 'Manpower management', in *The Management of Construction Firms, Aspects of Theory*, P.M. Hillebrandt and J. Cannon (eds), pp. 108–127, Macmillan Press, Hampshire.
CIOB (Chartered Institute of Building) (1995), *Time for Real Improvement: Learning from Best Practice in Japanese Construction R&D*, Report of the DTI Overseas Science and Technology Expert Mission to Japan, December 1994, CIOB Publications, Ascot.
Coleman, T. (1965), *The Railway Navvies, A History of the Men Who Made the Railways*, Hutchinson, London.
Cooney, E. (1955), 'The origins of the Victorian master builders', *Economic History Review*, VIII, pp. 167–176.
Construction Industry Task Force (1998), *Rethinking Construction*, Department of the Environment, Transport and the Regions. London, DETR.
Cowley, U. (2001), *The Men Who Built Britain, A History of the Irish Navvy*, Wolfhound Press, Dublin.
Hiraoka, L.S. (2000), *Global Alliances in the Motor Vehicle Industry*, Westport, CT, USA.
Helps, A. (1872), *The Life and Labours of Mr. Brassey*, An unpublished booklet, London.
Hillebrandt, P.M. (1984), *Analysis of the British Construction Industry*, Macmillan Publishers, London.
Hobhouse, H. (1971), *Thomas Cubitt, Master Builder*, Macmillan, London.
Klein, N. (2001), *No Logo*, Flamingo, London.
Lawrence, F. (2004), *Not on the Label: What Really Goes into the Food on Your Plate*, Penguin, London.
Murray, M. and Langford, D. (eds) (2003), *Construction Reports 1944–98*, Blackwell Science, Oxford.
Murray, M. (2003), 'Rethinking Construction: The Egan Report (1998)', in *Construction Reports 1944–98* (M. Murray and D. Langford), pp. 178–195.
Phelps Brown, E. (1968), *Report of the Committee of Inquiry into Certain Matters concerning Labour in Building and Civil Engineering*, Cmnd 3714, HMSO, London.
Postgate, R. (1923), *The Builder's History*, Labour Publishing, London.
Price, R. (1980), *Masters, Unions and Men*, Cambridge University Press, Cambridge.
Reid, A.J. (2004), *United We Stand, A History of Britain's Trade Unions*, Penguin Books, London.

Respect for People Working Group (2000), *A Commitment to People 'Our Biggest Asset'* (interim report), Rethinking Construction Limited, London.

Respect for People Working Group (2002), *Respect for People, A Framework for Action* (final report), Rethinking Construction Limited, London.

Sako, M. (1992), *Price, Quality and Trust: Inter-firm Relations in Britain and Japan*, Cambridge Studies in Management, Cambridge.

Tomlinson, R. (2003), *Ricky*, Time Warner Books, London.

Wood, L. (1979), *A Union to Build, The Story of UCATT*, Lawrence and Wishart, London.

Index

eBooks – at www.eBookstore.tandf.co.uk

A library at your fingertips!

eBooks are electronic versions of printed books. You can store them on your PC/laptop or browse them online.

They have advantages for anyone needing rapid access to a wide variety of published, copyright information.

eBooks can help your research by enabling you to bookmark chapters, annotate text and use instant searches to find specific words or phrases. Several eBook files would fit on even a small laptop or PDA.

NEW: Save money by eSubscribing: cheap, online access to any eBook for as long as you need it.

Annual subscription packages

We now offer special low-cost bulk subscriptions to packages of eBooks in certain subject areas. These are available to libraries or to individuals.

For more information please contact webmaster.ebooks@tandf.co.uk

We're continually developing the eBook concept, so keep up to date by visiting the website.

www.eBookstore.tandf.co.uk

Milton Keynes UK
Ingram Content Group UK Ltd.
UKHW031141141024
449569UK00024B/1150